Water Purification - Present and Future

Edited by Magdy M.M. Elnashar and Selcan Karakuş

Published in London, United Kingdom

Water Purification - Present and Future
http://dx.doi.org/10.5772/intechopen.111206
Edited by Magdy M. M. Elnashar and Selcan Karakuş

Contributors
Alejandro Ruiz, Baik-Ho Kim, Carlos Montalvo, Claudia A. Alejandra Aguilar, Enrique O. Martínez, Eric Houbron, Fahad S. Aljohani, Hafsa Muzammil, Hesham Mohamed Abdal-Salam Yehia, Hwan-Seok Choi, Juan C. Robles, Juhani Pylkkanen, Khagani Farzulla Mammadov, Lekshmi Gangadhar, Lina María Agudelo-Escobar, Magdy M. M. Elnashar, Majid Alruwaili, Manikandan Paramasivam, María Dolores Gil García, María Jesús Martínez Bueno, María Martínez Galera, Methaq H. Lafta, Muhammad Adnan Shahid, Muhammad Danish Majeed, Muhammad Safdar, Muhammad Zaman, Raed Al-Wasify, Rosa A. Martinez H., Rosa M. Cerón, Said Mahmoud Said, Samar Ragab, Santiago Erazo Cabrera, Saranyadevi Subburaj, Selcan Karakuş, Shimaa Hamed, Tayyaba Majeed, Zahid Maqbool

First published in London, United Kingdom, 2024 by IntechOpen
IntechOpen is the global imprint of INTECHOPEN LIMITED, registered in England and Wales, registration number: 11086078, 167-169 Great Portland Street, London, W1W 5PF, United Kingdom

British Library Cataloguing-in-Publication Data
A catalogue record for this book is available from the British Library

Additional hard and PDF copies can be obtained from orders@intechopen.com

Water Purification - Present and Future
Edited by Magdy M. M. Elnashar and Selcan Karakuş
p. cm.

This title is part of the Environmental Sciences Book Series, Volume 16
Topic: Water Science
Series Editor: J. Kevin Summers
Topic Editor: Yizi Shang

Print ISBN 978-1-83769-219-4
Online ISBN 978-1-83769-220-0
eBook (PDF) ISBN 978-1-83769-221-7
ISSN 2754-6713

For EU product safety concerns:
IN TECH d.o.o., Prolaz Marije Krucifikse Kozulić 3, 51000 Rijeka, Croatia,
info@intechopen.com or visit our website at intechopen.com.

IntechOpen Book Series
Environmental Sciences
Volume 16

Aims and Scope of the Series

Scientists have long researched to understand the environment and man's place in it. The search for this knowledge grows in importance as rapid increases in population and economic development intensify humans' stresses on ecosystems. Fortunately, rapid increases in multiple scientific areas are advancing our understanding of environmental sciences. Breakthroughs in computing, molecular biology, ecology, and sustainability science are enhancing our ability to utilize environmental sciences to address real-world problems.

The four topics of this book series - Pollution; Environmental Resilience and Management; Ecosystems and Biodiversity; and Water Science - will address important areas of advancement in the environmental sciences. They will represent an excellent initial grouping of published works on these critical topics.

Meet the Series Editor

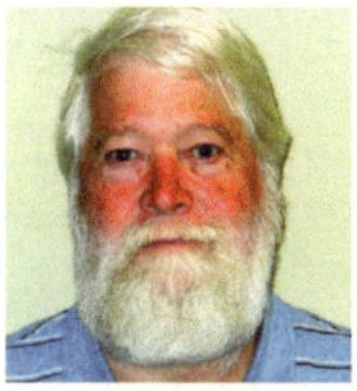

J. Kevin Summers is a Senior Research Ecologist at the Environmental Protection Agency's (EPA) Gulf Ecosystem Measurement and Modeling Division. He is currently working with colleagues in the Sustainable and Healthy Communities Program to develop an index of community resilience to natural hazards, an index of human well-being that can be linked to changes in the ecosystem, social and economic services, and a community sustainability tool for communities with populations under 40,000. He leads research efforts for indicator and indices development. Dr. Summers is a systems ecologist and began his career at the EPA in 1989 and has worked in various programs and capacities. This includes leading the National Coastal Assessment in collaboration with the Office of Water which culminated in the award-winning National Coastal Condition Report series (four volumes between 2001 and 2012), and which integrates water quality, sediment quality, habitat, and biological data to assess the ecosystem condition of the United States estuaries. He was acting National Program Director for Ecology for the EPA between 2004 and 2006. He has authored approximately 150 peer-reviewed journal articles, book chapters, and reports and has received many awards for technical accomplishments from the EPA and from outside of the agency. Dr. Summers holds a BA in Zoology and Psychology, an MA in Ecology, and Ph.D. in Systems Ecology/Biology.

Meet the Volume Editors

Professor Magdy M.M. Elnashar is employed at the School of Medicine, Curtin University, Australia, as well as the Centre of Excellence for Advanced Sciences, Egypt. Dr. Elnashar has a multidisciplinary background in chemistry, chemical engineering, and biochemistry. He received his Ph.D. in Biochemistry in 2005 from Leeds University (top 75 in the world), UK. Dr. Elnashar is the recipient of many awards in research and commercialization from Australia, the United Kingdom, the United States, and Egypt. He also received the National 2019 Australian Award for University Teaching. He has six patents and several research articles, books, and book chapters to his credit. His research expertise is in preparing/grafting hydrogels and nanocomposites for immobilization of enzymes, foods, and drugs.

Associate Professor Selcan Karakuş is currently working in the Department of Chemistry at Istanbul University-Cerrahpaşa (IUC), Türkiye. She received her MSc in Physical Chemistry from Istanbul University (IU) in 2006. She received her Ph.D. in Physical Chemistry from IU in 2011. She worked as a visiting researcher in the Department of Polymer Science and Engineering, University of Massachusetts, USA. Dr. Karakuş has research experience in drug carrier systems, nanoparticles, nanocomposites, nanoemulsion self-assembled polymeric nanostructures, biosensors, and copolymer blends. She has worked on different projects funded by IUC and has published several research articles and book chapters.

Contents

Preface

Water holds the key to life on Earth. Every cell, every ecosystem, and every human society thrives on its availability. Yet, access to clean, safe water is far from universal. Globally, 2.7 billion people live in water-stressed countries, facing scarcity that worsens with climate change and population growth.

Water Purification – Present and Future presents a comprehensive review of the different types of water, the elements that negatively affect their quality, and the industrial processes for water purification, desalination, and energy production. It includes twelve chapters organized into five sections:

- Section 1: "Drinking Water, Freshwater, and Rainwater"
- Section 2: "Sorbents and Adsorbents"
- Section 3: "Colors, Dyes, and Antibiotics"
- Section 4: "Biological Decontamination and Nanomembranes"
- Section 5: "Industrial Process and Energy Generation"

We would like to thank all the authors for their contributions, and we hope that this book will help save our planet from the water crisis, as without water, there will be no living organisms, including us!

Magdy M.M. Elnashar
Professor,
Faculty of Health Sciences,
School of Medicine,
Curtin University,
Perth, Australia

National Research Center,
Center of Excellence for Advanced Sciences,
Cairo, Egypt

Selcan Karakuş
Associate Professor,
Faculty of Engineering,
Department of Chemistry,
Istanbul University-Cerrahpaşa,
Istanbul, Turkey

Section 1

Drinking Water, Freshwater, and Rainwater

Chapter 1

Study of Sources of Drinking Water and Processes of Water's Purification from Pollutants

Khagani Mammadov

Abstract

This chapter presents an analysis of the sensory, physical-chemical, and microbiological aspects of water samples from drinking water sources in Azerbaijan. In the investigation, the presence of heavy metals, trace natural radionuclides, and other inorganic components is assessed. Results show that the chemical composition and microbiological parameters of waters transported to Baku, Azerbaijan through the Shollar and Oguz-Gabala pipelines comply with GOST 2874-82, AZS 216-2006, and AZS 282-2007 standards for drinking water, making them suitable for consumption. The study also found that the radioactive radon concentration in the thermal springs of the Istisu sanatorium in the Kalbajar region of Azerbaijan falls below the guideline indicator, rendering the springs safe for medical use under physician supervision. However, the radon levels in cold and thermal springs situated at an altitude of 2385 m above sea level exceed the recommended threshold by 100 times, making them unfit for drinking. The study also investigated the polluted water of the Araz River as well as measured background radiation along its watercourse. Finally, the study determined the optimal parameters for the chemical and radiological purification of water contaminated with harmful emissions and pathogenic microorganisms. This chapter also delves into the mechanisms underlying chemical and radiation-chemical processes involved in purifying polluted water.

Keywords: drinking water samples, radioactive radon, heavy metals, harmful emissions, microorganisms, radiological disinfection

1. Introduction

In 2015, we began carrying out sensory, chemical, physical-chemical, and microbiological analyses of water samples taken from polluted sources of water in Azerbaijan to study the degree of pollution of water reservoirs in the country. Determining the amounts of pollutants in water sources and methods for purification of contaminated water is important for preventing pollution of human habitats [1–14].

Ensuring a consistent supply of safe drinking water to people worldwide is one of the most pressing challenges of our time. The escalating ecological crisis, exacerbated by human activities, underscores the critical need to safeguard our natural habitats and resources, particularly drinking water sources. It is imperative that we address these challenges promptly and efficiently to ensure the sustainable and economical utilization of these vital resources [13–18].

The problem of providing drinking water to large cities in Azerbaijan was solved by using the flow of two large intercrossing rivers: Kur and Araz [14, 16]. The provision of drinking water to Baku, a city housing a significant portion of the country's population, has always been a focal point. However, relying solely on the water flow from the Sabirabad district into the Caspian Sea in the Neftchala region proved insufficient to meet the agricultural and drinking water needs of the country. To tap into alternative underground water sources, the Shollar drinking water pipeline was constructed, spanning from the Khachmaz district to Baku. Though operational since the early twentieth century, this pipeline struggled to keep up with the city's growing demand. In response, the Oguz-Gabala drinking water pipeline was completed at the beginning of the twenty-first century.

Today, the Jeyranbatan, Shollar, and Oguz-Gabala water pipelines supply drinking water to all areas within the city of Baku. Additionally, modern drinking water treatment facilities meeting contemporary standards, along with the Hovsan aeration station for wastewater treatment, have been established and are operational in the city.

Addressing ecological safety concerns surrounding agricultural products involves studying the accumulation of toxic elements in water, soil, and vegetation. The accumulation of harmful substances in the soil poses a risk of these elements migrating into the human body through water pathways [14–16].

The pollution of soil and major rivers in the country by anthropogenic emissions contributes to the increase in xenobiotic levels in the environment. This increasing technogenic pressure on natural habitats, coupled with the utilization of natural reserves, has led to critical ecological conditions. Therefore, there is a pressing need to conduct radiological, chemical, and biological analyses to gather objective data about the state of the environment, which allows for predicting trends in the occurrence of environmental crises.

Electron accelerators, as well as radioactive sources such as Co60, have proven effective in diverse applications within radiation-chemical technology, including in the purification and disinfection of water sources contaminated by anthropogenic emissions [1, 4, 5, 16].

Our investigation of pollution levels caused by anthropogenic organic and inorganic emissions (xenobiotics) in one of Azerbaijan's major rivers yielded valuable insights that offer opportunities for strategizing preventive and regulatory measures, as well as implementing chemical and radiological purification and disinfection processes for water treatment.

2. Methodical part and discussion of results

2.1 Instruments and methods for analysis

For radiometric measurements and analysis of water samples, we used İnSpector 1000 and Radiagem 2000 radiometers, a gamma spectrometer featuring an HPGe detector from Canberra, and the identiFİNDER radionuclide identification device (RID) from Thermo Scientific. Additionally, we conducted atomic absorption using the AA-6800 spectrometer by Shimadzu. We also used Expert-3 L and XRF X-ray fluorescence spectrometers [14–21].

We carried out microbiological analyses by growing colonies of microorganisms on selective standard nutrient media and microbiological tests from HiMedia in India, Condalab in Spain, and R-Biopharm in Germany [7–10, 14, 16].

We determined components in solutions using physical-chemical methods, such as liquid chromatography, gas chromatography, and mass spectroscopy (LC-10AVP, GC-2010, GCMS-QP 2010).

We used traditional analytical-chemical and modern research physical-chemical methods to identify and determine the quantity of reactants and organic and inorganic (harmful chemicals, heavy metals, radionuclides) emissions.

We used presterilized dishes for water sampling, which was carried out in accordance with the requirements of GOST 24481-80 and 18968-73, as were transportation and storage of the samples.

Express analyses with microbiological test napkins and sensory assays were carried out in all possible cases.

We conducted the comprehensive sensory, analytical-chemical, physical-chemical, and bacteriological analyses in accordance with the requirements of GOST standards 2761-84, 3351-74, 2874-82, 18164-72, 4151-72, 4011-72, 4245-72, 4386-81, 18963-73, 4595-49, 18826-73, and 18190-72 to determine the conformity of water samples to standards for drinking water (GOST 2874-82, AZS 216-2006, AZS 282-2007) [14, 16].

We used the AlphaGUARD professional radon monitor (Frankfurt, Germany) to determine levels of radioactive radon in the investigated water sources. The radioactive radon gas was carried into the ionization chamber of the measuring device by bubbling 3 liters of atmospheric air from the water sample for 10 minutes [6].

Analytical-chemical and physical-chemical laboratory installations and equipment were used for conducting comprehensive analyses of the water and its minerals.

We measured the energy of ionizing rays or particles using a detector and this measured energy value was then used to identify the specific element or its isotope that emits these rays or particles.

We used LDZH-30FBS and Tengor steam sterilizers to sterilize chemical glassware and glass jars for water sampling.

The GFL-2304 distiller produced double-distilled water for rinsing chemical glassware and for making solutions.

We used TDL-5 M and TD5A-WS centrifuges for centrifugation and sedimentation of solid impurities in water samples in ampoules with different volumes. Membrane filters were used to isolate solid impurities from water samples and for accelerated filtration.

The surfaces and structure of water minerals were comprehensively examined and studied under a scanning electron microscope (SEM, Carl Zeiss) at various magnifications. X-ray spectra and the composition of the content of chemical elements in the studied water minerals were obtained using an additional stand for the SEM equipped with an X-ray tube.

Weighing obtained by evaporation of water samples of mineral masses was carried out on a laboratory electronic balance. Active reactions of aqueous samples were determined by universal indicator tests and pH meters. Evaporation of aqueous samples from laboratory heat-resistant glass beakers was carried out on laboratory tiles.

To determine the types and number of colonies of microorganisms in water samples, we used both express test napkins and nutrient media. Nutrient media were applied in a thin layer on the inner surface of Petri dishes stored in a thermostat for the reproduction of bacterial colonies for 1 day at 37°C and for the reproduction of microscopic fungi colonies for 2 days at 28°C. After this time, the total number of colonies formed was determined by multiplying the total number of cells of colony counter by the number of colonies per cell. A detailed description of the rules for carrying out these procedures is given in the instructions of these test napkins and nutrient media.

Irradiation of water samples was carried out at radiation research facilities (absorbed dose rate 0.0045, 0.18 Gy/s or $2.81 \cdot 10^{13}$, $1.12 \cdot 10^{15}$ eV/g s) [16–22].

2.2 Discussion of results

In this study, we determined the mineral content of water taken from reservoirs, rivers, springs, and water pipes. The conducted analyses show that the mineral balance of the water samples varies in the range of 0.3–3.5 g/l. **Tables 1** and **2** present the results of the water sample analysis.

When contaminated water is exposed to gamma rays and supplemented with small wood chips, water molecules and organic waste break down into active particles and physical and chemical sorption takes place. This entails the adhesion of molecules and active particles onto the surface of the chips, enhancing the purification process. **Table 3** shows the mechanism of this process [15, 17–22].

The elementary reactions (4), (5), (8), (9), (11), (12), (17), (22), (23), (24), (28), (29), (33), and (34) are the chain termination reactions. Reactions (7), (8), (9), (11), (17), (20), (21), (22), (23), and (24) are reactions for the formation of new products of the process. Reactions (1) and (2) are initiation reactions of the process.

Regions	Components, mg/l									
	Sulfates	Chlorides	Na; K	Ca/all carbonates/	J	Sr	NO_3; NO_2	Fe; Mn	Zn	As
Khachmaz/ water of spring/	24	22	19; 11	138	0.04	0.6	3; 0.001	0.2; 0.2	0.01	0
Quba/Qudiyal River/	42	35	25; 12	96	0.05	0.4	4; 0	0.2; 0.03	0.1	0.001

Table 1.
Concentrations of the chemical components in samples taken from water sources of the Azerbaijan regions.

Regions	Isotopes, Bq/l									
	$_{11}Na^{22}$	$_{19}K^{40}$	$_{26}Fe^{60}$	$_{27}Co^{57}$	$_{30}Zn^{65}$	$_{38}Sr^{91}$	$_{50}Sn^{113}$, $_{50}Sn^{126}$	$_{63}Eu^{152}$, $_{63}Eu^{154}$	$_{88}Ra^{226}$	$_{90}Th^{228}$
Khachmaz/ water of spring/	0.4	0.12	0.24	0.24	0.05	0.22	0.04; 0.1	0.16; 0.2	0.24	0.03
Quba / Qudiyal River/	0.5	0.16	0.24	0.26	0.05	0.08	0.06; 0.06	0.12; 0.12	0.24	0.02

Table 2.
Activity of radionuclides in water samples taken from the water sources of region of the Azerbaijan.

N:	Reactions	W_n/W_3
(1)	$RH \rightsquigarrow H^{\bullet} + R^{\bullet}$ $(ROH \rightsquigarrow H^{\bullet} + RO^{\bullet})$	10^{-8}
(2)	$H_2O \rightsquigarrow H, {}^{\bullet}OH, e^{-}_{aq}, H^{+}_{aq}, H^{-}_{aq}$	0.03–0.3
(3)	$H^{\bullet} + RH \leftrightarrow R^{\bullet} + H_2$ $(H^{\bullet}\,RH)$	1
(4)	$H^{\bullet} + R^{\bullet} \rightarrow RH$	10^{-5}–10^{-6}
(5)	$H^{\bullet} + H^{\bullet} \rightarrow H_2$	1–10
(6)	$H^{\bullet} + H_2O \rightarrow {}^{\bullet}OH + H_2$	2–20
(7)	$R^{\bullet} + RH \rightarrow$ products	10^{-8}–10^{-11}
(8)	$R^{\bullet} + H^{\bullet}\,RH \rightarrow$ products	10^{-18}–10^{-20}
(9)	$R^{\bullet} + R^{\bullet} \rightarrow$ products	10^{-14}–10^{-17}
(10)	${}^{\bullet}OH + RH \rightarrow R^{\bullet} + H_2O$	100
(11)	${}^{\bullet}OH + R^{\bullet}$ $(RH) \rightarrow HORH$ $(R^{\bullet})$	10^{-4}–10^{-5}
(12)	$H^{\bullet} + OH^{\bullet} \rightarrow H_2O$	10–100
(13)	$H^{\bullet} + O_2 \rightarrow HO_2^{\bullet}$	10^{5}–10^{8}
(14)	$RH + HO_2^{\bullet} \rightarrow R^{\bullet} + H_2O_2$	10^{-5}–10^{-6}
(15)	$2HO_2^{\bullet} \rightarrow 2{}^{\bullet}OH + O_2$ $(H_2O_2 + O_2)$	10^{-3}–10^{-4}
(16)	$H_2O_2 \rightsquigarrow 2{}^{\bullet}OH$	10^{-9}–10^{-10}
(17)	$R^{\bullet} + HO_2^{\bullet} \rightarrow HOOR$ $(O_2 + RH)$	10^{-7}
(18)	$H_2O_2 + RH \rightarrow {}^{\bullet}OH + R^{\bullet} + H_2O$	10^{-8}
(19)	$RH + O_2 \rightarrow R^{\bullet} + HO_2^{\bullet}$	10^{-13}–10^{-15}
(20)	$R^{\bullet} + O_2 \rightarrow RO_2^{\bullet}$	10^{4}–10^{6}
(21)	$RH + RO_2^{\bullet} \rightarrow R^{\bullet} + ROOH$	10^{-7}–10^{-8}
(22)	$R^{\bullet} + RO_2^{\bullet} \rightarrow ROOR$	10^{-8}–10^{-10}
(23)	$RO_2^{\bullet} + RO_2^{\bullet} \rightarrow$ products	10^{-9}–10^{-12}
(24)	$RO_2^{\bullet} + HO_2^{\bullet} \rightarrow ROOH + O_2$	10^{-4}–10^{-6}
(25)	$H_2O_2 + R^{\bullet} \rightarrow RH + HO_2^{\bullet}$	10^{-9}–10^{-10}
(26)	$H^{\bullet} + HO_2^{\bullet} \rightarrow H_2O_2$ $(2{}^{\bullet}OH)$	1–10
(27)	$H^{\bullet} + H_2O_2 \rightarrow {}^{\bullet}OH + H_2O$	0.1–0.01
(28)	${}^{\bullet}OH + HO_2^{\bullet} \rightarrow H_2O + O_2$	1–10
(29)	${}^{\bullet}OH + {}^{\bullet}OH \rightarrow H_2O_2$	10–100
(30)	${}^{\bullet}OH + H_2O_2 \rightarrow HO_2^{\bullet} + H_2O$	0.1–0.01
(31)	$e\text{-}_{aq} + RH \rightarrow$ products	10^{-1}–10^{2}
(32)	$e\text{-}_{aq} + H_2O \rightarrow$ products	10^{1}–10^{2}
(33)	$e\text{-}_{aq} + H^{+}_{aq} \rightarrow H^{\bullet} + H_2O$	10^{2}–10^{3}
(34)	$OH^{-}_{aq} + H^{+}_{aq} \rightarrow H_2O$	10^{2}–10^{3}
(35)	$e\text{-}_{aq} + e\text{-}_{aq} \rightarrow H_2 + 2OH^{-}_{aq}$	10^{2}–10^{3}
(36)	$e\text{-}_{aq} + {}^{\bullet}OH \rightarrow OH^{-}_{aq} + H_2O$	10^{2}–10^{3}
(37)	$e\text{-}_{aq} + {}^{\bullet}H \rightarrow H_2 + OH^{-}_{aq}$	10^{1}–10^{2}
(38)	${}^{\bullet}OH + OH^{-}_{aq} \rightarrow O^{-} + H_2O$	10^{1}–10^{2}

N:	Reactions	W_n/W_3
(39)	$O^{-} + RH \rightarrow R^{\bullet} + OH^{-}_{aq}$	10^{-2}–1
(40)	$H^{+}_{aq} + RH \rightarrow$ products	10^2–10^3
(41)	$OH^{-}_{aq} + RH \rightarrow$ products	1–100
(42)	$H_2O^{+} + H_2O \rightarrow H_3O^{+} (H^{+}_{aq}) + {}^{\bullet}OH$	1–10
$W_3 = 5 \cdot (10^{-6}–10^{-7})$ M/s, $W_1 = 3 \cdot (10^{-13}–10^{-15})$ M/s [7–12]		

Table 3.
Comparison of the rates of elementary reactions for radiolysis of oxygen-containing water contaminated with organic substances, oil products (RH, ROH) in volume to which was added the mass of cut wood chips.

Table 3 clearly shows that the macroradicals formed by ionizing rays on the surface of wood chips (RH, R) are actively involved in fast reactions 10, 11, and 14.

The principal reactions in the radiation-induced conversion of organic waste in water solutions involve the generation of hydroxyl-substituted derivatives and the formation of products resulting from the mutual recombination of organic peroxides, linked by an oxygen bridge. In the presence of small chips, the hydroxyl-substituted and peroxide radicals of organic compounds interact with other hydroxyl-substituted radicals and macro radicals present on the surface of the chips.

The notable increase in the weight of removed organic waste during reservoir irradiation suggests chemical sorption, in addition to physical adsorption, occurring on the surface of wood chips. This dual sorption effect holds promise for enhancing water purification processes aimed at removing oil products and phenol.

After centrifugation, filtration, and evaporation of different samples of drinking water, it was determined that the compositions of the obtained mineral residues are not identical. **Table 4** presents the results of the organoleptic, physical-chemical, and microbiological analyses of drinking water samples.

2.2.1 Shollar drinking water

As shown in **Table 4**, the total salinity of the Shollar drinking water is 450 mg/l. Of this amount, sulfates account for 23–33 mg/l is sulfates, chlorides account for 20 mg/l, calcium carbonates accounts for 142 mg/l, nitrates account for 2 mg/l, and nitrite compounds amount to practically zero.

The samples of Shollar drinking water are characterized by the same indicator, with the following trace element concentrations: Na 18 mg/l, K 10 mg/l, I 0.04 mg/l, Sr 0.4 mg/l, Fe 0.2 mg/l, Mn 0.1 mg/l, Zn 0.01 mg/l, and As 0 mg/l.

The Shollar drinking water samples exhibited consistent trace amounts of radionuclides, with the following concentrations:

- ${}_{11}Na^{22}$: 0.30 Bq/l
- ${}_{19}K^{40}$: 0.10 Bq/l
- ${}_{26}Fe^{60}$: 0.14 Bq/l

№	Parameters	Requirement of standards (2874-82, AZS216-2006)	Factual data					
1	2	3	4	5	6	7	8	9
1.	Transparency, sm	>30*	>30*	>30*				
		>10**			>30	>30	>30	>30
2.	Turbidity, °	≤1.5	0	0	0	0	0	0
3.	Sediment	0 or traces	0	0	0	0	0	0
4.	Color, °	≤20*	0	0				
		≤40**			0	0	0	0
5.	Smell, in points at 20°C	≤2*	0	0				
		≤3**			1	0	1	0
6.	Taste and flavors, in points at 20°C	≤2*	0	0				
		≤2**			1	0	1	0
7.	Active reaction (pH)	6.0–9.0	6.9	6.8	7.7	7.6	7.6 (7.7)	7.6
8.	Dry residue, mg/l	100–1000*	350	450				
		>1000***			950	370	470 (800)	570
9.	Total rigidity, mg-ekv/l	7.0*	2.9	3.7	8.7	3.1	4.1 (7.1)	5.0
10.	Disposable rigidity, mg-ekv /l	<7*	0.8	1.0				
		Not standardized**			2.5	1.2	2.3	1.4
11.	Zn, mg/l	5	0.2	0.01	0	0	0	0
12.	Phenol, mg/l	0.001	0	0	0	0	0	0
13.	Sulfide, mg/l	0.05	0	0	0	0	0	0
14.	I, mg/l	0.001–0.05*	0.05	0.04				
15.	Residual chlorine, mg/l	0–0.5*	0.01	0.01	0	0	0	0
16.	Chlorides, mg/l	350*	18	20	90	33	78 (90)	80
17.	Nitrates, mg/l	45*	4.5	2.0	7.0	4.5	5 (7)	2
18	Cd, mg/l	0*	0	0	0	0	0	0
19	As, mg/l	0.05*	0	0	0.003	0	0	0
20	Fe, mg/l	0.3*	0.14	0.2				
		1**			0.01	0.01	0.01	0.01
21	Pb, mg/l	0.03*	0	0	0	0	0	0
22	Hg, mg/l	0*; 0.001**	0	0	0	0	0	0
23	Nitrites, mg/l	0–0.1*	0	0				
		0.1**			0.3	0.05	0.05 (0.07)	0.02
24	Sulfates, mg/l	500*	10	33	155	40	111 (148)	50

№	Parameters	Requirement of standards (2874-82, AZS216-2006)	Factual data					
1	2	3	4	5	6	7	8	9
25	*Ecoli* number, in 1 liter	≤3*	1	1				
		≤9**			30–40	14–19	22 (35)	15
26	Coli-titr, quantity water in ml wherein found *E. coli*	>300*	1000	1000				
		>100**			30	67	50 (30)	67
27	Pathogens (Salmonella, Yersinia), in 1 liter	0*	0	0	6–10 (*Salm.*)	0 (*Salm.*)	1 (3) (*Salm.*)	1 (*Salm.*)
28	Vibrio cholerae, in 1 liter	0*	0	0	0	0	0	0

**Standard for drinking water [14, 16].*
***Water for technical use [14, 16].*
Note: 4—water taken from the water pipeline, supplying the city of Baku with "Oguz-Gabala" drinking water; 5—water taken from the water pipeline, supplying the city of B aku with "Shollar" drinking water; 6—water taken from Araz River at the crossing of the Beylagan district's territory; 7—water taken from Kur River at the crossing of the Mingaçevir district; 8—water taken from of Rivers stream on the Suqovushan village after intercrossing Kur and Araz Rivers: sample taken from surface part of stream; (sample taken from bottom part of stream); 9—water taken from of Rivers stream after intercrossing (1 km) of Kur and Araz Rivers.

Table 4.
The results of the analysis of water samples taken from the Araz and Kur Rivers before and after their discharge into a single stream (in the Sabirabad region) and from the old and new water supply systems supplying residential areas of Baku with "Shollar" and "Oguz-Gabala" drinking water.

- ${}_{27}Co^{57}$: 0.12 Bq/l
- ${}_{30}Zn^{65}$: 0.05 Bq/l
- ${}_{38}Sr^{91}$: 0.12 Bq/l
- ${}_{63}Eu^{152,154}$: 0.12 Bq/l
- ${}_{88}Ra^{226}$: 0.12 Bq/l

2.2.2 Oguz-Gabala drinking water

As also shown in **Table 4**, the total salinity of the Oguz-Gabala drinking water is 350 mg/l. Of this amount, sulfates account for 10–24 mg/l, chlorides account for 18–19 mg/l, calcium carbonates account for 76 mg/l, nitrates account for 1.3–4.5 mg/l, and nitrites account for 0–0.01 mg/l.

The samples of Oguz-Gabala drinking water are characterized by the same indicator, with the following trace element concentrations: Na 20 mg/l, K 10 mg/l, I 0.05 mg/l, Sr 0.9 mg/l, Fe 0.14 mg/l, Mn 0.03 mg/l, Zn 0.16 mg/l, and As 0.001 mg/l.

The Oguz-Gabala drinking water samples exhibited consistent trace amounts of radionuclides, with the following concentrations:

- $_{11}Na^{22}$: 1.0 Bq/l
- $_{19}K^{40}$: 0.15 Bq/l
- $_{26}Fe^{60}$: 0.48 Bq/l
- $_{27}Co^{57}$: 0.28 Bq/l
- $_{30}Zn^{65}$: 0.05 Bq/l
- $_{38}Sr^{91}$: 0.16 Bq/l
- $_{63}Eu^{152,154}$: 0.2 Bq/l (each)
- $_{88}Ra^{226}$: 0.28 Bq/l

2.2.3 Kur River water

Table 4 also shows the total salinity of Kur River water, which is 370 mg/l. Of this amount, sulfates account for 40 mg/l, chlorides account for 33 mg/l, calcium carbonates account for 165 mg/l, nitrates account for 4.5 mg/l, and nitrite compounds account for 0–0.05 mg/l.

The samples of Kur River water are characterized by the same indicator, with the following trace element concentrations: Na 20 mg/l, K 8 mg/l, I 0.02 mg/l, Sr 1.1 mg/l, Fe 0.01 mg/l, Mn 0 mg/l, Zn 0 mg/l, and As 0 mg/l.

The Kur River water samples exhibited consistent trace amounts of radionuclides, with the following concentrations:

- $_{11}Na^{22}$ 0.48 Bq/l
- $_{19}K^{40}$ 0.16 Bq/l
- $_{26}Fe^{60}$ 0.24 Bq/l
- $_{27}Co^{57}$ 0.16 Bq/l
- $_{30}Zn^{65}$ 0.02 Bq/l
- $_{38}Sr^{91}$ 0.08 Bq/l
- $_{63}Eu^{152,154}$ 0.15, 0.12 Bq/l
- $_{88}Ra^{226}$ 0.12 Bq/l

2.2.4 Araz River water

As shown in **Table 4**, the total salinity of the Araz River water is 950 mg/l. Of this amount, sulfates account for 155 mg/l, chlorides account for 55–90 mg/l, calcium carbonates account for 540 mg/l, nitrates account for 7.0 mg/l, and nitrite compounds account for 0.3 mg/l.

The samples of Araz River water are characterized by the same indicator, with the following trace element concentrations: Na 38 mg/l, K 9 mg/l, I 0.02 mg/l, Sr 4 mg/l, Fe 0.01 mg/l, Mn 0.1 mg/l, Zn 0 mg/l, and As 0.003 mg/l.

The samples of Araz River water exhibited consistent trace amounts of radionuclides, with the following concentrations:

- ${}_{11}Na^{22}$ 0.50 Bq/l
- ${}_{19}K^{40}$ 0.22 Bq/l
- ${}_{26}Fe^{60}$ 0.21 Bq/l
- ${}_{27}Co^{57}$ 0.38 Bq/l
- ${}_{30}Zn^{65}$ 0.04 Bq/l
- ${}_{38}Sr^{91}$ 0.2 Bq/l
- ${}_{63}Eu^{152,154}$ 0.2, 0.2 Bq/l
- ${}_{88}Ra^{226}$ 0.3 Bq/l

The results of the complex organoleptic, physical-chemical, and microbiological analyses show that the waters of the Araz and Kura Rivers, without special treatment, are not potable.

However, the chemical composition and microbiological indicators of the waters transported to Baku through the Shollar and Oguz-Gabala pipelines are in accordance with the requirements of drinking water standards GOST 2874-82, AZS 216-2006, and AZS 282-2007. The pathogenic bacteria *Salmonella*, *Yersinia*, and *Cholera vibrio* are absent in these samples and the amount of the bacteria *Escherichia coli* in these samples is less than the permissible amount in drinking water; thus, these waters are drinkable.

The concentration of radioactive radon in the thermal springs in the foothills along the Dalidag mountain range of the Kalbajar region of Azerbaijan, as well as in the Lower Istisu thermal spring and in the waters of the thermal springs of the Istisu sanatorium in the Kalbajar region, is below the guideline indicator (60 Bq/l). As such, these waters are suitable for use in medical procedures.

The concentration of radioactive radon in the waters of cold and thermal springs located at an altitude of 2385 m above sea level in the western territories of the region exceeds the directive indicator by 100 times, and thus, these waters are undrinkable.

Table 5 lists concentrations of chemical trace elements determined by analytic-chemical and physical-chemical analyses in soil samples taken from the surrounding Araz River's landscape in Nakhchivan city, Sadarak and Ordubad regions of the Nakhchivan Autonomous Republic (AR), and the Beylagan district.

Table 6 lists concentrations of chemical trace elements determined by analytic-chemical and physical-chemical analyses in vegetation samples (trees, leaves, and small branches of bushes, flowers, and grass) taken from the surrounding Araz River's landscape in Nakhchivan city, Sadarak and Ordubad regions of the Nakhchivan AR, and the Beylagan district.

№	Trace elements	In the soil of the Sadarak region, mq/kq	In the soil of the Nakhchivan sity, mq/kq	In the soil of the Ordubad region, mq/kq	In the soil of the Beylagan district, mq/kq
1.	Sulfates	1440	148	180	113
2.	Chlorides	260	106	127	111
3.	Na; K	220; 120	207; 109	193; 500	35; 71
4.	$CaCO_3$	6587	707	1408	920
5.	Br	0.2	0.02	0.05	0.02
6.	Sr	18	12	10	8
7.	Ti; Zr	0; 0	0.1; 0.1	0.2; 0.1	0; 0.03
8.	Fe; P	5.7; 0.7	3.4; 0.1	5.5; 0.3	1.7; 0.05
9.	Zn	0.05	0.02	0.05	0.02
10.	As	0.03	0.005	0.005	0.004

Table 5.
Concentrations of trace elements in soil samples taken from the surrounding Araz landscape.

№	Trace elements	In the soil of the Sadarak region, mq/kq	In the soil of the Nakhchivan sity, mq/kq	In the soil of the Ordubad region, mq/kq	In the soil of the Beylagan district, mq/kq
1.	Sulfates	477	366	380	342
2.	K	6462	5567	9075	5885
3.	Ca	7645	933	1675	1755
4.	Sr	104	97	93	90
5.	Ti; Zr	0.1; 0	0.1; 0.1	0.2; 0.1	0.1; 0.03
6	Fe; P	14; 1.7	8; 0.7	1; 0.9	3; 0.4
7.	Zn	1.1	0.6	1.1	0.6
8.	As	0.007	0	0	0
9.	Mn	18	6.7	15	14

Table 6.
Concentrations of trace elements in vegetation taken from the surrounding Araz River landscape.

The radioactive background and the types of radioactive radiation were detected in the territory of the Nakhchivan AR and the degree of contamination with radionuclides of the Araz River along the territory of the Azerbaijan was determined.

Alpha and beta radiation were not detected from water samples taken from the Araz River.

The measured dose rate from the natural radioactive background (soil) varies in the range of 0.03–0.08 μSv/h (the maximum permissible value is 0.12 μSv/h), and the alpha radiation intensity is within 0–0.11 Bq_{eq}/sm^2.

The measured dose rate from the natural radioactive background detected in the surrounding Araz River's landscape in the territory of the Beylagan and İmishli regions varies in the range of 0.03–0.04 μSv/h and the alpha radiation intensity is 0 Bq_{eq}/sm^2.

The activity (radiation intensity) of Na^{22} and K^{40} isotopes in the water of the Araz River flowing through the territory of the Nakhchivan AR is 0.54 and 0.40 Bq/l, respectively.

The activity of Na^{22} and K^{40} isotopes in the water of the Araz River flowing through the territory of the Beylagan-İmishli regions is 0.35 and 0.20 Bq/l, respectively.

For comparison, the concentration of the Na^{22} isotope (radiation intensity) in the samples of drinking water supplied to enterprises and the population in the cities of Nakhchivan and Baku, respectively, is 0.32 Bq/l and 0.28 Bq/l. Similarly, the concentration of the K40 isotope is 0.18 Bq/l and 0.16 Bq/l, respectively.

Additionally, the dose rate of gamma radiation in the natural radioactive background in these cities varies, with intervals of 0.04–0.07 μSv/h for Nakhchivan and 0.01–0.04 μSv/h for Baku. The intensity of alpha radiation also varies, with intervals of 0–0.01 Bq_{eq}/sm^2 for Nakhchivan and 0–0.04 Bq_{eq}/sm^2 for Baku.

The results of microbiological analyses, as presented in **Table 7**, indicate that the count of pathogenic microorganisms in each liter of water samples taken from the

№	Parameters	Requirements of standards (2874-82)	Actual results				
1	2	3	4	5	6	7	8
1.	Transparency, sm	>30*	>30*				
		>10**	—	>30	>30	>30	>30
2.	Turbidity, °	≤1.5	0	0	0	0	0
3.	Sludge	0 or traces	0	0	0	0	0
4.	Color, °	≤20*	0	0	0	0	0
1	2	3	4	5	6	7	8
5.	Odors, at 20°C, points	≤2*	0	1–2	1–2	1	1
		≤3**					
1	2	3	4	5	6	7	8
6.	Tastes, at 20°C points	≤2*	0	2	2	1	1
		≤2**					
7.	Activity (pH)	6.0–9.0	6.8	7.1	7.2	7.2	7.7
8.	Dry residue, mg/l	100–1000*	166	1645	1495	595	950
9.	Total hardness, mg-eq/l	7.0*	1.6	15	14	5.4	8.7
10.	Hydrocarbonate hardness, mg-eq/l	<7*	—	—	—	—	—
		Are not standardized**	0.7	7.1	6.7	1.4	2.5
11.	Zinc, mg/l	5	0	0.01	0	0.01	0
12.	Phenol, mg/l	0.001	0	0	0	0	0
13.	Sulfides, mg/l	0.05	0	0	0.01	0	0
14.	Iodine, (bromine), mg/l	0.001–0. standardized**	0.02 (Br)	0.15 (Br)	0.2 (Br)	0.05 (Br)	0.02 (Br)
15.	Residual chlorine, mg/l	0–0.5*	0–0.01	0	0	0	0
16.	Chlorides, mg/l	350*	36	55	238	90	90

№	Parameters	Requirements of standards (2874-82)	Actual results				
17.	Nitrates, mg/l	45*	4.4	12.8	4.4	3.4	4.9
18	Cadmium, mg/l	0*	0	0	0	0	0
19	Silver, mq/l	Are not standardized**	0	0-0.01	0	0	0
20	Strontium, mg/l	7.0*	0.7	9	2.5	2.3	7
21	Arsenic, mg/l	0.05*	0	0.005	0.005	0.005	0.003
22	İron, mg/l	0.3* 1**	0.2	0.2	1.0	1.0	0.01
23	Lead, mg/l	0.03*	0	0	0	0	0
24	Mercury, mg/l	0* 0.001**	0	0	0	0	0
25	Nitrites, mg/l	0–0.1* 0.1**	0.05	0.1	0.2	0.1	0.3
26	Sulfates, mg/l	500	18	318	230	68	155
1	2	3	4	5	6	7	8
27	Na, K, W, mg/l	Are not standardized**	23; 9; 0.03	56; 12; 0.1	54; 18; 0.1	58; 22; 0.05	38; 9; 0
28	E.coli, the number of bacteria in 1 l	≤3* ≤9**	3 3	20–26	82–90	20	30–40
29	Coli-titr, amount of water in which 1 bacterium was detected, ml	>300* >100**	333 333	50–40	12–10	50	30
30	The amount of Saprophytic microorganisms in 1 ml	100*	0	8	10	6	10
	Aspergillus (in 1 ml)		0	0	1	0	0-1
	Candida al. (in 1 ml)		0	6-8	8-9	6	6-9
31	Staphylococcus, Strept., (in 1 1);	0	0 (St_{aur}), 0 (St_{epid})	0 (St_{aur}), 16 (St_{epid})	30 (St_{aur}), 42 (St_{epid})	0–1 (St_{aur}), 7–8 (St_{epid})	8 (St_{aur}), 20(St_{epid})
32	Pathogenic microorganisms, including Salmonella (in 1 l)	0	0	0–1	10–13	3–6	6–10

**Standard for drinking water.*
***Spring water and water for technical use.*
Notes: 4—water taken from the water pipeline, supplying the city of Nakhchivan and its population with drinking water; 5—water taken from Araz river at the crossing of the Sadarak region of the Nakhchivan Autonomous Republic with the borders of Turkey and Armenia; 6—water taken from Araz river at the territory of the Sadarak region after the Customs Transition Gate; 7—water taken from Araz river at the intersection of the territory of the Ordubad region with the borders of the IİR and Armenia; 8—water taken from Araz river at the crossing of the Beylagan district's territory.

Table 7.
The results of organoleptic, analytical-chemical, physico-chemical and microbiological analyzes of water samples taken from the Araz River in the Beylagan district, in the territories of the Sadarak and Ordubad districts of the Nakhchivan AR and drinking water samples, supplied to enterprises and population of the city of Nakhchivan.

Araz River at the crossing of the Sadarak region of the Nakhchivan AR exceeds the maximum permissible norm. However, at the intersection of the Ordubad region, there is a significant reduction in the microorganism count in the water of the Araz River. Conversely, in the territory of the Beylagan region, the count of pathogenic microorganisms is observed to increase. This suggests that pollution of the Araz River occurs at the intermediate area situated between Ordubad and Beylagan.

The results of analytic-chemical, physical-chemical, and microbiological analyses show contamination by organic and inorganic emissions of the Araz River when it reaches the territory of the Sadarak region and its further contamination through parts of the territory of neighboring countries located between the Ordubad region of Nakhchivan AR and Beylagan district.

In addition to the increased number of microorganisms, the concentration of chlorides and other halides, nitrites, sulfides, and alkaline-underground metals in the water of the Araz River in the Sadarak region is also relatively increased up to the permissible limits.

The results of the chemical and physical-chemical analyses presented in **Table 7** show that total quantity of inorganic substances in water samples taken from Araz River at the crossing of the Sadarak region of the Nakhchivan AR is three times greater than the maximum permissible limit for drinking water. The total quantity of inorganic compounds decreases along the path of the river to the Ordubad region, naturally diminishing by a factor of 3 due to partial adsorption on silt soils. However, in the Beylagan-Imishli territory, it increases again by 50%. Upon entering the Sadarak region, the concentration of strontium (Sr88) in the river water exceeds the maximum permissible norm by 30%. This concentration naturally decreases along the path of the river to the Ordubad region, due to partial adsorption on the silt soils, reducing by 3–4 times. However, in the Beylagan-Imishli territory, it increases again by 2–3 times.

Moreover, the relative amounts of sulfates and nitrates decline from Sadarak to Ordubad by 4–5 times, but in the Beylagan-Imishli territory, they increase again by 2 times.

Nitrite-containing compounds were detected in the water samples taken from the Araz River, even though the probability of finding nitrite-containing compounds in current waters is generally negligible.

The formation of nitrites from nitrate-containing salts of alkaline earth metals occurs at sufficiently high temperatures (400–500):

$$2KNO_3 - -T, K \rightarrow 2KNO_2 + O_2 \qquad (1)$$

The disproportionation of nitrogen dioxide/nitrogen (IV) oxide/in alkaline media (reaction (2)) also occurs in water (reaction (3)):

$$2NO_2 + 2NaOH \rightarrow NaNO_3 + NaNO_2 + H_2O \qquad (2)$$

$$2NO_2 + H_2O \rightarrow HNO_3 + HNO_2 \qquad (3)$$

However, the absorption of nitrogen dioxide on the surface of river water and reaction (3) proceed at low rates. Therefore, the observed excessive levels of nitrites cannot be solely explained by reaction (3). As mentioned, pollution of the Araz River by anthropogenic emissions, along with high concentrations of inorganic compounds, including chlorine-containing compounds, in the river water and chlorine-containing salts in the soil of Nakhchivan AR, indicate that the formation of nitrites in the water of the Araz River is accompanied by processes involving the mechanism of nitrite formation in chlorine-containing waters [18]:

$$NO_2 + ClO \rightarrow ClNO_3 \quad (4)$$

$$ClNO_3 + H_2O \rightarrow HOCl + HNO_3 \quad (5)$$

$$4Mg\ (Ca) + 10HNO_3/\text{slightly acidic}/ \rightarrow 4(Ca)\ Mg(NO_3)_2 + N_2O\ (NO) + 5H_2O \quad (6)$$

$$3Cu + 8HNO_3/\text{slightly acidic}/ \rightarrow 3Cu(NO_3)_2 + 2NO_2 + 4H_2O \quad (7)$$

$$N_2O + NO \rightarrow N_2 + NO_2 \quad (8)$$

$$NO_2 + NO_2 \rightarrow N_2O_4 \quad (9)$$

$$N_2O_4 + H_2O \rightarrow HNO_2 + HNO_3 \quad (10)$$

$$2HNO_2 + Mg(HCO_3)_2 \rightarrow Mg(NO_2)_2 + 2H_2CO_3 \quad (11)$$

As evident from the outlined mechanism, molecules of atmospheric nitrogen dioxide (NO_2) and $ClNO_3$, NO_2, N_2O_4 molecules, formed in reactions (4) and (6)-(9), react with water to produce nitric acid (HNO_3) molecules in a slightly acidic environment. Additionally, the minerals present in water contribute to the formation of nitrogen oxides through slow reactions (6) and (7).

Nitrite acid molecules, resulting from reactions (3) and (10) of nitrogen oxides with water, facilitate the formation of small concentrations of stable nitrite compounds in the water of the Araz River via reaction (11) with hydrocarbons.

The chemical composition and the sensory, physical-chemical, and microbiological characteristics of water samples taken from Araz River in Sadarak and Ordubad districts of the Nakhchivan AR and in the Beylagan-Imishli region do not meet the requirements of the GOST 2874-82, AZS 216-2006, and AZS 282-2007 standards for drinking water.

The results of experiments involving the disinfection of water samples from the Araz River using various concentrations of calcium perchlorate show that treatment with 1 mg of calcium perchlorate (a solution containing 1 mg of calcium perchlorate with 60% active chlorine in 5 ml of distilled water added to 1 liter of river water) resulted in a reduction of approximately 10% in the microorganism count after 1 hour. No residual chlorine was detected in the river water sample.

Increasing the concentration to 2 mg of calcium perchlorate led to a more significant decrease in the microorganism count, approximately by 2–3 times, after 1 hour. Similarly, no residual chlorine was observed in the river water sample.

Sterilization experiments using 5 mg of calcium perchlorate resulted in the complete elimination of microorganisms and the absence of residual chlorine after 1 hour in the river water samples.

Lastly, experiments with 10 mg of calcium perchlorate demonstrated the absence of microorganisms after 1 hour, with residual chlorine concentrations below the maximum permissible value (<0.5 mg/l) for drinking water.

Higher amounts of calcium perchlorate along with the total clearance of all microorganisms in contaminated river water leads to the accumulation of residual chlorine in quantities greater than the maximum permissible values for drinking water. The amount of calcium perchlorate in the range of 5–10 mg (optimal value) is sufficient for the total clearance of all microorganisms in contaminated water taken from Araz River and the accumulated amount of residual chlorine does not exceed the maximum permissible value.

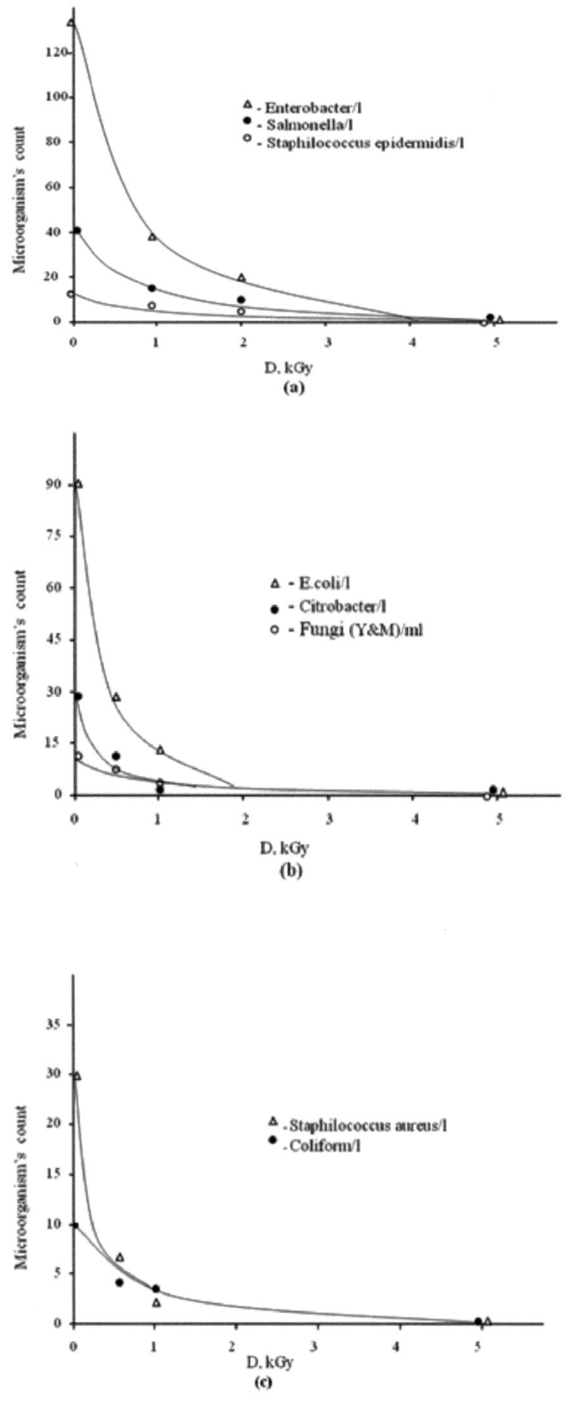

Figure 1.
Reducing the microorganism's count in water samples taken from the Araz River, depending on the value of the absorbed dose of ionizing radiation.

Figure 1 shows the results of disinfection experiments involving ionizing radiation of water samples taken from the Araz River.

As shown in **Figure 1**, the irradiation by absorbed dose 5 kGy of ionizing radiation ensures complete extermination of all microorganisms in water samples taken from the Araz River, which were contaminated with organic emissions and microorganisms.

Thus, both chemical (chlorination with perchlorates) and radiolytic (irradiation with ionizing radiation) methods of disinfection of the contaminated water of the Araz River are highly efficient methods of purification.

Based on the experimentally obtained results and considering the observed process of slow natural self-purification of water, particularly the partial adsorption of inorganic pollutants on the silt soils along the Araz River on its way to the territory of the Beylagan-Imishli region, it becomes evident that there is potential to implement deep purification of the Araz River water in Nakhchivan AR.

This can be achieved by passing the river water through a thermally regenerated and periodically updated large adsorbent sand filter after chemical or radiolytic disinfection of the river's water. This approach would help enhance the quality of the water supply in Nakhchivan AR, ensuring safer and cleaner drinking water for the population.

3. Conclusions

In this chapter, we analyzed the sensory, physical-chemical, and microbiological properties of water samples collected from drinking water sources to determine the distributions of heavy metals, trace natural radionuclides, and other inorganic components. The chemical composition and microbiological parameters of waters transported to Baku through the Shollar and Oguz-Gabala pipelines adhere to the drinking water standards 2874-82, AZS 216-2006, and AZS 282-2007. Thus, these waters meet the criteria for safe consumption.

The radioactive radon concentration in the thermal springs of the Istisu sanatorium, located in the Kalbajar region of Azerbaijan, remains below the recommended threshold (60 Bq/l). These waters are deemed suitable for medical applications as prescribed by healthcare professionals. Conversely, the radioactive radon levels in both cold and thermal springs situated at an altitude of 2385 m above sea level in the western territories of the region exceed the directive indicator by 100 times, rendering them unsuitable for drinking.

In conjunction with measuring background radiation in adjacent areas along the Araz River watercourse, we conducted an extensive analysis of the river's polluted water.

During the irradiation of contaminated water reservoirs, notably high values of removed petroleum products have been observed. This observation suggests the presence of chemical sorption alongside physical adsorption on the surface of wood chips added to the reservoir. This crucial phenomenon should be considered in the radiation purification of water contaminated with various organic compounds, crude oil, and phenol. Moreover, we identified the optimal parameters for chemical and radiological purification of water polluted by harmful emissions and pathogenic microorganisms. Additionally, we discussed the mechanisms of chemical and radiation-chemical processes involved in water purification.

Conflict of interest

The author declared no conflict of interest.

Author details

Khagani Mammadov
Institute of Radiation Problems of Azerbaijan National Academy of Sciences, Baku, Azerbaijan

*Address all correspondence to: xagani06@mail.ru

References

[1] Piskarev IM. Purification water of open reservoirs by chain reactions of hydroxyl radicals [thesis]. M: MGU-NIIYAF-7; 2004. p. 4

[2] Roginskiy VA. Phenolic Antioxidants. Reactivity and Efficiency. M: Nauka; 1988. 247 p

[3] Neta P, Grodkowski J. Rate constants for reactions of phenoxyl radicals in solution. The Journal of Physical Chemistry. 2005;**34**(1): 109-199

[4] Kaushansky DA, Kuzin AM. Radiation-Biological Technology. M: Energoatomizdat; 1984. 152 p

[5] Lebedeva NE, Gorbatova EN, Golovkina TV. Screening method for substances acting in ultra-low concentrations. Radiation Biology. Radioecology. 2003;**43**(3): 282-286

[6] Levin MN, Negrobov OP, Gitlin VR, Selivanova OV, Ivanova OA, Radon. Tutorial. Voronej: Publishing and Printing Center of Voronezh University; 2008. 40 p

[7] Mogilevich MM, Pliss EM. Oxidation and Oxidative Polymerization of Unsaturated Compounds. Moscow: Khimiya; 1990. p. 240

[8] Mamedov K, Aliyev A, Mamedova N, Badalova A. Radiolytic decomposition of TCDD in water solutions and in damp plant products. Problems of Atomic Science and Technology. 2015;**96**(2):48-51. EID: 2-s2.0-84928675513

[9] Orlov DS, Sadovnikova LK, Sukhanova NI. Chemistry of Soils. M: Higher School; 2005. 558 p

[10] Mukha VD, Kartamyshev NI, Mukha DV. Agricultural soil science. M: Kolos; 2003. 528 p

[11] Mammadov K. Effect of natural radionuclides on the processes occur in the environmental objects. Studying the cleaning methods of water contaminated with organic emissions and the methods for cleaning of soil contaminated with heavy metals and radionuclides. International Journal of Applied Chemistry. 2020;**7**(01):18-24. DOI: 10.14445/23939133/IJAC-V7I1P105

[12] Piskunov AS. Methods of Agrochemical Research. Moscow: Kolos; 2004. 312 p

[13] Pikaev A.K. Modern radiation chemistry. Radiolysis of gases and liquids. M.: Nauka; 1986. 440 p.

[14] Mamedov KF. Research of Distribution of Heavy Metals and Radioactive Elements in Sources of Drinking Water, Soil and Grass Cover of Azerbaijan Regions. Petrozavodsk: Tsifra; 2020. 43 p. ISBN: 978-5-901619-31-5

[15] Mamedov KF. Radiolysis and photolysis of water solutions of phenol. European Researcher. Series A. 2014;**7-1**: 1216-1236. DOI: 10.13187/issn.2219-8229

[16] Mammadov KF. Determining the sources of contamination of the waters discharged into the Araz river with inorganic xenobiotics and pathogenic microorganisms. Pedagogy University News. Mathematics and Natural Sciences Series. 2020;**3**:168-185. ISSN: 2520-2049

[17] Mamedov KF. Radiolytic conversion of natural toxins in contaminated plant products and aqueous solutions.

Scientific Journal "Science Rise". 2014; **4**(2):116-121. DOI: 10.15587/2313-8416.2014.29402

[18] Mammadov XF. Kinetics of formation of nitrites in chlorinated waters of closed water sources. Scientific Works – Fundamental Sciences. Chemistry. 2012;**1**(41):163-165. ISSN: 1815-1779

[19] Mahmudov HM, Kuliyeva UA, Kerimov VK, Kurbanov MA. Water radiolysis on the surface of Al_2O_3 nano-catalyst. European Journal of Analytical and Applied Chemistry. 2015;**1**:58-62. DOI: 10.1155/2021/9493765

[20] Mammadov K, Shiraliyeva H, Aliyeva U, Gulamirov A. Research into causes of rise in the level of beta radiation in atmospheric air on the territory of Azerbaijan. Chemical Problems. 2018;**3**:376-380. DOI: 10.32737/2221-8688/-2018-3-376-380/

[21] Mamedov KF, Aliev AG, Mamedova NA, Badalova AR. Radiolysis of 2,3,7,8-tetrachlorodibenzo-p-dioxin in water solutions. Journal of Physical Chemistry A. 2015;**89**(9):1707-1709. DOI: 10.1134/S0036024415090228

[22] Sakumoto A, Miyata T. Treatment of waste water by a combined technique of radiation and conventional method. Radiation Physics and Chemistry. 1984; **24-1**:99-115

Chapter 2

Improving the Quality of Drinking Water by Raising the pH Levels Using a Natural Na_2SiO_3 Physical Field

Hesham Mohamed Abdal-Salam Yehia
and Said Mahmoud Said

Abstract

The sustainable development goals (SDGs) established by the United Nations aim to ensure access to clean water and sanitation for all. One way to improve the quality of drinking water is by using a natural substance made up of environmental elements. Therefore, it was necessary to maintain the degree of alkalinity of the water, starting from 7 pH level. While previous studies have focused on raising the pH levels of water by adding alkaline compounds, recent research has suggested using baking soda, which has a pH of approximately 9. However, there are several unfavorable health effects associated with adding baking soda to water, including hypokalemia, hypochloremia, high blood salt levels, deterioration of kidney illness, deterioration of heart failure. In this study, a substance made from silica using the thermal fusion method has been developed by the authors. This substance raises the pH of water without the need for additional components, using a physical field of 80 cm. The outcomes of the study demonstrated the efficiency of this natural chemical in improving water quality. Using a natural silica physical field to raise the pH level of water is a safe and effective method without the need for additional components that can have unfavorable health effects. The results recommended that water at 7 degrees is completely pure distilled water suitable for drinking in normal ways and indicated that alkaline water with a pH between 8.5 and 10 degrees is beneficial for people suffering from irritable bowel syndrome. Moreover, improving drinking water quality is essential for achieving the SDGs, specifically SDG 6, which aims to ensure access to clean water and sanitation for all. By utilizing natural methods to improve water quality, we can work toward achieving a more sustainable and equitable future.

Keywords: SDGs, drinking water, pH, silica, biotechnology, water purification, Na_2SiO_3

1. Introduction

The sustainable development goals (SDGs) established by the United Nations are a set of 17 goals aimed at promoting sustainable development worldwide. One of the key goals of the SDGs is to ensure access to clean water and sanitation for all [1]. Access to clean water and sanitation is a fundamental human right and is essential for human health and well-being. However, millions of people worldwide still lack access to clean water and sanitation facilities, which can lead to the spread of disease and other health problems [2].

To address this issue, the SDGs include a specific goal, SDG 6, which aims to ensure availability and sustainable management of water and sanitation for all. This goal includes targets such as achieving universal and equitable access to safe and affordable drinking water and sanitation facilities, improving water quality, and increasing water-use efficiency. Achieving SDG 6 requires a comprehensive approach that involves improving infrastructure and technology, promoting sustainable water management practices, and increasing awareness and education about the importance of clean water and sanitation [3].

Governments, businesses, and individuals all have a role to play in achieving SDG 6. Governments can invest in water infrastructure projects and implement policies to promote sustainable water management practices. Businesses can incorporate sustainable water management practices into their operations and products. Individuals can conserve water and support initiatives that promote clean water and sanitation for all [4].

Water with a pH of approximately 7 is considered neutral, whereas water with a pH level higher than 7 is considered alkaline, as shown in **Figure 1**. The ability of water to neutralize acids is referred to as its alkalinity [5]. The pH scale ranges from 0 to 14, with 7 being neutral and everything in between being either acidic or alkaline. While previous studies have focused on raising the pH levels of water by adding alkaline compounds, recent research has suggested using baking soda, which has a pH of approximately 9. However, there are several unfavorable health effects associated with adding baking soda to water, including hypokalemia, hypochloremia, high blood salt levels, deterioration of kidney illness, deterioration of heart failure, muscle weakness and spasm, and increased production of stomach acid [6]. Studies also determine that alkaline water reduces acid in the stomach, which helps kill pathogens and bacteria in the stomach. Moreover, an increase in alkalinity in the body may irritate

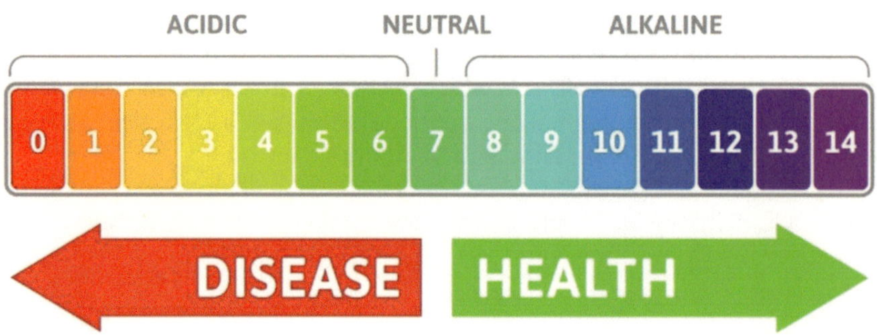

Figure 1.
Water pH scale.

the digestive capacity of the stomach, which leads to stomach-related health problems such as upset stomach, bloating, and acidity [6].

The authors of a recent study have developed a substance made from silica using the thermal fusion method, which can effectively raise the pH of water without requiring additional components. This natural chemical utilizes a physical field of 80 cm to achieve a pH level of 8.5 from 80 cm. The study aimed to analyze the effectiveness of this substance in both its solid and liquid forms, with the primary objective of raising the pH of water without causing any negative health effects.

2. Literature review

Water is a vital component of life, as it is necessary for the survival of human beings [7]. However, not all water is created equal, and the quality of drinking water can vary significantly depending on its source. One way to improve the quality of drinking water is by raising its pH levels using a natural silica physical field. Silica is a naturally occurring mineral that is found in rocks, sand, and soil [8]. It is also found in certain types of water, such as spring water and artesian well water. Silica has been found to have a positive effect on the pH levels of water when it is exposed to a natural silica physical field [9]. A natural silica physical field is created when water flows through layers of silica-rich rocks. The movement of water through these rocks creates a natural physical field that can raise the pH levels of water. This process is called silicification, and it is a natural way to improve the quality of drinking water.

When water has a low pH level, it can be corrosive and cause damage to pipes and plumbing fixtures. It can also cause a metallic taste in the water [10]. On the other hand, water with a high pH level can taste bitter and have a soapy feel [11]. By raising the pH levels of drinking water using a natural silica physical field, the water can have a more balanced and pleasant taste.

In addition to improving the taste of drinking water, raising its pH levels using a natural silica physical field can also have health benefits. Alkaline water has been found to have antioxidant properties that can help neutralize free radicals in the body. Free radicals are unstable molecules that can cause damage to cells and contribute to the development of chronic diseases. Drinking water with a high pH level has also been found to have anti-inflammatory properties that can help reduce inflammation in the body [12]. Inflammation is a natural response of the immune system, but chronic inflammation can lead to a variety of health problems, including heart disease, diabetes, and cancer.

Raising the pH levels of drinking water using a natural silica physical field can also improve the absorption of minerals in the body. When water has a low pH level, it can interfere with the absorption of essential minerals such as calcium, magnesium, and potassium. Alkaline water, on the other hand, can help improve the absorption of these minerals, which are essential for bone health, muscle function, and overall health. In addition to its health benefits, raising the pH levels of drinking water using a natural silica physical field can also have environmental benefits [13]. Improving the quality of drinking water can reduce the need for bottled water, which can have a significant environmental impact. Bottled water production requires a lot of energy and resources, and plastic bottles can take hundreds of years to decompose [14].

Overall, raising the pH levels of drinking water using a natural silica physical field is a natural and effective way to improve the quality of drinking water. It can improve the taste and health benefits of the water, while also reducing the environmental impact of bottled water. While more research is needed to fully understand the health benefits of alkaline water, the evidence suggests that it is a safe and effective way to improve the quality of drinking water.

3. Materials and methods

To improve the quality of drinking water by raising its pH levels using a natural silica physical field, a substance made from silica using the thermal fusion method was utilized. This substance was developed by the authors of a recent study and was tested in both its solid and liquid forms. To conduct the study, the pH levels of the water samples were measured using a pH meter before and after exposure to the natural silica physical field. The distance between the water samples and the natural silica physical field was also recorded. The study included both laboratory experiments and field tests. In the laboratory, water samples were placed in containers and exposed to the natural silica physical field for a set amount of time. In the field, the natural silica physical field was created by placing the water source near layers of silica-rich [6].

The main component of this experiment is Na_2SiO_3 as, a chemical compound composed of two sodium (Na) atoms, one silicon (Si) atom, and three oxygen (O) atoms. It is also known as sodium silicate. Sodium silicate solutions can be acidic, neutral, or alkaline, depending on the ratio of sodium oxide (Na_2O) to silicon dioxide (SiO_2) in the compound, as depicted in **Table 1**.

The composite material was extracted based on **Table 1**, as illustrated in **Figure 2**, which displays the chemical makeup of the active elements in silica.

Item	Value
% Na_2O	10
% SiO_2	20
% H_2O	—
pH	pH ≈ 10.43
Density	Density 1000 g/m^3 liquid
Melting point	Around 1400°C 1700°C
Boiling point	2230°C

Table 1.
The properties of sodium silicate solutions Na_2SiO_3.

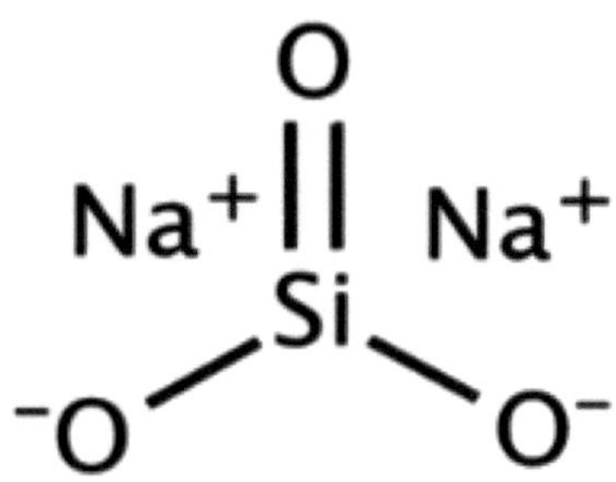

Figure 2.
Sodium silicate structure.

Figure 3.
Sodium silicate in liquid form.

Figure 3 shows the compound in its liquid form that was used in the current research experiment.

4. Experiment

The present research experiment was carried out in collaboration with the Drinking Water Authority in Sharkia Governorate, Egypt, as part of a cooperative agreement between the Nanotechnology Research Laboratory of HST Company and the Sharkia Drinking Water facility. The experiment spanned one month, during which samples were collected at regular intervals to verify and maintain the accuracy of the results. These samples were analyzed in a laboratory to measure the pH and physical characteristics of the water quality, ensuring its safety for human consumption.

4.1 ASTM E70 standard test method

After the natural silica physical field was created by passing water through layers of silica-rich rocks, the solution was examined for its ability to raise the pH value. The pH value was determined using the ASTM E70 standard test method, which was utilized to calibrate the instrument and measure the pH of the aqueous solution [15]. The ASTM E70 standard test method is a widely accepted and recognized protocol for measuring the pH of water and other aqueous solutions. The test was conducted under controlled conditions, with a temperature of 22°C and a humidity level of 73%. The standard test method involved using a pH meter to measure the hydrogen ion concentration in the solution [16]. The pH meter was calibrated using buffer solutions with known pH values before the measurements were taken. The pH meter was then used to measure the pH of the aqueous solution, and the results were recorded. By utilizing the ASTM E70 standard test method, the researchers were able to accurately

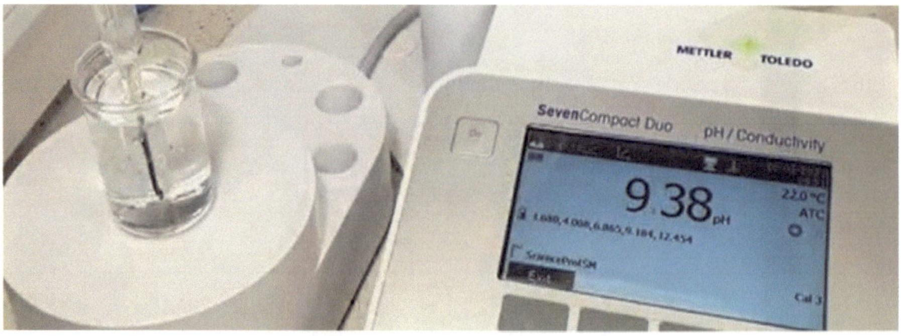

Figure 4.
The ASTM E70 standard test.

measure the pH of the solution and verify its ability to increase the pH value of water to the desired level, as seen in **Figure 4**. The controlled test conditions ensured the accuracy and reliability of the results.

5. Results

The results presented in this section highlight the promising potential of the material extracted from silica in raising the pH levels of water and suggest that further research in this area is warranted.

5.1 Determine whether the pH values

To determine whether the pH values were successfully increased, the following evaluation criteria were implemented:

1. A 100 ml beaker was filled with 50 ml of tap water and covered with a stretch cover.

2. The pH meter was calibrated using buffer solutions with known pH values (4, 7.02, and 10 pH units) before each pH measurement.

3. The water temperature and pH were measured.

4. The 100 ml beaker containing the 50 ml water was placed on the sample shown in **Figure 2** to measure the effect of the natural silica physical field on raising the pH value.

5. The temperature and pH of the water were measured every hour for a total of five hours. The beaker was re-covered with the stretch cover after each measurement, and no other interference with the water was made except for placing it on the base that contained the manufactured material in this study. The results of this experiment are illustrated in **Figure 5**.

The results shown in **Figure 5** indicate a clear and significant effect of the material extracted from silica in raising the pH levels of water. The data presented in the

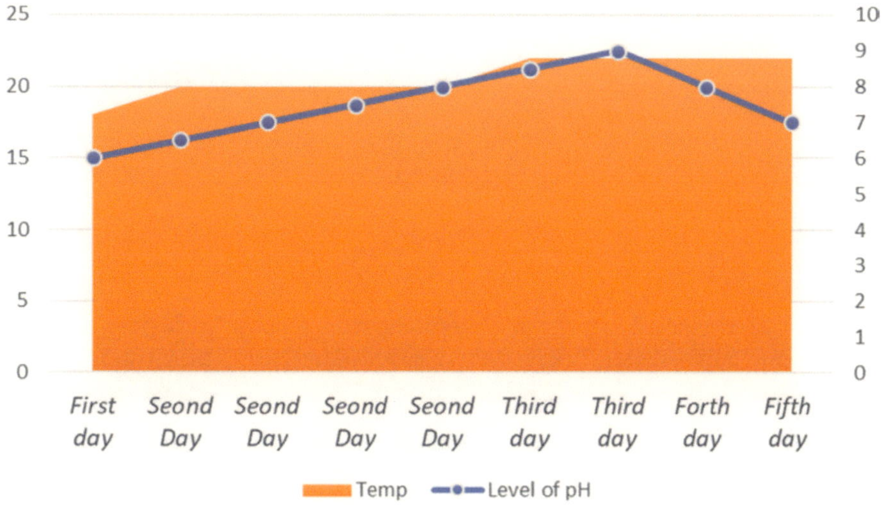

Figure 5.
The results of the pH value over time (Trial 1).

Figure 5 demonstrate a consistent increase in pH levels over time, indicating that the substance has a long-lasting effect on the pH of water. The effectiveness of the extracted material in raising the pH levels can be attributed to the natural silica physical field created by passing water through layers of silica-rich. This physical field is believed to alter the ionization state of the water, resulting in an increase in the concentration of hydroxide ions, and hence, a rise in the pH levels. The findings of this study are particularly significant for water treatment applications, as they demonstrate the potential of natural silica physical fields to increase the pH levels of water, thereby making it safer for human consumption. The use of natural silica physical fields could provide a more sustainable and cost-effective alternative to traditional water treatment methods that involve the use of chemicals and artificial pH stabilizers.

To ensure the accuracy and validity of the results, the tests were repeated under the same conditions as the first experiment in the laboratories of the Drinking Water and Wastewater Authority in Egypt after a 2-week interval. The results of the second experiment confirmed that the pH levels increased when the substance was used in a natural silica physical field, as depicted in **Figure 6**. The repetition of the experiment under the same conditions helped to confirm the reliability of the initial results and validate the effectiveness of the natural silica physical field in raising the pH of water.

The results presented in **Figures 5** and **6** provide promising evidence of the potential of the material extracted from silica to raise the pH levels of water. Further research in this area could lead to the development of more sustainable and cost-effective water treatment methods that leverage the unique properties of silica to improve the quality and safety of our water resources. These results agree with recent scientific studies that have confirmed the effectiveness of natural silica physical fields in raising the pH levels of water [17]. These studies have shown that the dissolved silica in water can form complexes with metal ions, resulting in the precipitation of metal hydroxides. This process results in the removal of heavy metals and other impurities from water and the increase in the pH levels [18].

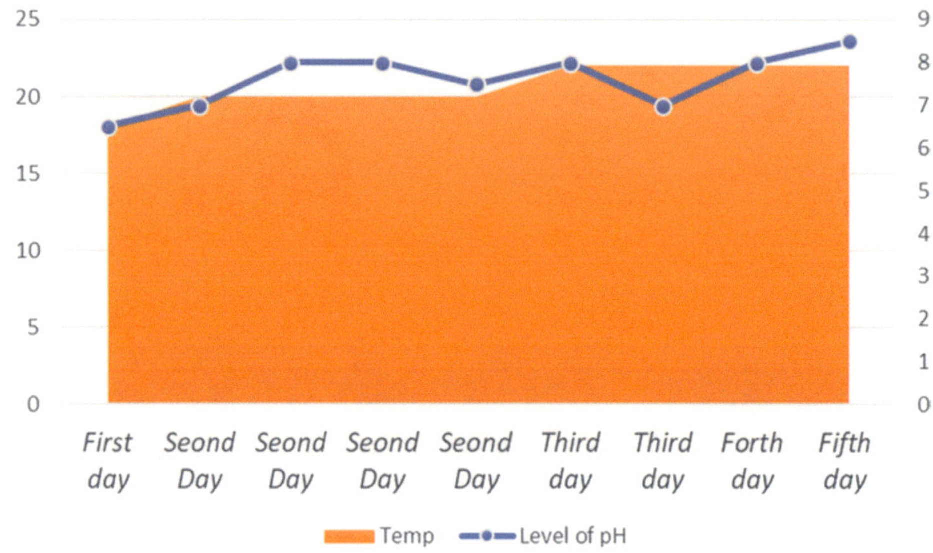

Figure 6.
The results of the pH value over time (Trial 2).

5.2 Radiation test

Radiation testing is a process used to detect, measure, and evaluate the levels of ionizing radiation in a particular material or environment. **Table 2** summarizes the evaluation method and procedures used to determine whether the manufactured material (Silica SiO_2) contains radioactive or nonradioactive elements. The goal of this evaluation was to ensure that the material was safe for human use and did not pose a risk of radiation exposure. The evaluation process involved several steps, including sample preparation, measurement of radioactivity, and analysis of the results. The samples were prepared by grinding the material into a fine powder and then placing it in a sealed container to prevent contamination. The radioactivity of the samples was measured using a gamma spectrometer, which is a device that can detect and measure gamma radiation emitted by radioactive isotopes. The gamma spectrometer was calibrated using a set of standard reference materials before the measurements were taken. The results of the radioactivity measurements were then analyzed to determine whether any radioactive isotopes were present in the material. If radioactive isotopes were detected, their concentrations were calculated and compared to the limits set by international safety standards to determine whether the material was safe for human use.

Device name	Type	Manufacturer
Detector	NaI (TI) scintillation detector 3 × 3 Model NAIS-3 × 3 sodium iodide Scintillation detector is a high-efficiency Scintillation detector featuring a 3 × 3 in. NaI(Tl) crystal	Canberra
Analyzer	Multi-channel analyzer	Ortec
Detector	Contamination monitor LB122	Berthold

Table 2.
Radiation evaluation method and procedures.

Based on the previous results, it is evident that Na_2SiO_3 can effectively raise the pH levels of water, and it was safe according to the radiation test results in **Table 2**. These findings are consistent with the conclusions of Du et al. [19], who demonstrated that precipitated silica could be produced using Na_2SiO_3 as a precursor and HCl as a precipitant under specific conditions. What sets our study apart from previous research is that we were able to raise the pH of the water using a natural silica physical field without any direct chemical interference with the water. This approach is unique because it helps to protect the water from any potential chemical effects that may arise from other pH-raising methods [20]. Our findings suggest that natural silica physical fields have significant potential for pH control and water treatment applications. The use of natural silica physical fields offers a more sustainable and environmentally friendly alternative to traditional chemical-based methods for pH control.

6. Conclusion

The use of natural silica physical fields for pH control and water treatment has gained increasing attention in recent years due to its potential for a more sustainable and cost-effective alternative to traditional chemical-based methods. The unique properties of silica, such as its high surface area, charge density, and porosity, make it an ideal material for water treatment applications. The objective of the current research was to investigate a natural substance derived from environmental elements. This substance, which was extracted from silica using the thermal fusion method, was developed by the authors to enhance the pH level of water with an 80 cm physical field. Notably, no additional components were added to the water during the experimentation process. The study included both laboratory experiments and field tests. In the laboratory, water samples were placed in containers and exposed to the natural silica physical field for a set amount of time. In the field, the natural silica physical field was created by placing the water source near layers of silica-rich. The results of our study provide further evidence of the potential of natural silica physical fields for pH control and water treatment. The unique approach taken in this study demonstrates the effectiveness of this method for raising the pH levels of water without any direct chemical interference. These findings have important implications for the development of more sustainable and environmentally friendly water treatment methods.

Acknowledgements

The authors express their gratitude and appreciation to the Drinking Water Company in Sharkia Governorate, Arab Republic of Egypt, for their generous support and assistance in conducting the required analyses and testing. Additionally, the authors are pleased to acknowledge the National Research Center and the National Institute for Standardization for conducting the necessary evaluations to verify the efficacy of the manufactured material in enhancing pH levels in water. The support and assistance provided by these entities were essential in enabling the successful execution of the study and the attainment of the research objectives. The authors are grateful for their diligent efforts and expertise, which contributed significantly to the study's reliability and validity.

Conflict of interest

The authors declare no conflict of interest.

Author details

Hesham Mohamed Abdal-Salam Yehia* and Said Mahmoud Said
Department of Biotechnology, HST Company, Cairo, Egypt

*Address all correspondence to: heshamyehia@gmail.com

References

[1] Hák T, Janoušková S, Moldan B. Sustainable development goals: a need for relevant indicators. Ecological Indicators. 2016;**60**:565-573

[2] Vohidovna OD, Olimovich KK. Medical and social significance of water supply, sanitation and hygiene in human activity. Nexus: Journal of Advances Studies of Engineering Science. 2022;**1**(1):20-24

[3] Yadav M, Kumar BS, editors. Can Asia accomplish the sustainable development goal for water and sanitation by 2030? In: Handbook of Research on Sustainable Consumption and Production for Greener Economies. IGI Global; 2023. pp. 198-212

[4] Ali KA, Hassan ME, Elnashar MMM. Development of functionalized carrageenan, chitosan and alginate as polymeric chelating ligands for water softening. International Journal of Environmental Science and Technology. 2017;**14**:2009-2014

[5] Yehia HMA-S, Said SM. Drinking water treatment: pH adjustment using natural physical field. Journal of Biosciences and Medicines. 2021;**9**(6):55-66

[6] Yehia HMA-S, Said SM, Ragb NGER. Silica nanoparticles in targeted human cancer therapy. Medical Journal of Cell Biology. 2022;**10**(1):7-17

[7] Remick KA, Helmann JD. The elements of life: a biocentric tour of the periodic table. In: Advances in Microbial Physiology. Vol. 82. Academic Press; 2023. pp. 1-127

[8] Akter T et al. Water quality index for measuring drinking water quality in rural Bangladesh: a cross-sectional study. Journal of Health, Population and Nutrition. 2016;**35**:1-12

[9] Prasad AN et al. Smart water quality monitoring system. In: 2015 2nd Asia-Pacific World Congress on Computer Science and Engineering (APWC on CSE). IEEE, 2015

[10] McFarland ML, Provin TL, Boellstorff DE. Drinking Water Problems: Corrosion. USA: Texas AgriLife Extension Service and USDA National Institute of Food and Agriculture; 2017

[11] Islam R et al. Assessment of pH and total dissolved substances (TDS) in the commercially available bottled drinking water. IOSR Journal of Nursing and health Science. 2017;**6**(5):35-40

[12] Siczek K et al. Synthesis and evaluation of anti-inflammatory properties of silver nanoparticle suspensions in experimental colitis in mice. Chemical Biology & Drug Design. 2017;**89**(4):538-547

[13] Larrosa M et al. Anti-inflammatory properties of a pomegranate extract and its metabolite urolithin-A in a colitis rat model and the effect of colon inflammation on phenolic metabolism. The Journal of Nutritional Biochemistry. 2010;**21**(8):717-725

[14] Ullah I et al. An osteogenic, antibacterial, and anti-inflammatory nanocomposite hydrogel platform to accelerate bone reconstruction. Journal of Materials Chemistry B. 2023;**11**:5830-5845

[15] Neumann J et al. Production and characterization of a new quality

pyrolysis oil, char and syngas from digestate – introducing the thermo-catalytic reforming process. Journal of Analytical and Applied Pyrolysis. 2015;**113**:137-142

[16] Jourabchi SA, Gan S, Ng HK. Comparison of conventional and fast pyrolysis for the production of Jatropha curcas bio-oil. Applied Thermal Engineering. 2016;**99**:160-168

[17] Lemessa F et al. Assessment of the impact of industrial wastewater on the water quality of rivers around the Bole Lemi Industrial Park (BLIP), Ethiopia. Sustainability. 2023;**15**(5):4290

[18] Mpewo M, Kizza-Nkambwe S Jr, Kasima S. Heavy metal and metalloid concentrations in agricultural communities around steel and iron industries in Uganda: implications for future food systems. Environmental Pollutants and Bioavailability. 2023;**35**(1):2226344

[19] Du C-M et al. Effects of acid and Na_2SiO_3 modification on the dissolution behavior of $2CaO \cdot SiO_2$-$3CaO \cdot P_2O_5$ solid solution in aqueous solutions. ISIJ International. 2016;**56**(8):1436-1444

[20] Suriyaprabha R et al. Silica nanoparticles for increased silica availability in maize (Zea mays. L) seeds under hydroponic conditions. Current Nanoscience. 2012;**8**(6):902-908

Chapter 3

Alternative Challenges to Water Quality Improvement and Conservation of Freshwater Bivalves

Hwan-Seok Choi and Baik-Ho Kim

Abstract

The response surface method has been used by researchers to optimize the conditions of the organic filtration reaction using shellfish. They analyzed the data by setting the size, flow rate, and filtration rate of shellfish as independent variables using the central composite design program in the software. The highest filtration rate was obtained at a flow rate of 24 L/h and a shell size of 10.6–11.4 cm with a residence time of 22.7 hr., while the lowest filtration rate was obtained at a flow rate of 48 L/h with a residence time of 10.4 hr. The excessively fast and low flow rate decreased the filtration rate owing to the increase in the residence time. The researchers suggest that for validation purposes, water velocities in a further range, along with an increase in retention time, should be assessed. They used a central composite design to optimize Sinanodonta woodiana filtration rates and feces production to identify the key factors and optimal conditions. The proposed response surface model illustrated the interactions among the variables on mussel filtration rate. The experimental filtration rate of 4.47 ± 1.82 L/mussel/h corresponded to the predicted value of 8.4 L/mussel/h, which validated the practicability of this optimization strategy.

Keywords: freshwater bivalves, water quality, filtration rates, ecological restoration, response surface methodology (RSM)

1. Introduction

Freshwater bivalve species and populations are declining globally due to multiple factors, such as habitat destruction, pollution, loss of host fish for larval development, and invasive species outbreaks [1, 2]. Stream ecologists aim to improve the water quality and conserve mussel species diversity, particularly in the context of climate change. To contribute to these efforts, our laboratory and field studies aimed to understand the ecological role of mussels, including their filtration rates, feces and pseudo-feces production, mortality, impact on water quality, and interactions with other biota [3–5]. However, our findings indicate that uncontrolled introduction of mussels to field conditions is not an effective method for improving water quality or conserving species diversity, as it may result in increased ammonia tolerance and mortality.

Freshwater unionid mussels, such as Anodonta and Unio species, play a vital role in freshwater ecosystems through the biodeposition of feces and pseudo-feces [6], filter feeding on seston, and linking pelagic and benthic food webs [7, 8]. And, the growth and reproduction of juvenile and adult mussels are influenced by the type and depth of stream sediment as well as the availability of host fish for larval development [9, 10].

In temperate regions, the nutrient content in water systems can quickly increase, leading to the growth of cyanobacteria and diatoms, causing water quality problems. Researchers have attempted various eco-friendly physicochemical methods to improve water quality, such as using food chains or nutritional interactions to promote the growth of animal plankton, which are avian predators, and controlling and managing the optimal growth conditions of predators or competitors. However, these methods have drawbacks, such as difficulties in rapid ecosystem transition and negative effects on water resource development and preservation. Biological manipulation for water quality improvement and ecological restoration is challenging, as it requires accurate prediction of the characteristics of the introduced organisms and anticipating expected changes after biological application while considering the risks associated with introducing exotic species.

Bivalve mollusks are recognized for their high biomass in freshwater and marine ecosystems because of their filter-feeding activity. This activity helps to remove phytoplankton and other suspended matter from the water column, leading to improved light penetration and facilitating the growth of macrophytes that provide habitats for other biota [11–13]. Studies have demonstrated that greater biological richness is associated with a greater abundance of unionid mussels in freshwater systems [14, 15]. Freshwater bivalves, such as zebra mussels, have been recognized as potential biofilters for drinking water treatment has been recognized [16–19]. The method of utilizing top-down effects through food chains additionally can be employed to indirectly induce the actions of other organisms. Filter-feeding bivalves, with their excellent filtration capabilities, have been extensively researched, primarily in Europe, for their potential application in water quality improvement. The utilization of these bivalves as unique and diverse biological filters holds significant promise for the removal of organic matter in aquatic environments.

However, the optimal conditions for effective biofiltration systems involving bivalves are still under investigation. Response Surface Methodology (RSM) is a statistical technique used to evaluate the effects of different factors and to identify the optimal conditions for desirable responses in experiments [20, 21]. In this study, we aimed to identify the optimal conditions for particulate material removal by the freshwater mussel *Sinanodonta woodiana* Lea, using RSM. This large mussel species is native to the Amur and Yangtze river basins, but has spread to other regions due to fish farming [22]. Researchers have investigated the effects of shell size, water flow rate, filtration rate (FR), and production of feces/PF on the removal of suspended material from the water column.

In this study, we aimed to identify the optimal conditions for particulate material removal from a water column using freshwater mussels. Specifically, we investigated the relative importance of shell size, water flow rate, filtration rate (FR), and the production of feces and pseudo-feces (PF) using RSM. We focused on the freshwater mussel *Sinanodonta woodiana* Lea, which is a large species native to the Amur and Yangtze River basins and has spread throughout Southeast Asia and South America due to fish farming. Our study proposes the use of domestic freshwater shellfish to control algae in highly polluted agricultural reservoirs, establishing a sustainable and ecologically friendly water improvement method using response surface analysis. We used a central composite design to efficiently create approximate secondary or higher response functions.

2. Methods and materials

2.1 Test animal collection

We collected adult *Anodonta woodiana* Lea and *Unio Douglasiae* Griffith et Pidgeon from tributaries of the Geumgang and Mangyeonggang Rivers to serve as spawning sites for fish. Shellfish are commonly found in domestic water systems and usually coexist in the same river; however, *Anodonta woodiana* Lea is abundant in organic matter, and *Unio Douglasiae* Griffith et Pidgeon is relatively common in the upper stream of the river in irrigation channels with developed sand or pearls. They primarily feed on plankton or bottom organic matter [22]. We immediately placed the shellfish collected in an icebox and transported them to the laboratory. After washing them two to three times with desalinated tap water, they were kept in a management tank for more than three months. We acclimated *S. woodiana* specimens in laboratory aquaria for at least three months which were collected directly from waterways and streams in Korea. We used stainless steel treatment baths (80 × 80 × 145 cm) with a working volume of 500 L to study filtration. We acclimated 30 individuals of test mussels to holding aquaria for 18 days before the commencement of the experiments. The experimental equipment used has been described in detail by Lee et al. [23]. An experiment was conducted involving filtration by *S. woodiana* under conditions including a water temperature of 19 ± 3°C, water flow rates ranging from 24 L/h (T1) to 48 L/h (T2), and a photoperiod of 12 D:12 L.

2.2 Removal of organic matter by freshwater bivalves

We measured the filtration rate and production of bivalve feces and pseudo-feces using a device with the same structure as Lee et al. [3] using the continuous organic material removal method with shellfish. The device consisted of a storage tank, treatment tank, and analysis tank. We made a treatment tank of stainless steel (80 × 80 × 145 cm) with a total treatment capacity of 500 L. We installed a lattice-type shellfish holder (75 × 75 cm) under 10 cm of the check surface in the treatment tank and placed the shellfish on it. The surface water of the eutrophic reservoir was passed through the tank at 24 and 48 L/h. The water temperature was adjusted to 18 ± 2°C and the optical instrument was maintained at 12:12 L. We supplemented shellfish of a similar size to those that died during the experiment. To understand the effect of shellfish treatment on water quality, we measured changes in the water quality environment daily and analyzed the samples collected from the analysis tank to observe changes in organic and nutritional salts. We also collected shellfish excrement every five days to measure the production of excrement.

We collected data on the filtration rates for each animal every 42 days (**Table 1**). The production of feces and pseudo-feces by each mussel was also measured simultaneously, with sediments from individual mussels being analyzed at three-day intervals for nine days during the 42-day experiment. The ash-free dry mass of each of the 30 mussels was measured at the end of the experimental period.

We harvested sedimented feces and pseudo-feces from the experimental aquaria and placed them in sterilized dishes. The weight of the feces was measured after drying

Characters	Operation conditions
Kind of mussels	*Anodonta woodiana*
Mussel density (mussel/L)	0.5
Total weight (kg)	10.5–30.1
Shell length (cm)	8.09–11.35
Shell width (cm)	4.74–6.71
Total AFDW (g)	1.33–3.42
Tank volume (L)	500
Flow rate (L/h)	22.0
Total capacity (m^3/d)	0.52
Filtration area (m^2)	0.64
Temperature (°C)	11.9–21.8
Exposure time (days)	42
Sampling time (every)	AM 11:00

Table 1.
The experimental design and operation conditions for the removal of organic matter by the freshwater bivalve Anodonta woodiana.

at 70°C for 1 h. The production of mussel feces and pseudo-feces was quantified by calculating the disparity in dry weights (mg/g AFDM/h) of sedimented particulate matter between reactor treatments with and without mussels, employing the subsequent equation: PFs were determined using the equation: PFs = (V/M) × ln(T/C)/t, where V represents the volume of the experimental chamber in liters, M signifies the total ash-free dry mass (AFDM) of the mussels, T and C denote the total dry weights of sedimented particulate matter in the reactor with and without mussels, respectively, and t stands for the duration of the experiment in hours. After detaching the entire mussel body from the shell and measuring its weight, the mussel was placed into a heat-resistant container. Subsequently, it was subjected to desiccation at 100°C for 20 minutes in a drying oven until a consistent mass was achieved. Following this, the mussel was incinerated in a muffle furnace at 500°C for 2 hours, as per the methodology outlined in APHA 1995. The ash-free dry mass (AFDM) of the mussel body was calculated by utilizing the disparity in dry weight before and after the incineration process. The filtration rate of each mussel (FR: L/mussel/h) was determined for each experiment and day using the equation established by Coughlan [24]: FR = (V/M) × ln(T/C)/t, where V represents the volume of the experimental reactor in liters, M signifies the total ash-free dry mass (AFDM) of the mussels, T and C denote the concentrations of suspended solids in water that passed through the reactor with and without mussels, respectively, and t (hours) stands for the duration of the experiment.

2.3 Experimental design and the modeling of filtration by mussels

We analyzed the data to optimize the organic filtration reaction conditions using shellfish by applying statistical methods and the response surface method (RSM) through Minitab software (MINITAB Release 14.12.1, Korea). We considered the size, flow rate, and filtration rate of shellfish as independent variables and utilized the central composite design (CCD) program in the software to set their high and low levels. The size of the shellfish was set to 2 cm and 20 cm for high and low levels, respectively, while the flow rate was set to 48 L/h and 24 L/h for high and low levels, respectively. **Table 2** presents the experimental design used in this study, including the trends, interactions, and experimental errors of the operating variables.

An experimental design was employed to determine the optimal levels of three variables—mussel size (×1), experimental time (×2), and water flow (×3)—regarding their effects on filtration rates and feces production. This design aimed to model mussel filtration conditions with respect to body size (**Table 3**). Using 23 central composite designs (CCD) with three factors, including six center points, a series of 30 experiments were conducted, with each experimental condition involving a single individual mussel. All variables were standardized to a central coded value of zero. The ranges encompassing the minimum and maximum values of the investigated variables, along with the comprehensive experimental plan detailing their actual and coded forms, are provided in **Table 4**.

After conducting the experiments, The average maximum filtration rate was employed as the dependent variable or response (Y), and a second-order polynomial equation was applied to the dataset through the utilization of the multiple regression procedure.

This led to the development of an empirical model that establishes a correlation between the responses and the independent variables of the experiment. The formulated model equation for the three-factor system is expressed as follows: $Y = \beta 0 + \beta 1\ x1 + \beta 2\ x2 + \beta 3\ x23 + \beta 11\ x12 + \beta 22\ x22 + \beta 33\ x32 + \beta 12 \times x1 \times x2 + \beta 13 \times x1 \times x3 + \beta 23 \times x\ 2 \times x3$, where Y represents the predicted response, $\beta 0$ denotes the intercept, $\beta 1$, $\beta 2$, and $\beta 3$ stand for the linear coefficients, $\beta 11$, $\beta 22$, and $\beta 33$ represent the squared coefficients, and $\beta 12$, $\beta 13$, and $\beta 23$ signify the interaction coefficients. The statistical analysis of the data was conducted using the Minitab software package (Minitab Release 14.12.1).

Variables	Range of levels					
	Actual	Coded	Actual	Coded	Actual	Coded
Mussel size (cm)	5.0	−1	8.0	0	12.0	+1
Water flow (L/h)	12.0	−1	24.0	0	48.0	+1
Filtration rate (L/mussel/h)	0.5	−1	1.0	0	2.0	+1

Table 2.
Experimental range and levels of the three independent variables used in response surface methodology in terms of actual and coded factors.

	Shell length (cm)	Flow rate (L/h)	Retention time (h)	Filtration rate (L/mussel/h) (mean ± SE)
Sinanodonta woodiana	8.5 ± 1.0	12	1.5	0.87 ± 0.17
Sinanodonta woodiana	11.4 ± 1.8	24	22.7	4.47 ± 1.82
Sinanodonta woodiana	11.4 ± 1.3	24	3.9	1.23 ± 0.20
Sinanodonta woodiana	10.6 ± 1.9	24	20.8	3.30 ± 1.03
Sinanodonta woodiana	10.8 ± 1.9	48	10.4	2.30 ± 0.97

Table 3.
Filtration rate of the mussel Anodonta woodiana according to differences in flow rate and retention time.

Run order	Shell size (×1)	Flow rate (×2)	Production of feces (×3)	Mean measured response	Predicted response
1	−1	−1	−1	2.36	3.18
2	−1	−1	1	3.40	4.59
3	−1	−1	−1	3.57	4.81
4	−1	−1	1	2.75	3.71
5	−1	1	−1	2.77	3.73
6	−1	1	1	4.06	5.48
7	−1	1	−1	3.21	4.32
8	−1	1	1	1.72	2.33
9	1	−1	−1	2.58	3.47
10	1	−1	1	2.62	3.53
11	1	−1	−1	6.13	8.27
12	1	−1	1	4.78	6.45
13	1	1	−1	2.81	3.79
14	1	1	1	2.47	3.33
15	1	1	−1	13.45	18.14
16	1	1	1	10.71	14.45
17	1	0	0	1.69	2.28
18	1	0	0	6.15	8.29
19	0	−1	0	1.43	1.93
20	0	1	0	4.78	6.45
21	0	0	0	3.97	5.35
22	0	0	0	2.38	3.21
23	0	0	−1	6.53	8.80
24	0	0	1	6.09	8.21
25	0	0	0	6.11	8.24
26	0	0	0	6.45	8.70
27	0	0	0	6.55	8.83
28	0	0	0	6.53	8.80
29	0	0	0	6.28	8.48
30	0	0	0	6.08	8.20

Table 4.
Experimental designs used in response surface methodology using three independent variables with the centre point showing measured and predicted values of the filtration rates of freshwater bivalve Anodonta woodiana.

2.4 Data analysis

We used a HORIBA U-22XD (HORIBA Ltd., Japan) to measure the temperature, pH, electrical conductivity, dissolved oxygen (DO), and turbidity of both the control and treatment groups daily at 11:00 AM. We collected samples from the analysis tank at the same time each day to observe the changes in organic and nutritional salts.

Parameters	Unit	Control	T1 (24 l/h)	T2 (48 l/h)	F	P
Temperature	°C	16.09 ± 3.41	15.75 ± 3.22	15.79 ± 3.19	0.03	P = 0.971
Conductivity	μS/cm	66.12 ± 88.83	67.66 ± 90.12	67.71 ± 89.36	0.001	P = 0.999
DO	mg/L	11.83 ± 2.53b	8.36 ± 1.88a	6.92 ± 1.54a	13.87	P < 0.001
Turbidity	NTU	26.37 ± 10.14b	8.85 ± 4.53a	6.84 ± 3.35a	23.13	P < 0.001
pH	—	8.77 ± 0.24	8.54 ± 0.19	8.53 ± 0.33	2.23	P = 0.128
SS	mg/L	22.58 ± 3.17b	5.96 ± 1.20a	3.93 ± 2.40a	163.27	P < 0.001
Chl-a	μg/L	24.77 ± 8.99b	7.42 ± 4.67a	3.98 ± 2.68a	30.50	P < 0.001
NO_2-N	μg/L	4.10 ± 2.05	8.68 ± 6.85	8.27 ± 2.43	3.03	P = 0.067
NO_3-N	mg/L	0.12 ± 0.04	0.17 ± 0.03	0.17 ± 0.04	3.67	P = 0.041
NH_4-N	μg/L	22.29 ± 23.98a	261.88 ± 94.86b	472.20 ± 92.38c	75.55	P < 0.001
TN	mg/L	1.49 ± 0.23b	1.30 ± 0.12a	1.67 ± 0.14b	9.70	P < 0.001
PO_4-P	μg/L	7.45 ± 2.09a	8.57 ± 4.39a	49.05 ± 22.57b	28.44	P < 0.001
TP	μg/L	98.83 ± 16.40b	58.83 ± 8.58a	115.41 ± 24.37b	24.38	P < 0.001

Control; only lake water without mussels, DO; dissolved oxygen, SS; suspended solids, Chl-a; chlorophyll-a, TN; Total nitrogen; and TP; total phosphorus, Letters (a, b and c) indicate significant differences by ANOVA and Tukey's THD test.

Table 5.
Summary of ANOVA on environmental quality by the stocking water of freshwater bivalve Anodonta woodiana.

We filtered the *Chl*-a sample with a GF/F filter (Whatman Inc., England), extracted it with 90% acetone for 24 h, and separated it by centrifugation for 20 min following the Water Quality Process Test Act [25]. For nutritional was measured NO_2-N by the Colorimetric method, NO_3-N using the cadmium reduction method, NH_3-N using the phenate method, TN using the cadmium reduction method, PO_4-P using the ascorbic acid method, and TP using the ascorbic acid method after persulfate decomposition [26]. We also monitored the water temperature and light intensity at 30-minute intervals during the experiment using a HOBO Pendant Temperature/Light Data Logger (UA-002-08). Light intensity was used as an indirect indicator to observe changes in the turbidity of organic matter in the water. We conducted ANOVA and Tukey's THD test using the SPSS package (version 12.0.1, SPSS Inc., 2004 release) to determine the correlation between the application of shellfish and changes in the water quality according to the flow rate; the significance level was set at $P < 0.05$ (**Table 5**).

3. Results and discussion

3.1 Effect of mussel size, flow rate, and retention time on Sinanodonta filtration rate

Table 2 shows that the mussel treatments caused changes in several water quality parameters. As expected, turbidity, suspended sediments, and algae (as reflected by chl a) were significantly reduced by filtration. The excretion of feces and pseudo-feces by mussels was caused in the significantly higher concentrations of ammonium and orthophosphate as phosphorus in those treatments. Across various mussel sizes (mean value ± SE: 8.5 ± 1.0 to 11.4 ± 1.8 cm), flow rates (12–48 L/h), and retention times (1.5–22.7 h), the filtration rate exhibited a range spanning from 0.87 ± 0.17 to 4.47 ± 1.82 L/mussel/h (with an average of 2.67 ± 1.00 L/mussel/h; see **Table 4**). Filtration rate displayed an increment with the enlargement of mussel size and a decrease in flow rate. Specifically, the group of larger mussels (11.4 ± 1.8 cm) exhibited higher filtration rates in comparison to the smaller size group (8.5 ± 1.0 cm).

3.2 Optimization by response surface methodology

Response surface methodology was used to optimize the filtration rate of *Sinanodonta woodiana* in response to three independent variables: mussel size, water flow rate, and retention time. **Table 4** presents the results of the CCD experiments and the mean predicted and observed responses. After analyzing the variance (ANOVA), the researchers obtained an R2 value of 0.7625, indicating that the quadratic model had a satisfactory adjustment to the experimental data and could explain 76% of the variability in the response. The coefficients of the regression equation was obtained: Y = 18.214–10.211x1 + 10.105x2 + 12.542x3–12.458x12–8.243x22–9.5 49x32 + 13.263x1 x2 + 17.671x1 x3–15.842 × 2 × 3.

We determined the interaction between the experimental components and the optimum of each component required for the maximum filtration rate by using three-dimensional response surface curves. **Figures 1** and **2** illustrate the comparative impacts of two variables, namely mussel size and flow rate, across different retention times. The researchers anticipated the mussels' highest attainable filtration rate to reach 8.4 L/mussel/h within these parameters, aligning with the maximum levels (+1) of mussel size (13.0 ± 0.2 cm) and flow rate (17.5 L/h; depicted in **Figure 1**). Nevertheless, the curve also illustrates the responsiveness altering in concurrence with the water flow velocity. As the flow rate escalates (surpassing 30 L/h) and the retention time diminishes, the feces production by mussels escalates further, reaching 11.1 g AFDM/ind./h (as depicted in **Figure 2**). The curves of the response surface exhibited a lack of curvature, adopting a flattened disposition. While mussel size exhibited a relatively marginal influence, higher flow rates were associated with amplified fecal production (as seen in **Figure 2**). **Figure 3** provides insight into the higher fecal production rates linked to elevated flow rates, while indicating a limited impact of mussel size on fecal production.

We recognized the necessity for optimizing conditions, including the size and filtration rate of organisms, particularly in eutrophic regions, when employing freshwater mussels for organic matter removal. The surface plots indicate a requirement for reducing the flow rate and extending the retention time to minimize fecal production. In light of these findings, researchers propose the implementation of a further reduction in water flow and an extension of retention time within the system as a means of validation. The researchers also suggested that a multifactorial analytical approach, which takes into account the interaction of independent variables, provides a basis for models designed to assess the nonlinear nature of the response under limited experimental conditions.

3.3 Nutrient changes

The shellfish treatment group caused a significant change in the concentration of nutrient salts depending on the flow rate, resulting in a higher value of NO2-N (C:0.03 ± 0.01, FL – 1:0.11 ± 0.01, FL-2:0.07 ± 0.01LL-1) and NH3-N (C:36.77 ± 6.87, FL-1:306.09 ± 16.31, FL-2:214.99 ± 14.57LL-1) compared to the high flow rate

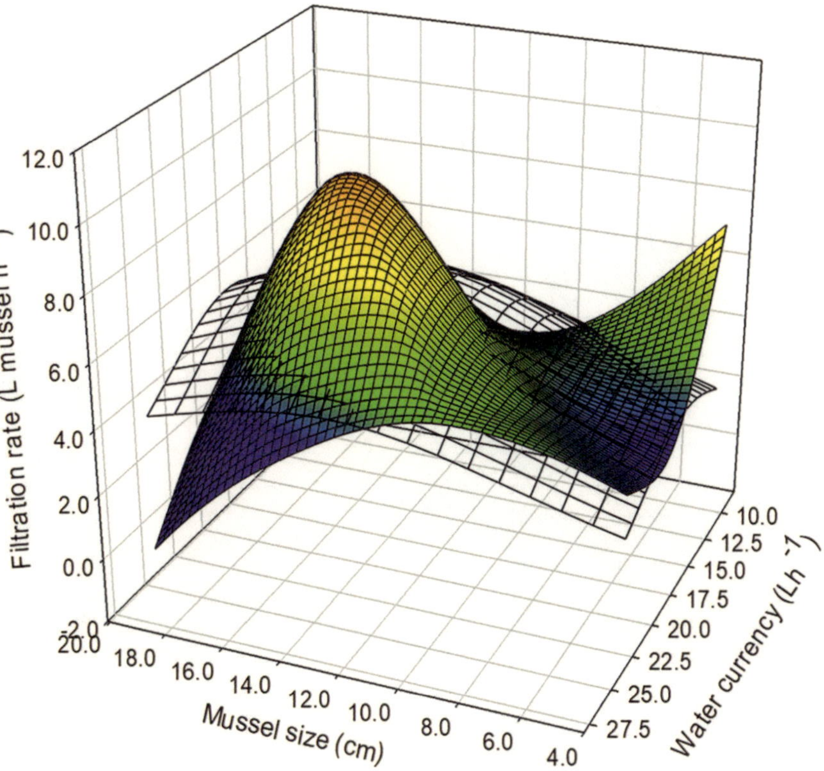

Figure 1.
Response surface curve of filtration rates by S. woodiana showing interaction between mussel size and water flow.

treatment group (FL-2) in the low flow rate treatment group (FL-1). Shellfish emit ammonia-type nitrogen during the eating process, and the amount of nitrogen emitted during the eating process is higher than that of phosphorus. Other experiments with various organisms have also shown an increase in underwater dissolved inorganic nutrients owing to shellfish treatment, which calls for further studies on the removal, concentration, and recovery of the generated nutritional salts for use by other organisms or plants. Increasing the transparency of water using shellfish creates a favorable environment for the growth of submerged plants or attached algae, thereby converting the basis of the food chain from plant plankton to attached algae or submerged plants. Fukushima et al. [27], Kim Deung [22, 28, 29], and Lee [23] conducted experiments that found that shellfish treatment increased underwater dissolved inorganic nutrients. Future studies are needed to determine how to remove, concentrate, and recover the generated nutritional salts so that they can be used by other organisms or plants.

3.4 Filtration rates

The shellfish treatment group had a higher average filtration rate (CR) than the control group (2.299 ± 0.97 L mussel−1d−1) at a flow rate of 48 L/h in the experiment to determine the effect of shellfish on organic material control according to flow rate. However, Kirby-Smith [30] and Jorgensen et al. [31] reported that shellfish's ability to filter particulate matter in water is reduced at high flow rates, and this study similarly showed low CR values at fast flow rates. The highest filtration rate was observed at a

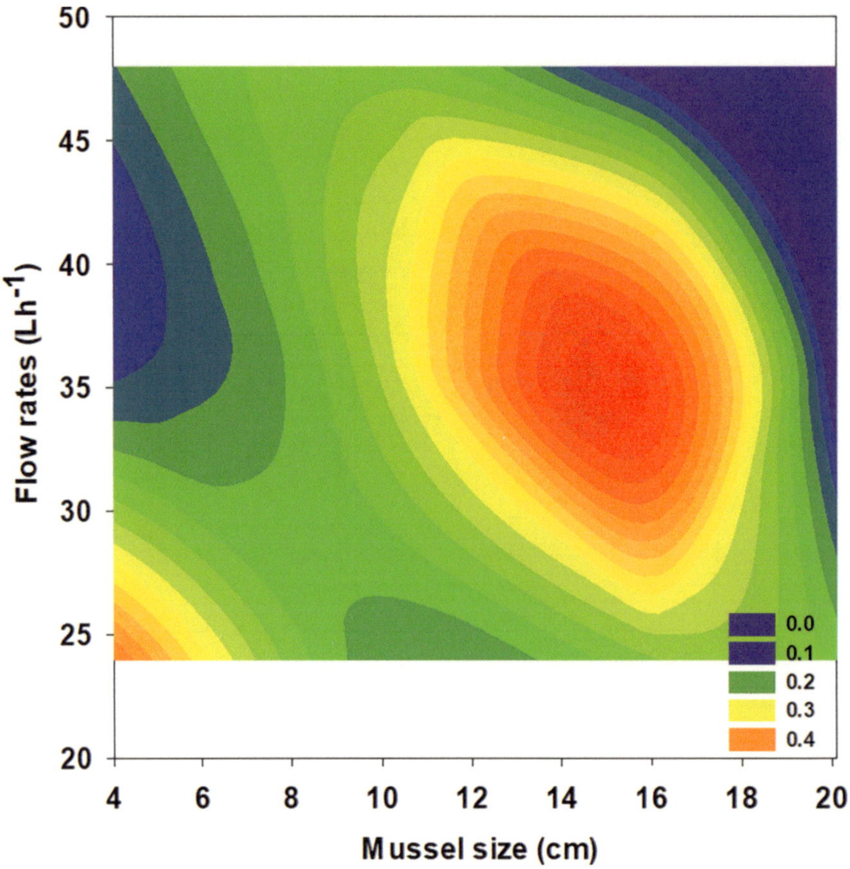

Figure 2.
Response surface contour plots of filtration rates by S. woodiana showing interaction between mussel size and water flow.

flow rate of 24 L/h and a shell size of 10.6–11.4 cm with a residence time of 22.7 hr., while the lowest filtration rate was observed at a flow rate of 48 L h − 1 with a residence time of 10.4 hr. The excessively high flow rate decreased the filtration rate, and the excessively low flow rate also decreased the filtration rate owing to the increase in residence time. Lee et al. [3] reported a lower filtration rate (0.87 ± 0.17 L /mussel/d) at a flow rate of 12 L/h, which was believed to be due to the long residence time affecting the reproduction of plant plankton. Previous studies have shown that the presence of shellfish can increase the growth of plankton. An optimal flow rate range for each shellfish exists, and for the continuous organic material control device using pearl shells, the optimal flow rate was found to be 24 L/h. Future studies can reveal the relationship between the flow rate and residence time and improve device efficiency.

3.5 Response surface analysis

We used the reaction surface analysis method to determine the optimal conditions for controlling organic material in shellfish. We determined the size and flow rate of the shellfish-dependent variables and conducted a multi-session and dispersion analysis using software. We also performed dynamic analysis and three-dimensional

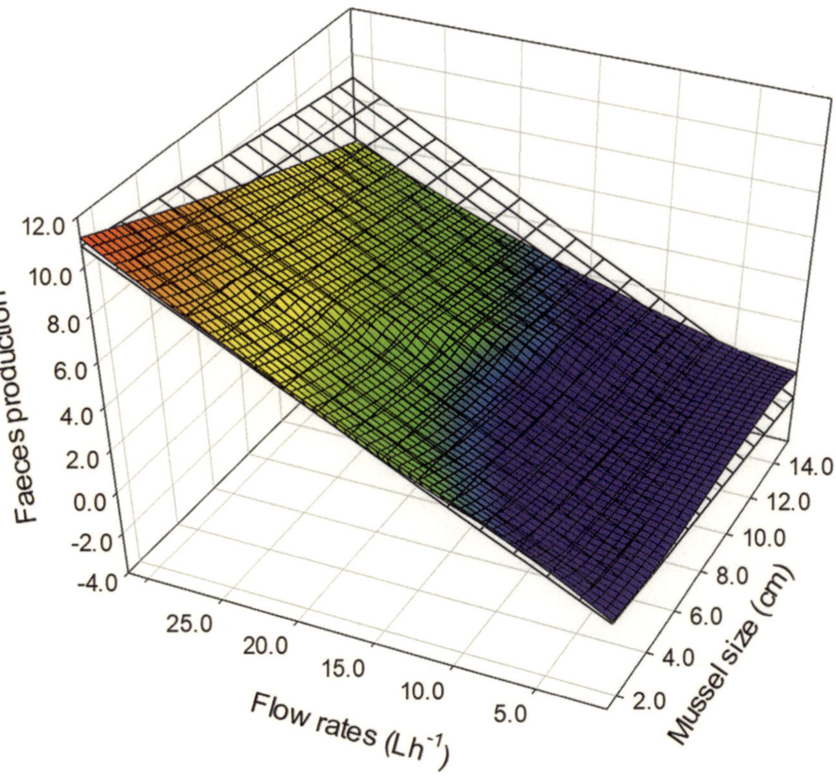

Figure 3.
Response surface contour plots of feces production by S. woodiana showing interaction between mussel size and water flow.

analysis. To design the experimental conditions, we used the central synthesis design method, setting the size of the shellfish to a high level of 2 cm, a low level of 20 cm, a flow rate of 48 L/h, and a low level of 24 L/h. The results in **Table 4** were obtained by performing the organic matter control reaction of shellfish, with the organic filtration rate as the dependent variable.

The results showed that the shellfish organic material control rate was higher when the size of the shellfish was larger and the flow rate was lower. We conducted an analysis in which we fixed the size (10 cm) and flow rate (36 L/h) as the independent variables. This analysis showed that the larger the size of the shellfish and the slower the flow rate, the higher the organic material control rate by the shellfish. However, environmental factors such as water temperature can affect the ability of shellfish to control irregular organic material. We also derived a contour plot of the relationship between the response factor and the response. Individual contour plots were obtained by fixing the size and flow rate of the shellfish to normal values. The results showed that both the size and flow rate of the shellfish are important factors in the organic material control reaction, and that stabilizing external environmental factors, such as water temperature, is also crucial for achieving stable organic material control.

Therefore, to increase organic material control using shellfish in a stable manner, we should consider not only the size, flow rate, and residence time of the shellfish but also the stabilization method of external environmental factors that affect the activity of shellfish.

3.6 Potential applications

Ismail et al. [32] highlighted that the utilization of environmentally relevant concentrations and treated wastewater can offer an initial insight into the potential effectiveness of bivalves in mitigating emerging contaminants and enhancing water quality. The application of the response surface methodology approach revealed optimal conditions, which could vary based on factors such as season, bivalve species, age, and prey competition, thereby necessitating additional research. While existing models reported in relation to eutrophication reduction using bivalves within lake systems can offer insights, their direct applicability for on-site organic matter removal is limited due to the absence of correlations between mussel filtration rates and rates of algae and organic matter removal.

Nonetheless, investigations have substantiated the noteworthy filtration efficiency and the utilization of bivalves for substantial water quality enhancement, thereby signifying the potential viability of bivalves in engineered systems or ecological rehabilitation efforts [19, 33]. The choice and upkeep of an apt bivalve species and population, alongside the fine-tuning of bivalve filtration rates and feces production, emerge as pivotal factors when evaluating the suitability of bivalve application for enhancing water environments, as advocated by these researchers.

The proliferation of dense algal blooms and the accumulation of excessive organic materials within the water column can degrade the habitat quality for various organisms and severely disrupt essential ecosystem services, including recreational activities, sanitation, irrigation, industrial cooling, and the supply of potable water. Consequently, numerous nations have endeavored to mitigate algal blooms and enhance the quality of eutrophic water bodies. Nonetheless, a significant drawback lies in the potential harm posed by various physical and chemical treatment approaches, which can trigger adverse environmental consequences, including secondary pollution. In response, the focus of this study is directed toward the management of organic matter levels and the population density of problematic algae, employing readily adaptable technologies. The native freshwater mussel *S. woodiana* possesses distinct advantages, owing to its robust resilience in challenging environments, including instances of harmful algal blooms (HABs). Notably, it can thrive at notably elevated densities, reaching levels as high as 60 individuals per square meter [34]. Additionally, with a lifespan of up to 12 years [35], this species proves conducive to environmental management strategies that require minimal maintenance.

From a pragmatic standpoint, our study's outcomes propose that the passage of nutrient-enriched water through mussel beds could establish effective biological filters. The optimal elimination of suspended sediments is attainable through the careful selection of appropriate mussel sizes, flow rates, and retention times. These systems have the potential to function as artificial conduits, seamlessly connecting to lentic or lotic water bodies. Furthermore, this technology lends itself to integration with other sustainable approaches for nutrient management, including the introduction of phytoplanktivorous fish [36] and the cultivation of rapidly proliferating macrophytes [37].

4. Conclusion

Using this response surface method, we optimized the conditions for the organic filtration reaction of shellfish. We analyzed the data using the central composite design program in the software, with the size, flow rate, and filtration rate of shellfish as independent variables.

- The experiment showed that the highest filtration rate was achieved by increasing the flow rate to 24 L/h and using shell sizes of 10.6–11.4 cm, with a residence time of 22.7 h. Conversely, the lowest filtration rate was observed with a flow rate of 48 L/h and a residence time of 10.4 h. Both excessively fast and excessively slow flow rates decreased the filtration rate, attributed to factors such as residence time. However, the model could only be validated with mussel size, and further validation is required with a broader range of water velocities and retention times.

- The utilization of response surface methodologies necessitates a limited number of individuals for the determination of optimal filtration parameters, rendering them a cost-effective resource for water management. The presented model equation vividly depicts the quantitative impact of variables and their interactions on mussel filtration rates. The conducted experiment validated the feasibility of the optimization strategy within the optimum conditions of mussel size (13.0 ± 0.2 cm) and flow rate (17.5 L/h), resulting in an observed filtration rate of 4.47 ± 1.82 L/mussel/h, aligning closely with the predicted value of 8.4 L/mussel/h.

From the above results, excessively fast or slow water flow induces a decrease in plankton filtration rates by bivlaves, and it was evident that such filtration rate reduction significantly depends on the residence time within the apparatus. Through further research, elucidating the relationship between water velocity and residence time, and incorporating it into device design, it is anticipated that a higher filtration efficiency can be achieved.

Thus, our research revealed that statistical design methodology is an effective and feasible approach to promote high filtration and low fecal production, indicating that indigenous bivalves offer significant potential to restore multiple ecosystem services provided by freshwater systems by removing organic matter from eutrophic systems.

Author details

Hwan-Seok Choi[1]* and Baik-Ho Kim[2]

1 Research Institute for Coastal Environment and Fishery-Policy, Kwangju, South Korea

2 Department of Life Science, Hanyang University, Seoul, South Korea

*Address all correspondence to: choihs1014@naver.com

References

[1] Bogan AE. Freshwater bivalve extinctions (Mollusca: Unionoida): A search for causes. American Zoologist. 1993;**33**:599-609

[2] McIvor AL, Aldridge DC. The reproductive biology of the depressed river mussel, Pseudoandonta complanata (Bivalvia: Unionidae), with implications for its conservation. Journal of Molluscan Studies. 2007;**73**:259-266

[3] Lee JH, Hwang SJ, Park SG, Hwang SO, Yu CM, Kim BH. Continuous removal of organic matters of eutrophic lake using freshwater bivalves: Inter-specific and intra-specific differences. Korean Journal of Limnology. 2009a;**42**:350-363

[4] Lee YJ, Kim BH, Kim NY, Um HY, Hwang SJ. Effects of temperature, food concentration, and shell size on filtering rate and pseudofeces production of Unio douglasiae on Microcystis aeruginosa. Korean Journal of Limnology. 2008;**41**:61-67

[5] Park GS, Kim BH, Um HY, Hwang SJ. Effects of dissolved oxygen and depth on the survival and filtering rate and pseudofeces production of a filter-feeding bivalve (*Unio douglasiae*) in the cyanobacterial bloom. Korean Journal of Limnology. 2008;**41**:50-60

[6] Jack JD, Thorp JH. Effects of the benthos suspension feeder Dreissena polymorpha on zooplankton on a large river. Freshwater Biology. 2000;**44**:569-579

[7] Dionisio Pires LM, Ibelings BW, Brehm M, Van Donk E. Comparing grazing on lake seston by Dreissena and Daphnia: Lessons for biomanipulation. Microbial Ecology. 2005;**50**:242-252

[8] Li XN, Song HL, Li W, Lu XW, Nishimura O. An integrated ecological floating-bed employing plant, freshwater clam and biofilm carrier for purification of eutrophic water. Ecological Engineering. 2010;**36**:382-390

[9] Jones JW, Neves RJ. Life history and propagation of the endangered fanshell pearly mussel, Cyprogenia stegaria Rafinesque (Bivalvia: Unionidae). Journal of North America Bentholic Society. 2002;**21**:76-88

[10] McIvor AL. Freshwater Mussels as Biofilters. Cambridge, UK: University of Cambridge; 2004. Available from: http://www.ourearth.co.uk/annamcivor/thesis.html

[11] Davenport J, Smith RJJW, Packer M. Mussels Mytilus edulis: Significant consumers and destroyers of mesozooplankton. Marine Ecology Progress Series. 2000;**198**:131-137

[12] Fanslow DL, Nalepa TF, Lang GA. Filtration rates of the zebra mussel (*Dreissena polymorpha*) on natural seston from Saginaw Bay, Lake Huron. Journal of Great Lakes Research. 1995a;**21**:489-500

[13] Newell RIE, Koch EW. Modelling seagrass density and distribution in response to changes in turbidity stemming from bivalve filtration and seagrass sediment stabilization. Estuaries. 2004;**27**:793-806

[14] Aldridge DC, Fayle TM, Jackson N. Freshwater mussel abundance predicts biodiversity in UK lowland rivers. Aquaatic Conservation Marine Freshwater Ecosystem. 2007;**17**:554-564

[15] Chowdhury GW, Zieritz A, Aldridge DC. Ecosystem engineering

by mussels supports biodiversity and water clarity in a heavily polluted lake in Dhaka, Bangladesh. Freshwater Science. 2016;**38**:188-199

[16] Elliott P, Aldridge DC, Moggridge GD. Zebra mussel filtration and its potential uses in industrial water treatment. Water Research. 2008;**42**:1664-1674

[17] Ibelings BW, Portielje R, Lammens EHRR, Noordhuis R, Van den Berg MS, Joosse W, et al. Resilience of alternative stable states during the recovery of shallow lakes from eutrophication: Lake Veluwe as a case study. Ecosystems. 2007;**10**:4-16

[18] Lammens EHRR, Van Nes EH, Mooij WM. Differences in the exploitation of bream in three shallow lake systems and their relation to water quality. Freshwater Biology. 2002;**47**:2435-2442

[19] McLaughlan C, Aldridge DC. Cultivation of zebra mussels (*Dreissena polymorpha*) within their invaded range to improve water quality in reservoirs. Water Research. 2013;**47**:4357-4369

[20] Jeong YS, Kim JW, Lee ES, Gil NY, Kim SS, Hong ST. Optimization of alkali extraction for preparing oat protein concentrates from oat groats by response surface methodology. Journal of Korean Society of Food Science and Nutrition. 2014;**43**:1462-1466

[21] Wang ZW, Liu XL. Medium optimization for antifungal active substances production from a newly isolated Paenibacillus sp. using response surface methodology. Bioresource Technology. 2008;**99**:8245-8251

[22] Kim BH, Baik SK, Hwang SO, Hwang SJ. Operation of CROM system and its effects of on the removal of Seston in a eutrophic reservoir using a native freshwater bivalve (*Anodonta woodiana*) in Korea. Korean Journal of Limnology. 2009;**42**:161-171

[23] Lee SH, Baik SK, Hwang SJ, Kim BH. Comparison of grazing characteristics of a freshwater bivalve Unio douglasiae (Unionidae) on the cold and warm phytoplankton communities in eutrophic lake. Korean Journal of Limnology. 2009b;**42**:115-123

[24] Coughlan J. The estimation of filtration rates from the clearance of suspensions. Marine Biology. 1969;**2**:356-358

[25] Ministry of Environment. The Water Pollution Standard Method of Korea. 1995. pp. 195-214

[26] American Public Health Association (APHA). Standard Methods for the Examination of Water and Wastewater. nineteenth ed. Washington, DC: American Public Health Association; 1995

[27] Fukushima M, Takamura N, Sun L, Nakagawa M, Matsushige K, Xie P. Changes in plankton community following introduction of filter-feeding planktivorous fish. Freshwater Biology. 1999;**42**:719-736

[28] Kim BH, Lee JH, Kim YJ, Hwang SO, Hwang SJ. Water quality improvement of Pocheon stream using freshwater bivalves: Development and operation of continuous removal of organic matters in stream (S-CROM). Korean Journal of Limnology. 2009;**42**:317-330

[29] Kim HS, Hwang SJ, Ko KM. Evaluation of water quality variation and sediment of a shallow artificial Lake (Lake llgam) in located the metropolitan area. Korean Journal of Limnology. 2003;**36**:161-171

[30] Kirby-Smith WW. Growth of the bay scallop: The influence of experimental water currents. Journal of Experimental Marine Biology and Ecology. 1972;**8**:7-18

[31] Orgensen BC, Famme P, Kristensen HS, Larsen PS, Mohlenberg F, Riisgard HU. The bivalve pump. Marine Ecology Progress Series. 1986;**34**:69-77

[32] Ismail NS, Muller CE, Morgan RR, Luthy RG. Uptake of contaminants of emerging concern by the bivalves Anodonta californiensis and Corbicula fluminea. Environmental Science Technology. 2014;**48**:9211-9219

[33] Gifford S, Dunstan RH, O'Connor W, Koller CE, MacFarlane GR. Aquatic zooremediation: Deploying animals to remediate contaminated aquatic environments. Trends in Biotechnology. 2007;**25**:60-65

[34] Kraszewski A, Zdanowski D. Sinandonta woodiana (Lea, 1834) (Mollusca) – A new mussel species in Poland: Occurrence and habitat preferences in a heated lake system. Polish Journal of Ecology. 2007;**55**:337-356

[35] Dudgeon D, Morton B. The population dynamics and sexual strategy of Anodonta woodiana (Bivalvia: Unionacea) in plover cove reservoir, Hong Kong. Journal of Zoology. 1983;**201**:161-183

[36] Ke Z, Xie P, Guo L, Liu Y, Yang H. In situ study on the control of toxic Microcystis blooms using phytoplanktivorous fish in the subtropical Lake Taihu of China: A large fish pen experiment. Aquaculture. 2007;**265**:127-138

[37] Srivastava J, Gupta A, Chandra H. Managing water quality with aquatic macrophytes. Reviews in Environmental Science and Biotechnology. 2008;**7**:255-266

Chapter 4

Purification of Rainwater Using a Photocatalysis Technique to be Applied to Communities in Ciudad del Carmen, Campeche, Mexico

Carlos Montalvo, Claudia A. Aguilar, Rosa A. Martínez, Rosa M. Cerón, Alejandro Ruiz, Eric Houbron and Juan C. Robles

Abstract

Small communities far from the municipal seat do not have access to drinking water, so many children suffer from various gastrointestinal diseases, which cause these children to grow up with nutritional deficiencies. In the state of Campeche, there are 300 sunny days. This energy can be used to install water treatment systems to make it drinkable. Therefore, a treatment system with heterogeneous photocatalysis was proposed using a zinc oxide catalyst doped with silver nanoparticles. The reactor has a metal structure with a flat plate where clay plates support the catalyst. Samples were taken every 2 h to carry out the corresponding analyses and in a period of 8 h of reaction. For the characterization of rainwater adhered to Mexican regulations. The results showed that there was 6400 NMP/100 mL for fecal coliforms at the beginning, and after 4 h, this parameter goes to <2 NMP/100 mL. Initially, the same happened for fecal coliforms; 9200 NMP/100 mL was determined. After 4 h, this parameter drops to <2 NMP/100 mL. The same behavior was observed with chlorides, hardness, and total alkalinity, which showed a tendency to decrease significantly. This confirms that the system works properly to eliminate organic compounds and purify rainwater.

Keywords: rainwater, zinc oxide, photocatalysis, drinking water, falling film photocatalytic reactor (FFPR)

1. Introduction

Water is one of the essential natural resources and, together with air, land, and energy, constitutes the four essential resources on which development is based [1].

Water, as a natural resource, is manipulated by man, thus altering its cycle. Water is extracted from ecosystems for use. But a greater supply of water to the cities brings with it a greater discharge of wastewater, which damages the vegetation and the quality of the subsequent discharge. It is here where the importance of sustainable development must be recorded, which is the one that makes it possible to make the use of resources compatible with the conservation of ecosystems.

The amount of water we have on Earth is neither increasing nor decreasing, but the human population has grown drastically, and therefore, our need for this resource has also grown. Also, while the amount of water is constant, the way it is distributed over time is not: it is irregular throughout the year. It varies differently depending on global climatic conditions. In the same way, the diverse ecosystems, such as humid forests, pine forests, scrublands, grasslands, or deserts, influence the way and the amount of water that enters the aquifer systems, its conservation in the soil, or its passage into the atmosphere, which causes the availability of this resource to be variable in each region of the planet [2].

Access to good quality water is directly linked to human health and development. At present, the use of water for human consumption is limited since various sources contaminate most of the available fresh water.

In many parts of the world with high or medium rainfall, the quantity and quality of water necessary for human consumption is unavailable. Rainwater is used as a source of supply.

The observation of this acute and critical problem induced to seek and promote some practical methods that provide the small communities of the Municipality of Carmen, Campeche, with water of a certain quality for their consumption; therefore, in this investigation, the design and construction of a falling film photocatalytic reactor, in which a ZnO-Ag photocatalyst is fixed utilizing clay plates, with the purpose of degrading organic components through the incidence of solar radiation, is proposed.

The methodology considers the design, dimensions, and materials for constructing the photocatalytic reactor: the processes for preparing the solutions to be treated, the process of adequacy and photo deposition of the catalyst, and the optimal operating parameters of the falling film system.

2. Collection and storage

Rainwater harvesting and storage have been practiced for over 4000 years.

In the case of Mexico, the “aguadas” (artificial deposits) were used in pre-Columbian times to irrigate crops in small areas.

In archeological zones of the Yucatan peninsula, from the year 300 A.D., collection systems known as “chultus” were used, whose function is to collect rainwater from the patios and conduct it through channels to deposits built with stone to be utilized later [3]. The average annual rainfall for the Mexican territory is 1500 cubic kilometers of water. If 3% of this amount were used, it would be possible to supply 13 million Mexicans who currently do not have drinking water; two supplemental irrigations would be given to 18 million hectares of rainfed land; 50 million animals would be supplied, and 100,000 hectares of greenhouses would be irrigated [4, 5].

The collection, treatment, and use of rainwater is an essential source of water supply for human, livestock, and agricultural use and consumption for rural communities or populations of less than 2500 inhabitants, which have difficulties with their collection due to their topography, isolation, dispersion of hamlets, or lack of supply sources, whether surface or subway [4].

2.1 Water use in Campeche

In the state of Campeche, despite having several sources of fresh water and annual rainfall reaching an average of 1681.4 mm, there are no storage and treatment systems to use for human consumption, livestock, aquaculture, and vegetable irrigation, since there is no adequate technology to ensure that the water meets the specifications for these purposes [6].

The municipality of Carmen is located southwest of the state of Campeche, bordered to the north by the Gulf of Mexico and the city of Champotón, to the south by the state of Tabasco, to the east by the towns of Escárcega and Candelaria, and to the west by the municipality of Palizada. It is located between parallels 17°52′ and 19°01′ north latitude and meridians 90°29′ and 92°28′ west longitude of Greenwich. It has a territorial extension of 9720.09 km^2, representing 17.09% of the state's surface [4].

The municipality is in the Grijalva-Usumacinta hydrological region, the most crucial hydrological system in the state; due to its rainfall, periods of drought, and the topography of the terrain, it maintains a regime of irregular flows throughout the year, with the highest flows during the rainy season in summer and autumn, which decreases in winter and spring.

Today's world depends without exception on chemicals, whether to increase food production, protect health, or facilitate daily life [7, 8].

2.2 Water pollution

Three-quarters of the planet's surface is covered by water; approximately, there are 1385, 000, 000 km^3 of water, of which 97.3% is salt water, 2.08% is frozen at the poles, and only 0.62% is available for the development and sustainment of human life [9].

Water pollution is defined as the presence of foreign substances or organisms in a body of water in such quantities and with such characteristics that they prevent its use for specific purposes [4]. Such contamination can be of natural or anthropogenic origin, a direct consequence of the development of water resources [4].

Such pollution can be of natural or anthropogenic origin, directly from sewage or industrial runoff (point sources) or indirectly from air pollution or agricultural or urban runoff (non-point sources) [10].

There is a great variety of components in water, which their nature or size can classify. Their physical or biological nature can distinguish them by size; they are classified as suspended matter, colloidal matter, and dissolved matter.

Chemical pollutants include organic and inorganic components, each with specific characteristics in contaminated water. The presence of organic compounds decreases the amount of oxygen dissolved due to their biodegradation by aerobic microorganisms. On the other hand, the presence of inorganic compounds is related to toxicity, especially in the case of heavy metals. Some inorganic compounds, such as sulfites and nitrites, oxidize, generating a demand for dissolved oxygen [11, 12].

Physical pollutants include thermal pollution due to the discharge of slightly hot water used in heat exchangers, turbidity (caused by suspended solids), color, foaming, and radioactivity [11].

Biological contaminants include remains of plants, animals, and microorganisms. These contaminants are responsible for transmitting diseases, some of which are transmitted by biological pollutants, such as cholera, typhoid, paratyphoid, and so on.

2.3 Water decontamination

The use of water is limited by its quantity and quality. Water quality used to be qualified only by some physical parameters such as appearance, color, taste, and odor, and now, with the scientific and technological advances that have had an impact on the development of analytical techniques and processes capable of identifying a wide range of compounds, giving way to chemical and biological parameters, to such an extent that it is possible to make drinking water by purifying wastewater.

Physical parameters are the characteristics that the senses can perceive, such as suspended solids, turbidity, color, taste, odor, and temperature.

In water, there are biological species of different sizes and complexity for which there are biological parameters; the presence of specific organisms, bacteria, viruses, and protozoa, can be used as an indicator of a contaminant.

2.4 Heterogeneous photocatalysis

Heterogeneous photocatalysis is a discipline whose main objective is to decrease the energy activation of a photochemical reaction. Heterogeneous photocatalysis includes various reactions: moderate or total oxidation, dehydrogenation, metal deposition, water detoxification, removal of gaseous pollutants, and so on. It can be considered one of the new "advanced oxidation technologies" for water and air purification treatments [13].

Photocatalysis differs from conventional heterogeneous catalysis by activating the solid catalyst through photo absorption instead of thermal activation [13, 14].

This type of activation requires a semiconductor material as a catalyst, which must be provided with radiation energy higher than its forbidden band.

The overall process is like a conventional catalysis, which can be broken down into five independent steps:

1. Transfer of reactants from the fluid phase to the surface 2

2. Adsorption of at least one of the reactants

3. Reaction of the adsorbed phase

4. Desorption of the products

5. Removal of the products from the interface region

Three components are fundamental for a heterogeneous photocatalytic reaction: a photon source (with appropriate wavelength), a catalytic surface (usually a semiconductor material), and an electron acceptor, which in many cases is oxygen.

Charge carriers are transferred to the photocatalytic surface, allowing REDOX reactions with the adsorbed reactants. On the catalytic surface, REDOX reactions are separated into oxidation and reduction processes, involving, on the one hand, conduction band electrons and electron acceptors, for example, oxygen molecules ($e_{CB}^{-} + A \rightarrow A\bullet^{-}$), and, on the other hand, valence band holes and adsorbed electron donors, such as organic molecules or generally specific pollutants ($h_{VB}^{+} + D \rightarrow D\bullet^{+}$). Indirect oxidation reactions also occur by forming the highly oxidizing hydroxyl radical (HO•) generated by water oxidation by the voids [15–17].

Each ion formed then reacts to form an intermediate and end product. Because of the reaction, photonic excitation of the catalyst appears as the initial step of the entire catalytic system. Hence, the photon efficiency must be considered a reactant, and the photon flows as a special fluid, the "electromagnetic" phase. The photon energy is adapted to the adsorption of the catalyst, not to that of the reactants. The activation of the process goes through the excitation of the solid but not through the reactants. As demonstrated below, the adsorbed phase has no photochemical process, only a proper heterogeneous catalysis regime [13].

Whenever different semiconducting materials have been tested under comparable conditions for the degradation of the same compounds, TiO_2 has generally been shown to be the most active. This semiconductor is of particular interest as it can utilize natural (solar) UV because it has an appropriate energy separation between its valence and conduction bands, which can be overcome by the energy content of a solar photon (390 nm > γ > 300 nm) [18, 19].

Systems based on disinfection by heterogeneous photocatalysis using solar radiation have several advantages, such as [19–22]:

Non-consumption of high-value oxidizing agents and high production of TiO_2.

Since the ultraviolet radiation necessary for catalyst activation can be obtained from solar radiation, this system consumes minimal maintenance energy and no external energy consumption during operation.

The oxidants produced are high-powered and non-discriminating, with the advantage of eliminating most microorganisms and degrading or mineralizing most organic pollutants.

It can be applied in rural areas or areas of difficult access since other similar technologies, such as UV irradiation or ozone application, require an external energy source.

2.5 Catalysts

Zinc oxide (ZnO) is an economically accessible metal oxide, easily detectable. Low energy light can excite it, absorbing part of the solar radiation incident on the earth's surface. It has received interest in removing and destroying recalcitrant organic compounds within the photocatalysis technology [23, 24].

ZnO exhibits a broad spectrum of biocidal activity toward different bacteria, fungi, and viruses given to the production of reactive oxygen species and the release of Zinc ions (Zn^{2+}); it is also demonstrated that the synthesis of this metal oxide by green chemistry routes has a higher bioactivity, which is possibly attributed to the greater surface area, higher absorption capacity, crystallinity, and transmission [24–26].

Ag (Silver) is the most studied agent, and therefore, more information is available regarding its mechanism of antimicrobial activity; it is active against Gram-negative (*Escherichia coli*) and Gram-positive (*Staphylococcus aureus*) bacteria [26].

3. Methodology

3.1 Reagents

The reagents used throughout the different processes for the development of the research work in the treatment of rainwater disinfection are:

Zinc oxide of the J. T. Baker brand, silver sulfate of the Fermon brand, pyridine of the Aldrich brand, silver nitrate of the J.T. Baker brand, sodium chloride of Baker, sodium chloride of the Zeus brand, potassium chromate of the Fermon brand, ammonium chloride of the Baker brand, and ammonium hydroxide of the Baker brand, among others, especially in the analysis of rainwater samples.

3.2 Equipment

During the realization of this project, equipment located in the Catalysis Laboratory of the Faculty of Chemistry UNACAR was used, as described in the following **Table 1**.

3.3 Falling film photocatalytic reactor (FFPR)

The falling film photocatalytic reactor (FFPR) is designed with a metallic structure base that provides a surface area of 1.00 m^2; with the impregnated catalyst, the contact area is increased to 852 m^2, with a height of 80 cm; the photoreactor presents an inclination that can be positioned at an angle of 15, 30, and 45°; the feeding works with a recirculation system by gravity and pressure action making use of a centrifugal electric pump with magnetic dragging that is supported by distributed hoses the water through channels that at the beginning circulate the water on the surface of the photoreactor and at the end lead the water to a tank where the experimental water is kept in a container, thus completing the connection of the recirculation system; the water flow is controlled with a ball valve as shown in **Figure 1**.

The materials used for the construction of this reactor are listed in **Table 2**. It is made up of a metallic structure, a ceramic bed where the doped catalyst is found, and a pumping system.

3.3.1 Catalyst support

Red clay plates support the catalyst on a porous surface due to their adsorption capacity, stability, and non-toxicity. To cover the area of 1.00 m^2, ten plates are selected, and the minimum necessary cuts are made to fit the reactor. They are leveled with sandpaper on one side to homogenize the surface and washed to eliminate loose dust. Finally, they are placed in the muffle for 1 hour at 550°C to eradicate moisture and organic matter.

Nombre	
Analytical balance	Brand: VELAB, Model: VE-204
Mufla	Brand: Fisher Scientific, Model: 550-14
UV-vis spectrophotometer	Brand: Agilent Technologies, Model: CARY – 60
Multiparameter meter	Brand: Hach Co, Model: HQ40d
Gas extraction hood	Brand: Formacero, Model: 120 Serie 2
Centrifugal electric pump	Brand: Laing Thermotech, Model: LMB0510196
Electric stove	Brand: Aceq Lab. Model: HRN-1003 de 110 Lts
Vacuum pump	Brand: Rocker Lab, Model: Rocker 300

Table 1.
Equipment used.

Figure 1.
Schematic diagram of the FFPR.

Quantity	Material
10 units	Red clay plates of 30 × 30 × 2.5 cm
1 m^2	Galvanized stainless steel mirror finish caliber sheet 28
23 mL	Tubular profile de 1 × 1″
4 mL	Screed angle de 90° × 1 × 1/8 "
1 mL	Carbon steel pipe 2."
1 unit	Carbon steel coupling 1"
3 units	Butt hinge of 1"
1 unit	Electro centrifugal pump ½ Hp
3 m	water hose ½"
1 unit	ball valve for the water of ½"
6 units	UV lamps 365 nm 15 w

Table 2.
List of materials for the construction of the RFPD.

3.3.2 Catalyst synthesis

To carry out the dry impregnation synthesis of the catalyst, a solution of distilled water and zinc oxide at non-fixed concentrations is prepared, preparing a total of 33 g of ZnO for 1600 mL of distilled water, after which it is applied on the surface face of each clay plate to cover the entire reactor and is left to stand for 90 minutes. After this step, to dry, adhere, and activate the catalyst, the plates are introduced into the muffle for 2 hours at 550°C.

3.3.3 Catalyst doping

Once the catalyst is supported on the plates, the catalyst is doped with silver nanoparticles, starting with the preparation of a solution of distilled water and avocado sulfate at a concentration of 400 ppm for 8 liters of distilled water, which is carried out with the photo deposition method on the impregnated catalyst, and the recirculation of the silver sulfate solution is started. At the same time, it is irradiated with six ultraviolet light lamps of 365 nm and 15 watts of power for a time of 6 hours and is left to stand for 2 hours; the next step is activation by calcination, and for this, each plate is introduced to the muffle for 2 hours at 550°C. Ultimately, the plates are placed on the reactor, ready to test its functionality and disinfect the rainwater.

3.3.4 Reactor start-up testing

Four water absorption tests were performed before using the FFPR to treat rainwater. The results of this absorption are shown in **Table 3**.

It was shown that before incorporating the solution to be treated over the reactor, it was necessary to include at least 3 L of water and consider that the clay tiles absorb this amount.

To ensure reliability in the reaction system, initial tests are made with the degradation of the pyridine compound based on previous research results [27]. Initial concentrations of 50 parts per million (ppm) of the pyridine compounds were used as the model compound, which was treated for 8 h, taking samples of 50 mL every 2 hours, which were analyzed in a Cary-60 Uv–vis spectrophotometer. In this equipment, a sweep was performed from 200 to 600 nm.

3.4 Rainwater samples

Collecting, preserving, and storing samples for subsequent characterization and microbiological decontamination are necessary for the disinfection treatment of rainwater over the photocatalytic reactor.

The collection of 25 samples of 200 mL of rainwater was carried out during the rainy season in a period from June to September in Ciudad del Carmen, Campeche, at location 18.642522, −91.816223, with a collection system built to capture water directly from the source, in the open air; they were preserved and stored under refrigeration until they were handled for the corresponding characterization.

3.5 Characterization of rainwater

The physical and chemical analyses are carried out by the Mexican Official Standard NOM-127-SSA1–1994, "Environmental Health, water for human use and consumption-permissible quality limits and treatments to which water must be subjected for its potabilization." In this case, 20 L of rainwater was treated in batches and treated in the reactor for 8 hours with exposure to sunlight; samples were taken every hour to see the progress of the disinfection or treatment given to this water.

Test number	Incorporated water (L)	Exposure time (h)	Water absorbed (L)
1	10	8	4
2	10	8	3.5
3	20	8	3
4	20	8	3

Table 3.
Analysis of absorption tests.

In the Teaching, Research and Services, Management and Environmental Control Laboratory of the Faculty of Chemical Sciences of the Veracruzana University Orizaba campus, Veracruz., samples were analyzed, according to the official Mexican regulations.

Table 4 shows the parameters analyzed and the regulations applied; each parameter is analyzed utilizing an official Mexican standard.

4. Results of the project

The results obtained throughout this project of constructing a photocatalytic reaction system to treat rainwater and be used in small communities of Carmen, Campeche, are shown.

The construction of the photocatalytic reaction system in its different stages of development and the tests for using the reactor are shown. And in the same way, results regarding the collection, handling, and physical, chemical, and microbiological analyses of rainwater samples are presented.

4.1 Construction of the falling film photocatalytic reactor

For the construction of the falling film photocatalytic reactor, as the design was presented in the methodology, the materials listed in **Table 2** were used.

As shown in **Figure 2**, its metallic structure provided an occupation area of 1 m^2 to support the clay plates that, at the same time, kept the catalyst for its subsequent impregnation and doping process, working at an angle of 30° inclination (**Figure 2a**).

Test carried out	Mexican regulations
Physical Characterization	
Temperature	NMX-AA-007-SCFI-2013 [28]
Hydrogen Potential	NMX-AA-008-SCFI-2016 [29]
Total Solids	NMX-AA-034-SCFI-2001 [30]
Volatile Total Solids	NMX-AA-034-SCFI-2001 [30]
Total Suspended Solids	NMX-AA-034-SCFI-2001 [30]
Volatile Suspended Solids	NMX-AA-034-SCFI-2001 [30]
Total Dissolved Salts	NMX-AA-034-SCFI-2001 [30]
Chemical Characterization	
Total Chlorides	NMX-AA-073-SCFI-2001 [31]
Total hardness	NMX-AA-072-SCFI-2001 [32]
Total Acidity and Alkalinity	NMX-AA-036-SCFI-2001 [33]
Chemical Oxygen Demand	NMX-AA-030-SCFI-2001 [34]
Microbiological analysis	
Determination of fecal and total coliforms	NMX-AA-042-SCFI-2015 [35]

Table 4.
Tests are carried out according to the corresponding regulations.

Figure 2.
Construction stages of the FFPR. (a) Metallic structure, (b) recirculation system, and (c) FFPR operating in previous degradation tests.

We also proceeded to incorporate the recirculation system (**Figure 2b**), which was supported by the electric pump to supply the liquid through hoses that lead it to the distribution channel, descending through the surface of the reactor to be collected directly in the collection channel that leads the liquid to an outlet with a hose so that the liquid reaches the collection tank again.

4.2 Catalyst support

For the catalyst support, clay plates were used that were subjected to a selection process during the process (**Figure 3a**). In addition, they were adapted to the reactor to support the total area (**Figure 3a**); and finally, they were subjected to a calcination process (**Figure 3b**).

4.3 Catalyst impregnation

For the impregnation of the support, the solution prepared with the ZnO catalyst and distilled water was used (**Figure 4a**); later it was dried to remove the liquid trapped in the pores (**Figure 4b**) then it's calcined at 550 °C for 1 h, and the activation was achieved of the catalyst. Calcination increases the surface concentration and generates a catalyst with uniform distribution (**Figure 4c** and **d**).

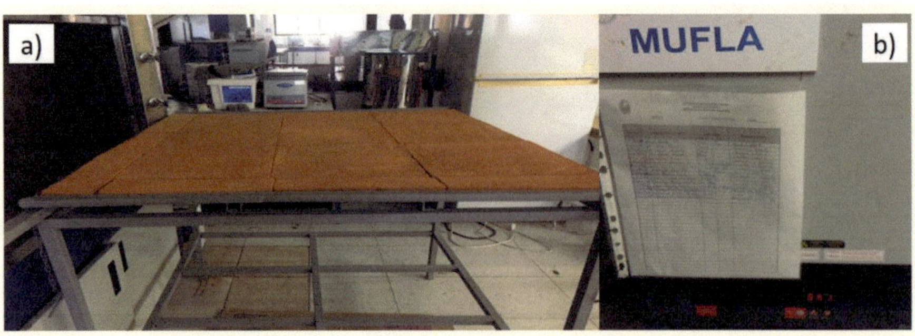

Figure 3.
(a) Catalyst support and placing clay plates on the FFPR, (b) calcination process, observing the porosity and homogeneity on the surface of the plate.

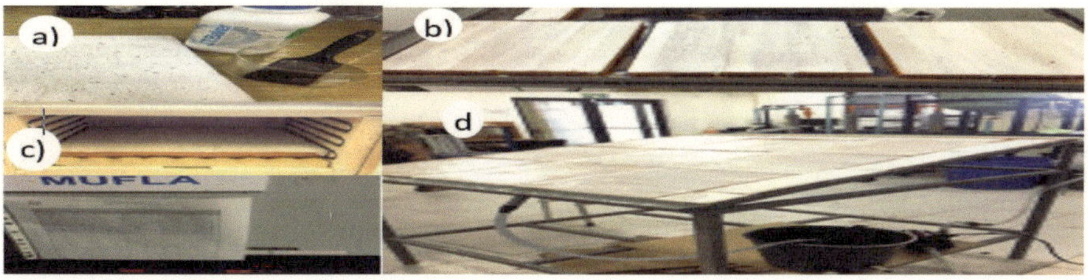

Figure 4.
ZnO catalyst impregnation process (a) and (b); impregnated support plates and fixation of the support on the FFPR and calcination of a plate in muffle at 550°C (c), completing the impregnation with the catalyst on the FFPR (d).

4.4 Catalyst doping

For the doping of the catalyst with Ag^+ nanoparticles, the solution prepared with Ag_2SO_4 and distilled water (**Figure 5a**) was used to initiate recirculation under exposure to UV light (**Figure 5b**) and thus achieve the photo deposition of the silver (**Figure 5c**). Finally, it was calcined to reach the activation of the bactericidal metal.

4.5 RFPD tests

The falling film photocatalytic reactor demonstrated its feasibility in the photodegradation of the organic pyridine compound in an initial concentration of 50 ppm using a ZnO – Ag^+-doped catalyst.

Figure 6 shows the monitoring of the photodegradation of the organic compound utilizing absorbance graphs at a wavelength of 200 to 600 nm. However, more excellent activity of the compound was observed in a range of 200 to 300 nm.

The organic compound presented a characteristic maximum peak of the same at a wavelength of 256 nm and simultaneously showed distinct intermediate peaks of the pyridine [36].

The pyridine photodegradation test was carried out with the same initial concentration for 6 h, taking a sample every 2 h to observe its behavior.

In the degradation of the pyridine, as shown in **Figure 7**, it can be observed how the behavior of the pyridine changes for the reaction time.

It is observed that in the time of zero hours (t = 0 h), the natural behavior of the compound is shown, which ensures that the organic compound is present in the solution, having an initial concentration of 50 ppm and showing absorbance of 2351

Figure 5.
Catalyst doping process with Ag^+ nanoparticles. (a) Preparation of the solution with silver sulfate, (b) doping process by photo deposition on the support impregnated with the catalyst; (c) support of the FFPR finishing the photo deposition.

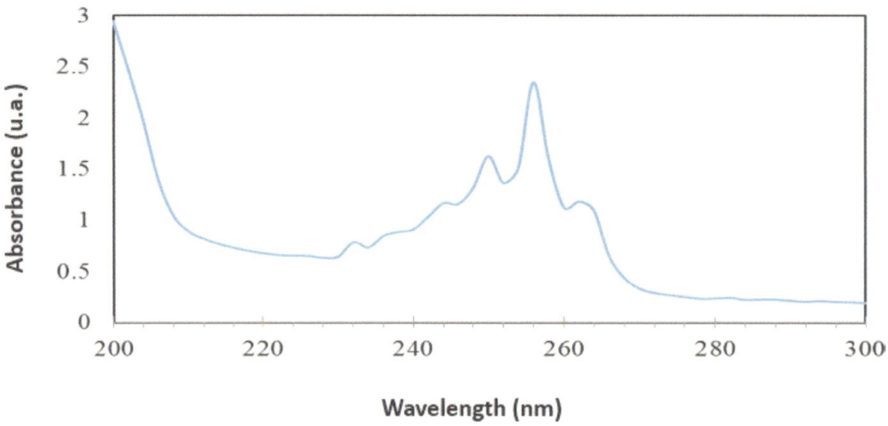

Figure 6.
A spectrum of organic compound pyridine with a concentration of 50 ppm.

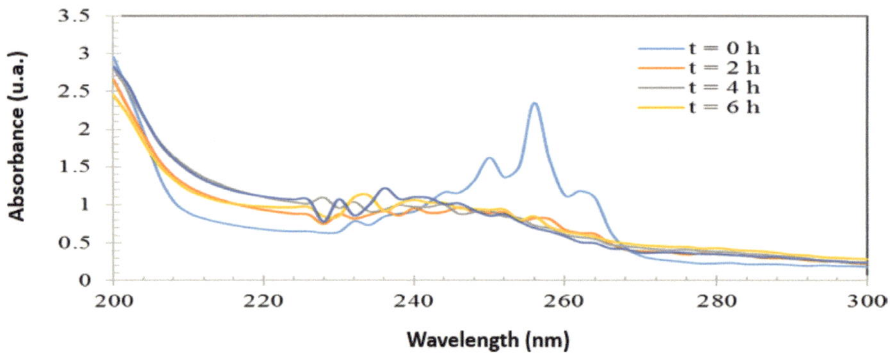

Figure 7.
The organic compound pyridine was degraded on the falling film photocatalytic reactor in the presence of the $ZnO - Ag^+$-doped catalyst for 6 h.

arbitrary units (u.a.); then, in the elapsed time of 2 hours (t = 2 h) of the degradation on the photoreactor, it can be observed that the concentration drops to 0.825 u.a., presenting a degradation of 65%; for the 4 hours of reaction (t = 4 h), it shows a decrease in the concentration to 0.731 u.a. resulting in a 69% degradation; later, after 6 hours of reaction (t = 6 h), it presents a slight increase in concentration to 0.845 u.a.; however, after 8 hours of reaction (t = 8 h) in the system, it is possible to drop to 0.695 u.a., reaching a final degradation of 70% [22].

It should be noted that randomly generated peaks appear from the tempo of 2 h to the time of 8 h because the main compound is degrading. Still, alternately, there is the presence of other intermediate compounds being formed.

Figure 8 shows the spectrophotometer reading carried out in a degradation of the same compound at initial concentrations of 50 ppm. In this image, you can better appreciate the formation of intermediate compounds after the first 2 h of photodegradation; such shapes can be attributed to the presence of 2 hydroxy pyridines, as reported by Montalvo (2009) [36]; however, at 4 h, the pyridine peak disappears.

In **Figure 9**, the trend line shows a reliability of 95.5% concerning the average data represented that oscillate in the initial concentration from 50 ppm to 16.22 ppm,

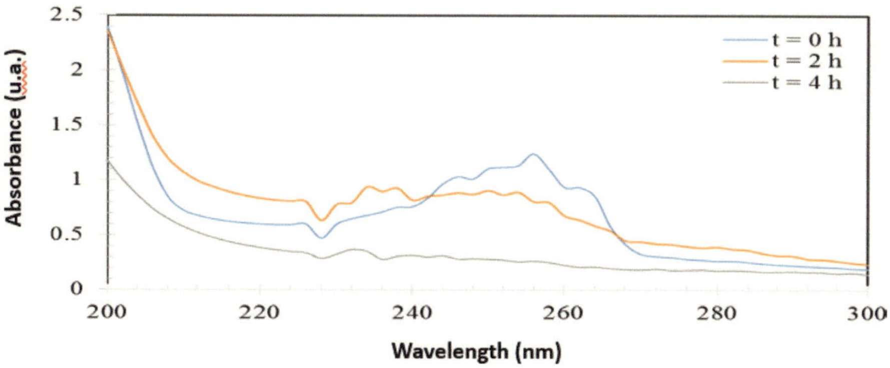

Figure 8.
Degradation of the organic compound pyridine during a period of 4 h.

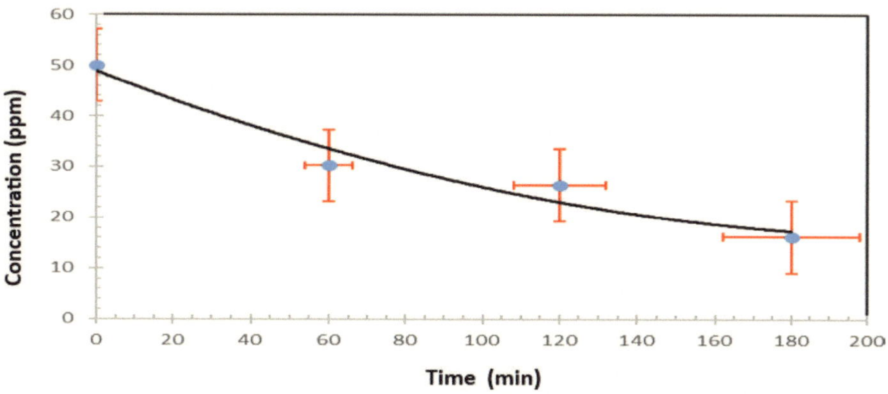

Figure 9.
Line of tendency and standard deviation of the degradation of the organic compound PYR in a period of 3 h.

considering a time of 0 to 180 minutes, which was the most representative time in the photodegradation tests.

Likewise, the standard deviation representing the variations of the average obtained in the data of the different photodegradation tests carried out to get these results is evidenced.

Figure 9 describes the degradation of the organic compound pyridine; it starts from a concentration of 50 ppm in its initial time. After 60 min of the photodegradation test, it degrades to 30.2599 ppm; after 120 min of degradation, the concentration of the compound drops to 26.3962; for a time of 180 min, the concentration is lowered to 16.2276 ppm.

In **Figure 10**, you can satisfactorily see the increase in the photodegradation of the organic compound; at time zero, there is no result since the presence of the contaminant is intact in the solution; for a time lapse of 60 min of photodegradation, there is a reduction from 50 ppm to 30.2599 ppm, achieving a percentage of photodegradation of 39% concerning the initial concentration of the original compound; it can also be observed that at 120 min, the photodegradation increases by 20% and after 180 minutes, a photodegradation of 68% [27].

Such results obtained are adequate due to the conditions managed, referring to the volume worked under direct exposure to sunlight.

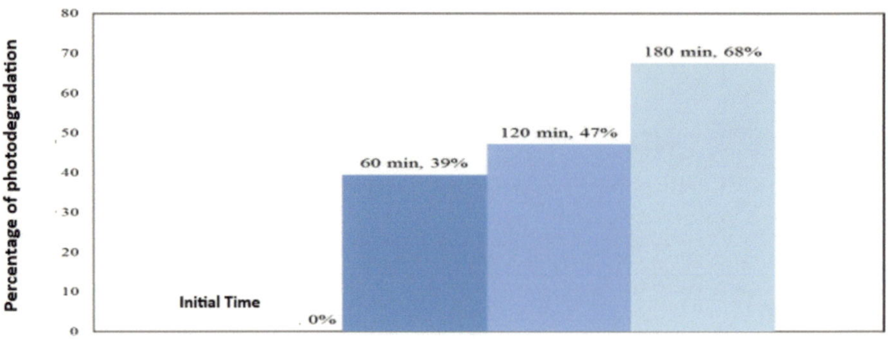

Figure 10.
Average percentage of degradation of the organic compound pyridine during 3 h.

4.6 Obtaining kinetic parameters

Various studies show that the first step in the photocatalytic degradation of organic compounds follows first-order or zero-order kinetics. The results clearly show that the reaction rate depends fundamentally on the concentration of the reaction (**Figure 9**), so it can be said that it follows first-order kinetics or is like first-order kinetics:

$$r_a = \frac{dC}{dt} = K \tag{1}$$

By integrating the velocity equation, the following function is obtained:

$$In\frac{C}{Co} = -K_{\text{apparent}} \tag{2}$$

Which is equivalent to $Co - K_{\text{apparent.}}$

Figure 11 plots the logarithms of the normalized concentrations log(C/Co) vs. reaction time. The values of the apparent reaction constant were obtained by linear regression.

The data are necessary to obtain the $K_{apparent}$ kinetic constants from the line equation. **Figure 11** shows the photocatalytic degradation of pyridine, and this follows a pseudo-first-order reaction since the reaction rate is the same as the concentration increases. **Table 5** shows the values of this constant obtained as the same from **Figure 11**.

4.7 Sample collection

A collection system was built to capture rainwater directly from the source, in the open sky, and thus collect 25 samples that were preserved, stored, and refrigerated until their characterization and the final analysis (**Figure 12**). These samples were collected during the rainy season from June to September in the municipality of Carmen city, Campeche.

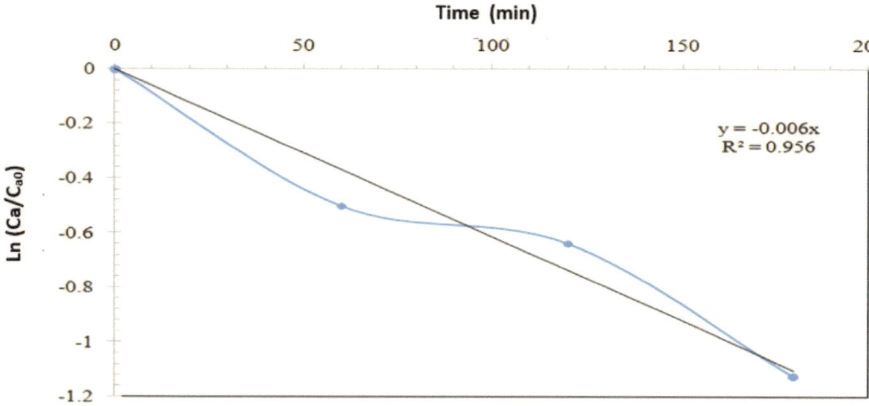

Figure 11.
Kinetics of the pyridine reaction using ZnO-Ag⁺ as a catalyst.

$C=C_o$ (ppm)	K_{apr} $(min)^{-1}$ irradiated with sunlight	R^2	r_o, (t = 0) (ppm/min)
50	0.006	0.956	0.3

Table 5.
The apparent velocity (min^{-1}) for the degradation tests utilizing a $ZnO\text{-}Ag^+$ reactor.

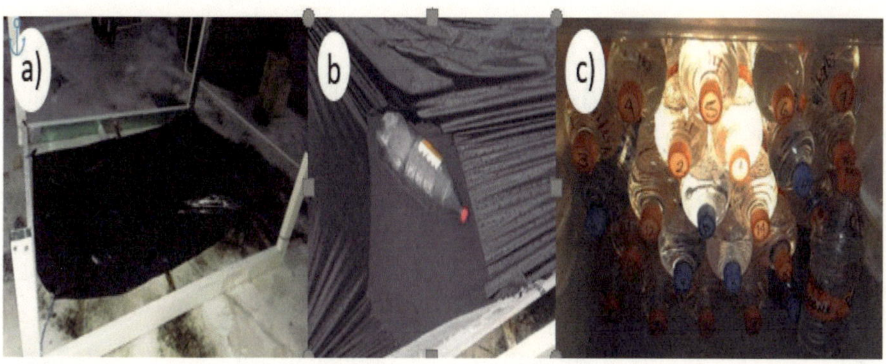

Figure 12.
Sample collection. a) Construction of the rainwater collection system, b) water collected during seasonal rain, and c) samples stored and refrigerated.

4.8 Characterization and analysis of rainwater samples

The Official Mexican Standard NOM-127-SSA1–1994 [37], referring to Environmental Health regarding water for human use and consumption, establishes the permissible limits of quality and treatments to which water must be subjected for purification. For this reason, it is considered to carry out the pertinent characterization tests on the water samples, following the criteria it establishes.

The average results obtained from the tests carried out during the characterization and physical and chemical analyses obtained are shown in **Tables 6** and 7, respectively, as well as their maximum permissible limit according to regulations.

Test carried out	Sample evaluated	Average value obtained	Maximum permissible limit	Unit of measurement
Temperature	01–18	26.2	40	°C
Total Solids	04, 05, 07, 12	28	200	mg/L de STT
Volatile Total Solids	04, 05, 07, 12	22	120	mg/L de STV
Total Suspended Solids	04, 05, 07, 12	8	100	mg/L de SST
Volatile Suspended Solids	04, 05, 07, 12	6	70	mg/L de SSV
Total Dissolved Salts	04, 05, 07, 12	20	1000	mg/L de SDT

Table 6.
Average results obtained from physical tests carried out on rainwater according to the corresponding regulations.

Test carried out	Sample evaluated	Average value obtained	Maximum permissible limit	Unit of measurement
Hydrogen Potential	01–18	7.33	6.5–8.5	Units of pH
Total Chlorides	10, 11, 18	0.0013	250.00	mg Cl−/L
Total hardness	01, 09, 14, 17	3.003	500	mg/L de $CaCO_3$
Alkalinity	02,03,08,15,16	13.2	—	Units of pH
Total Acidity	02,03,08,15,16	2.4	—	Units of pH
Chemical Oxygen Demand	06, 13	0.0002	<1	mg/L COD

Table 7.
Average results obtained from chemical tests carried out on rainwater according to the corresponding regulations.

The analyzed parameters show average results below the maximum permissible limit since, as previously mentioned, the rainwater was collected directly from the source in the open sky without having contact with any contaminated surface and was later stored in refrigeration, to preserve its properties until reaching the corresponding characterization analysis.

4.9 The treatment of collected rainwater

When carrying out the photodegradation treatment of rainwater collected in a cistern, the 20-L sample was recirculated in the RFPD to maintain contact of liquid with the surface of the reactor with the supported and impregnated catalyst and under the effect of sunlight for eight hours, a 1 L sample was taken every 4 h to carry out the tests on the main parameters and verify the effectiveness of the system when degrading the organic components present. The data shown in **Table 8** were obtained from the tests carried.

The data shown in **Table 8** show that the rainwater that was not promptly used for domestic use began to be collected through the roof, conducted by pipes to be stored in a typical cistern, where, by not having an immediate disinfection treatment, it became rich in microbiological contaminants (bacteria, fungi, algae, etc.) by carrying garbage, organic matter from the environment, bird droppings or other animals that circulate on the roof, and therefore, stored water becomes a cause of concern for health care.

Each microorganism grows individually. However, ambient temperature and an average pH of 6.69 are critical factors that increase the growth rate of

Parameter/Time (h)	0	4	8	Unit	Official Mexican Standard
Hydrogen Potential	6.44	7.07	6.55	pH	NMX-AA-008-SCFI-2006 [29]
Total Dissolved Salts	452	423.5	196.5	mg/L	NMX-AA-034-SCFI-2015 [30]
Chlorides	18.85	26.4	30.4	mg/L	NMX-AA-073-SCFI-2001 [31]
Hardness	68.79	158.52	220.33	mg/L	NMX-AA-072-SCFI-2001 [32]
Total alkalinity	121.45	120	125	mg/L	NMX-AA-036-SCFI-2001 [33]
Total coliforms	92,000	<2	<2	NMP/100 mL	NMX-AA-042-SCFI-2015 [35]
Fecal coliforms	6400	<2	<2	NMP/100 mL	NMX-AA-042-SCFI-2015 [35]

Table 8.
Results obtained from the analysis of rainwater collected according to the corresponding regulations.

microorganisms, mainly mesophilic bacteria, total coliforms, fecal [thermotolerant] bacteria, and *Escherichia coli*.

Regarding the dissolved salts, it can be noted that there was a decrease from 452 to 196.5 mg/L, which favors the water quality, an indicator of the effectiveness of the photocatalytic process of the water disinfection treatment.

Chloride ions are one of the most widespread ions in natural waters. It is not usually an ion that poses portability problems to drinking water, although it is an indicator of water contamination due to human action. The maximum permissible chloride concentration for human consumption is 250 mg/L. Therefore, an average of 25.22 mg/L is an acceptable average value for using water stored in a cistern.

Magnesium and calcium salt depend fundamentally on the geological formations that the water traverses before its collection. In this case, the hardness increases from a value of 68.79 to 220.33 mg/L; it is considered hard water, which means that it contains more calcium and magnesium minerals; as the hardness of water increases, more calcium and magnesium are dissolved. Magnesium and calcium are positively charged ions. Due to their presence, other positively charged ions will dissolve less easily in hard water than in water that does not contain calcium and magnesium.

Considering the total alkalinity parameter with an average value of 122.15 mg/L, which is an acquired value of the materials added in domestic uses due to the cistern water storage system, showing the presence of hydroxides, carbonates, and bicarbonates of elements such as calcium, magnesium, sodium, potassium, or ammonia, considered as biological nutrients.

Total coliforms and fecal coliforms are the most significant indicators of microbiological contamination; the presence and degree of fecal contamination are essential factors in evaluating water quality [38, 39]. The initial tests (t = 0 h) present a significant NMP of total and fecal coliforms, 92,000 and 6400 NMP/100 mL, respectively. In both cases, the effectiveness of the photocatalytic reaction is verified in the tests carried out at t = 4 h, eliminating thoroughly any coliform present in the medium.

5. Conclusions

Given the nature of the construction of the falling film photocatalytic reactor, it is concluded that it satisfactorily fulfilled the function for which it was designed when testing its functionality in the previous tests and the same degradation of rainwater collected in the cistern. It is an alternative for the development of future research with different contaminants and at larger scales since it did not show any complication in the mineralization of organic compounds and microorganisms using zinc oxide catalyst doped with Ag^+ nanoparticles, interacting under the incidence of sunlight.

The clay plates, due to their composition and their manufacturing, had deformations due to the preparation on the surface face and the cuts made to adapt the reactor to the size, creating pathways through which the water had greater ease in moving, and at the same time, crests were created where the passage of the flow of the treated liquid over the reactor occurred to a lesser extent.

However, for use as support, the clay was an excellent receptor of the catalyst since a uniform impregnation of ZnO on the clay was achieved, increasing the contact area of the catalyst from 1 m^2 of the surface area to 852.5 m^2.

The doping of the ZnO catalyst with the photo deposition of Ag^+ nanoparticles is a proven good alternative for the photocatalytic degradation in the mineralization of contaminating microorganisms, showing the breakdown of their cell wall in a period of 4 hours. At the same time, the degradation of the organic compound "pyridine" is verified, showing degradation efficiency of up to 68% in a time of 3 h.

The operating parameters of the falling film system were tested at an angle of 30°, at a flow rate of 8 L/min, the results obtained in the degradation of the organic compound pyridine being sufficient and optimal, using the low ZnO – Ag^+ doped catalyst—the incidence of sunlight.

The rainwater collected in a cistern was disinfected with the advanced oxidation treatment applied with the RFPD, complying with the NOM-127-SSA1–1994 regulations, considering the principal physical, chemical, and microbiological parameters for water use. It guarantees that this water can be used as drinking water, since the quantitative determination of each parameter was below the maximum permissible limits referenced in the Official Mexican Standard.

Conflict of interest

The authors declare no conflict of interest.

Author details

Carlos Montalvo[1]*, Claudia A. Aguilar[1], Rosa A. Martínez[1], Rosa M. Cerón[1], Alejandro Ruiz[1], Eric Houbron[2] and Juan C. Robles[1]

1 University Autonomous of Carmen, Carmen City, México

2 University Veracruzana, Veracruz, México

*Address all correspondence to: cmontalvo@pampano.uanacar.mx

References

[1] Vidales Contreras JA et al. Comportamiento de un sistema de pantanos construidos de flujo superficial para el tratamiento de efluente secundario y de retrolavado. Tropical and Subtropical Agroecosystems. 2011;**14**:375-384

[2] Aurazo, de ZM. Aspectos biológicos de la calidad del agua. In: Tratamiento de agua para consumo humano: Plantas de filtración rápida. Manual I: Teoría. Vol. 1. Lima, Perú. Centro Panamericano de ing. Sanitaria y ciencias ambientales; 2004. pp. 58-102

[3] Anaya GM. Manual de sistemas de captación y aprovechamiento del agua de lluvia para uso doméstico en América Latina y el Caribe [thesis]. Montecillo, México: IICA, Colegio de Posgraduados; 2004

[4] CONAGUA. Estadística del agua en México. Síntesis, secretaria de Medio ambiente y recursos Naturales. In Comisión Nacional del Agua. Sistema Nacional de información del Agua. [Internet]. 2022. Available from: https://sina.conagua.gob.mx/publicaciones/Numeragua_2022.pdf [Accessed: June 7, 2023].

[5] FAO. Captación y almacenamiento de agua de lluvia. Santiago de Chile: FAO; 2013. DOI: 10.1111/jce.13019

[6] Nava J, Arenas P, Cardoso F. Integrated coastal management in Campeche, Mexico; a review after the Mexican marine and coastal national policy. Ocean & Coastal Management. 2018;**154**:34-45. DOI: 10.1016/j.ocecoaman.2017.12.029

[7] Hassaan MA, El Nemr A. Pesticides pollution: Classifications, human health impact, extraction, and treatment techniques. The Egyptian Journal of Aquatic Research. 2020;**46**:207-2020. DOI: 10.1016/j.ejar.2020.08.007

[8] SEMABICC. Bitácora ambiental. Ordenamiento ecológico territorial del municipio de Carmen. Secretaría de medio ambiente, biodiversidad y cambio climático. Campeche, Mexico; 2013

[9] Bofill S, Casares P, Giménez N, De Porta C, Gonfa A, Llop R. Efectos sobre la salud de la contaminación de agua y alimentos por virus emergentes humanos. Revista Espanola de Salud Publica. 2005;**79**:253-269. DOI: 10.1590/s1135-57272005000200012

[10] Allegrini I, Ianniello A, Valentini F. Environmental air pollution: An anthropogenic or a natural issue. In: Current Trends and Future Developments on (Bio-) Membranes. Roma Italy: Elsevier; 2023. pp. 1-38. DOI: 10.1016/B978-0-12-824103-5.00007-3

[11] Rodrigues S, Römkens P. Chapter 9, human health risks and soil pollution. In: Soil Pollution. Human Health Risks and Soil Pollution. Portugal: Science Direct; 2018. pp. 217-250. DOI: 10.1016/C2016-0-02243-X

[12] Babu D, Srivastava V, Nidheesh P, Kumar M. Detoxification of water and wastewater by advanced oxidation processes. Science of the Total Environment. 2022;**696**:133961. DOI: 10.1016/j.scitotenv.2019.133961

[13] Herrmann J. Heterogeneous photocatalysis: Fundamentals and applications to the removal of various types of aqueous pollutants. Catalysis Today. 1999;**53**:115-129. DOI: 10.1016/S0920-5861(99)00107-8

[14] Gisbertz S, Pieber B. Heterogeneous Photocatalysis in organic synthesis. Chem. Photo. Chem. 2020;**4**:456-475. DOI: 10.1002/cptc.202000014

[15] Liu B, Zhao X, Parkin IP, Nakata K. Chapter 4 - charge carrier transfer in photocatalysis. In: Interface Science and Technology. Vol. 31. United Kingdom: Elsevier; 2020. DOI: 10.1016/B978-0-08-102890-2.00004-X

[16] Abdel Messih MF, Ahmed MA, Soltan A, Anis SS. Synthesis, and characterization of novel Ag/ZnO nanoparticles for photocatalytic degradation of methylene blue under UV and solar irradiation. Journal of Physics and Chemistry of Solids. 2019;**135**:109086-109104. DOI: 10.1016/j.jpcs.2019.109086

[17] Carbajo J. Aplicación de la fotocatálisis solar a la degradación de contaminantes orgánicos en fase acuosa con catalizadores nanoestructurados de TiO_2 [thesis]. Madrid, Spain: Facultad de Ciencias, Departamento de Química Física Aplicada. Universidad Autónoma de Madrid; 2013

[18] Malato S, Fernandez P, Maldonado MI, Blanco J, Gernjak W. Decontamination and disinfection of water by solar photocatalysis: Recent overview and trends. Catalysis Today. 2009;**147**:1-59

[19] El-Aassar A, Hameed M, Isawi H, El-Noss M, El-Kholy RA, Said MM, et al. Design and fabrication of continuous flow photoreactor using semiconductor oxides for degradation of organic pollutants. Journal of Water Process Engineering. 2019;**32**:100922. DOI: 10.1016/j.jwpe.2019.100922

[20] Hashim FS, Alkaim AF, Salim SJ, Alkhayatt AHO. Effect of (Ag, Pd) doping on structural, and optical properties of ZnO nanoparticles: As a model of photocatalytic activity for water pollution treatment. Chemical Physics Letters. 2019;**737**:136828. DOI: 10.1016/j.cplett.2019.136828

[21] Lu F, Wang J, Chang Z, Zeng J. Uniform deposition of Ag nanoparticles on ZnO nanorod arrays grown on polyimide/Ag nanofibers by electrospinning, hydrothermal, and photoreduction processes. Materials and Design. 2019;**181**:108069. DOI: 10.1016/j.matdes.2019.108069

[22] Meng X, Zong P, Wang L, Yang F, Hou W, Zhang S, et al. Au-nanoparticle-supported ZnO as highly efficient photocatalyst for H_2O_2 production. Catalysis Communications. 2020;**134**:10-13. DOI: 10.1016/j.catcom.2019.105860

[23] Chong MN, Bo J, Chow CK, Saint C. Recent developments in photocatalytic water treatment technology: A review. Water Research. 2010;**44**:2997-3010. DOI: DOI

[24] Wetchakun K, Wetchakun N, Sakulsermsuk S. An overview of solar/visible light-driven heterogeneous photocatalysis for water purification: TiO_2 and ZnO-based photocatalysts used in suspension photoreactors. Journal of Industrial and Engineering Chemistry. 2019;**71**:19-49. DOI: 10.1016/j.jiec.2018.11.025

[25] Ma W, Li L, Liu Y, Sun Y, Kim I, Ren X. Tailored assembly of vinylbenzyl N-halamine with end-activated ZnO to form hybrid nanoparticles for quick antibacterial response and enhanced UV stability. Journal of Alloys and Compounds. 2019;**797**:692-701. DOI: 10.1016/j.jallcom.2019.05.174

[26] Rosai C, Dilarri G, Froes C, Ruy V, Matos P, Bueno P, et al. Antibacterial action and target mechanisms of

zinc oxide nanoparticles against bacterial pathogens. Scientific Reports. 2020;**12**:2658-2670. DOI: 10.1038/s41598-022-06657-y

[27] Alcocer de la HR. Implementacion de un sistema de degradación de compuestos orgánicos (piridina) con luz natural y luz uv a diferentes condiciones de operación [thesis]. Campeche: UNACAR Ciudad del Carmen; 2018

[28] NMX-AA-007-SCFI-2013, Water analysis - determination of temperature in natural waters, wastewaters and treated wastewaters – Test method

[29] NMX-AA-008-SCFI-2016, Water analysis-pH measurement in natural, wastewater and treated wastewater" Test method- (CANCELS NMX-AA-008-SCFI-2011)

[30] NMX-AA-034-SCFI-2015. Water analysis - measurement of solids and dissolved salts in natural, wastewater and treated wastewater – Test method (CANCELS NMX-AA-034-SCFI-2001)

[31] NMX-AA-073-SCFI-2001. Water analysis - determination of total chloride in natural, wastewater and treated wastewater - Test method (CANCELS NMX-AA-073-1981)

[32] NMX-AA-072-SCFI-2001. Water analysis - determination of total hardness in natural, residual and treated waste water - Test method (CANCELS NMX-AA072-1981)

[33] NMX-AA-036-SCFI-2001. Water analysis - determination of acidity and alkalinity in natural, wastewater and treated wastewater - Test method (CANCELS NMX-AA-036-1980)

[34] NMX-AA-030-SCFI-2001. Water analysis - determination of the chemical oxygen demand in natural, wastewater and treated wastewater - Test method (CANCELS NMX-AA-030-1981)

[35] NMX-AA-042-SCFI-2015. Water analysis - enumeration of total coliform organisms, fecal coliform organisms (thermotolerant), and Escherichia coli - Most probable number method in multiple tubes (CANCELS NMX-AA-42-1987)

[36] Montalvo C. Degradación fotocatalítica de compuestos que aportan olor al agua residual y potable [thesis]. San Luis Potosí: Universidad Autónoma De San Luis Potosí; 2009

[37] NOM-127-SSA1-1994, “Environmental health, water for human use and consumption-permitted limits of quality and treatments to which the water must be subjected for its potabilization”

[38] Agrawal S, Mandal TK, Mohanty P. Ag^+ driven antimicrobial activity of Ag^+: ZnO nanowires immobilized on paper matrices. Materialia. 2019;**8**:100490. DOI: 10.1016/j.mtla.2019.100490

[39] Karami A, Xie Z, Zhang J, Kabir MS, Munroe P, Kidd S, et al. Insights into the antimicrobial mechanism of Ag and I incorporated ZnO nanoparticle derivatives under visible light. Materials Science and Engineering. 2020;**107**:110220. DOI: 10.1016/j.msec.2019.110220

Section 2

Sorbents and Adsorbents

Chapter 5

Green Adsorbents for Water Purification

Hafsa Muzammal, Muhammad Danish Majeed, Muhammad Zaman, Muhammad Safdar, Muhammad Adnan Shahid, Zahid Maqbool and Tayyaba Majeed

Abstract

Water purification is crucial for ensuring access to safe and clean drinking water. As a sustainable and effective solution, green adsorbents have gained significant attention in recent times. These adsorbents are composed of natural or waste-based materials that are biodegradable, renewable, and abundant, making them an eco-friendly substitute for traditional adsorbents. This chapter offers a comprehensive outline of the present and prospects of green adsorbents in water purification. It encompasses a comprehensive overview of the many forms of green adsorption agents comprising natural, agriculture waste-based, & industrial waste-based adsorbents, as well as their synthesis and modification methods. Furthermore, it also explores the potential applications of green adsorbents in removing heavy metals, organic pollutants, and inorganic contaminants from water. The challenges and future directions of green adsorbent research are also discussed, including the limitations of these adsorbents and the opportunities for enhancing their performance. Overall, this chapter offers valuable insights into the potential of green adsorbents as a sustainable and eco-friendly solution for water purification.

Keywords: green absorbent, water purification, green treatments, natural adsorbents, agricultural adsorbents, industrial adsorbents, heavy metals, organic and inorganic solutions

1. Introduction

Water conflicts are becoming more frequent because of the rising global need for safe drinking water [1]. The primary sources of drinking water are natural waters, such as groundwater, surface water (rivers and lakes), and rainwater, although desalination of brackish and seawater is becoming more important. More than 50% of the world's population uses groundwater for drinking, which accounts for 97% of all freshwaters [2]. Groundwater is the most economically viable option in many remote and underdeveloped regions where basic water delivery systems do not exist [3]. However, inorganic contaminants such as fluoride, uranium as arsenic, and boron are regularly detected in groundwater. Humans get negative health impacts from long-term exposure to such pollutants. For instance, chronic uranium intake from drinking water damages the kidneys, while excessive fluoride intake causes dental and skeletal fluoroscopic [4]. Inorganic pollutants are strongly influenced by geology. Therefore, creating proper and adaptable technology for local use is the most practical way to get removal of such ions, species, and toxins [5].

Adsorption, coagulation, flocculation, clarity, filtration, and disinfection are all combined in traditional water treatment techniques [6]. The main problem with conventional techniques is that they are usually inefficient at the removal of residual pollutants [5]. In comparison to traditional approaches, reverse osmosis (RO) and nanofiltration (NF) are particularly promising techniques, especially for applications involving drinking water. Because they use a variety of separation methods, NF/RO can produce high inorganic removal by processes that involve solution diffusion, size being excluded, charge repulsion, and adsorption. It is possible to choose a membrane with the proper properties for certain water quality. Bacteria, viruses, and micro-pollutants like pesticides are often present in groundwater along with inorganic contaminants, which are undesirable. Whereas elimination of micropollutants is dependent on specific properties, NF/RO can remove all those contaminants simultaneously. Additionally, NF/RO is a workable choice for rural settlements because of its modular architecture and flexible placement. However, the potential negative aspects of NF/RO, fouling of the membrane & scaling, concentrated disposal, & relatively high energy usage must all be avoided. Scaling is dependent on the likelihood of precipitate formation in each water, while fouling of membranes and scaling are inevitable throughout the separation process, they can be significantly reduced by optimizing the operating conditions. The implementation of renewable energy technology can offset high energy demand [7]. Given the high standard of water quality that NF/RO produces, using this water primarily for potable purposes is a suitable technique. Utilizing concentrates for non-potable purposes while having feed water quality permitting offers almost no opportunity for removal.

Several researchers have coupled adsorbent materials with further chemicals and substances to improve their ability to remove contaminants. Nanotechnology has recently been applied to almost every human-interest industry, especially soil and management of natural resources, particularly wastewater treatment [8]. Non-adsorbents are often more successful than traditional adsorbents at eliminating organic and inorganic pollutants from wastewater. These adsorbents are made of metallic oxide-based nanoparticles (NPs), nanotubes of carbon (CNTs), graphite, plant nanocomposites that (NCs), and other materials [9]. This study will examine the most common agricultural & non-agricultural material-based adsorbents. Their methods of removing pollutants are also discussed, along with potential modifications or treatments that could increase their effectiveness. Additionally, this evaluation outlines their chances of re-usability and safe disposal procedures [10].

2. Definition and importance of water purification

To make water safe, clean, and acceptable for use in a variety of purposes, such as drinking, irrigation, industrial processes, and recreational activities, water must first be purified. It entails the elimination of potential physical, chemical, and biological contaminants from water sources [11].

Due to the following factors, the significance of water purification cannot be overstated:

2.1 Keeping the public healthy

For the maintenance of the general welfare, access to clean and safe drinking water is essential. Waterborne illnesses brought on by bacteria, viruses, protozoa, and chemical pollutants can seriously harm your health. Purifying water helps keep people and communities healthy and stops the spread of disease.

Water purification techniques like filtration, disinfection, and chemical treatment efficiently remove or reduce the concentration of disease-causing microorganisms like bacteria (like *E. coli*, Salmonella), viruses (like Hepatitis A, norovirus), and parasites (like Giardia, Cryptosporidium), thereby reducing the risk of waterborne infections.

Getting rid of chemical pollutants Chemical contaminants such as heavy metals, pesticides, drugs, industrial chemicals, and organic molecules can all be found in water sources. These contaminants can be eliminated or reduced using purification techniques such e.g., clay, zeolites, biochar, activated carbon adsorption, oxidation, and ion exchange, protecting the environment and public health [12].

2.2 Enhancing flavor and odor

Water purification techniques can make it taste and smell better by removing or cutting down on contaminants that cause unpleasant tastes, odors, or discoloration. This improves the drinking experience overall and encourages taking in enough water to be hydrated.

2.3 Environmental protection

Water pollution can disrupt the ecological balance, degrade aquatic habitats, destroy aquatic species, and harm aquatic ecosystems. Excess nutrients, pollutants, and poisons, as well as other contaminants that might harm aquatic environments, can be eliminated by water filtration, protecting ecosystem health and biodiversity.

2.4 Supporting industrial and agricultural processes

Industrial processes and irrigation in agriculture both require water. Purification ensures that the water used in these industries follows quality requirements, preventing damage to machinery, maintaining product quality, and reducing the effects of pollution discharge on the environment.

2.5 Response to emergencies and disasters

Access to clean water becomes vital in times of emergency, such as during natural disasters or epidemics. To provide impacted communities with safe drinking water, avert further health emergencies, and ensure survival and well-being in difficult circumstances, water purification technologies and techniques are essential.

Overall, water purification is essential for maintaining ecosystems, promoting sustainable water usage, safeguarding public health, and facilitating a variety of human endeavors.

3. Overview of green adsorbents

This term refers to low-cost materials that come from (i) agricultural commodities as well as materials (fruits, vegetables, and foods); (ii) farming and waste material residues; and (iii) low-cost substances from that the most complicated adsorption agents will be manufactured (specifically, activated carbons produced after pyrolysis from plant sources). Whereas these "green adsorbents" are expected to have lesser adsorption capabilities than the super-adsorbents mentioned in the scientific literature (complex materials such as modified chitosan, activated carbon, highly complex inorganic material composites, and so on), they remain relevant due to their low cost. In this study, two major types of environmental pollutants were used: colors and heavy metals, as well as a few others: medicines, pesticides, and phenols. Despite their little studies, several contaminants were discovered in wastewater [12].

Green adsorbents are derived from a number of different sources, such as minerals from clay, natural polymers, organic substances, agricultural sources and by-products, agricultural residue and waste, and agricultural waste pyrolysis. Heavy metals must be removed from the aqueous phase, green adsorbents made from diverse agricultural waste products, by-products, and residues are becoming more and more popular today. Numerous agricultural by-products and residues, including rice bran, wheat bran, cottonseed hulls, maize corn cobs, dal husk, mango peel, sugar beet pulp, grape bagasse, and others, have been investigated as adsorbents. These agricultural by-products have the benefits of being affordable, widely accessible, and easily regenerated. They also have a high absorption capacity. They also have a high ability for adsorption as well as selective heavy metal adsorption [13].

4. Types of green adsorbents for water purification

Adsorbents are often divided into two categories: natural and synthetic, depending on where they were created. Minerals, charcoal, ores, clay, and zeolites are examples of natural adsorbents. While synthetic adsorbents are made from waste materials such as waste sludge, trash from agriculture, and industrial waste [14].

Figure 1 explain the flow chart of adsorbents for removal of pollutants. Its further divided in to five types include activated carbon adsorbents, non-conventional low-cost adsorbents, nanomaterial adsorbents, composite and nanocomposites adsorbents, and miscellaneous adsorbents. Activated Carbon adsorbents are further divided into two types including commercial activated carbon and AC prepared from water materials. Nano-conventional Low -cost Adsorbents are further divided into waste materials from Agriculture and industry, Natural Materials, and Bio-adsorbents. Nanomaterial Adsorbents are further divided into nanoparticles, nanotubes, nanowires, and nanorods. Waste materials from Agriculture further explain the agricultural solid waste and Industrial by-products. Natural materials are further divided into clays and zeolites and siliceous materials. Bio adsorbents further explain the Biomass and Biopolymers & Peal. Industrial by-products are divided into Metal hydroxide sludge, Fly Ash, Red Mud, Bio Solids, and waste slurry.

Chemicals coagulation, which precipitating, the decomposition of separation of membranes, sludge activation, flocculation, and ion exchange are just a few of the processes used for removing pollutants from contaminated water [15]. The demand for rapid and simple processes for removing harmful dyes and chemicals from dirty water is continually increasing. Utilizing green adsorption for water filtration is much

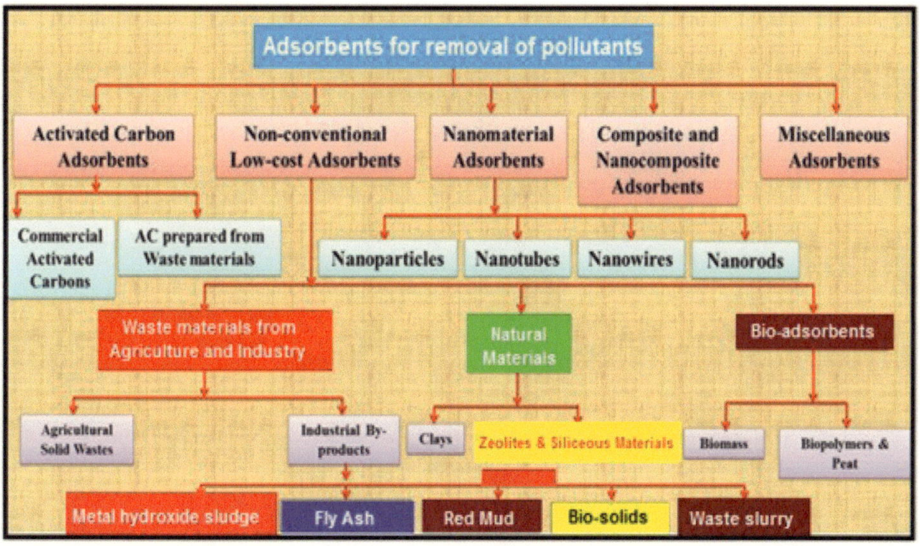

Figure 1.
Adsorbents for removal of pollutants.

more advantageous than utilizing conventional methods due to its ease of use, low price, high effectiveness, and environmental friendliness [16].

Natural green adsorbents such as orange peel, peel from bananas, husks of rice, bagasse from sugar cane, tea waste, avocado peel, Hami melons peel, the peel of dragon fruit peels, and other peels have been utilized to recover ions of heavy metals & organic dyes from contaminated water [17]. One of the most common applications for fruit peel debris is water treatment. Using solid waste as green adsorbents, harmful compounds such as metals, pigments, and pesticide can be eliminated from contaminated water [18].

5. Natural adsorbents (e.g., clay, zeolites, biochar, and activated carbon)

Natural resources are explained in **Figure 2** such as volcanic rocks, soil, plant biomass, industrial and agricultural waste, animal shells, microalgae, and fungal biomass are used to create biosorbents (natural adsorbents). These materials share the ability to physically retain arsenic ions or molecules due to their enormous specific surface area. In comparison to other technologies using synthetic membranes and materials requiring large chemical doses, natural adsorbents have a higher capacity to remove arsenic, are reusable, and are less expensive as well as having less of an impact on the environment. Agricultural wastes have the most functional groups that help keep heavy metals out of all the natural components, including proteins, extractives, hemicellulose, lignin, lipids, starch, and simple sugars [19].

Natural green adsorbents such as citrus peel, peel from bananas, rice husk, bagasse from sugar cane, tea waste, avocado peel, Hami melons peel, dragon fruit peels, as well as other peels have been utilized to recover heavy metal ions & organic dyes from contaminated water. One of the most common applications for fruit peels is water treatment. Using solid waste as green adsorbents, harmful compounds such as heavy metals, dyes, & pesticides can be eliminated from contaminated water.

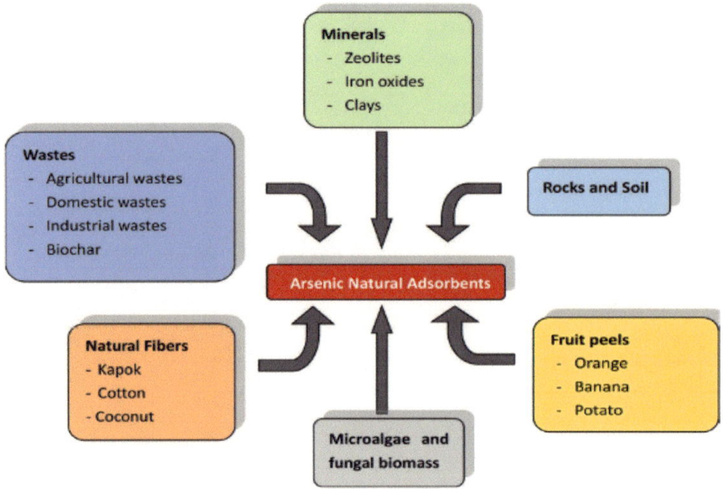

Figure 2.
Natural adsorbents.

5.1 Clay

Clay is a multilayered natural adsorbent that contains the minerals vermiculite, smectites (saponite and montmorillonite), pyrophyllite (talc), mica (illite), kaolinite, serpentine, & sepiolite. Adsorption is caused by the minerals' net-negative charge, which negative charge allows the clay material to absorb positively charged ions. Their high porosity and broad surface area account for the majority of their sorption capacities [19].

5.2 Siliceous

One of the most popular and fairly cost adsorbents is siliceous material. Alunite, dolomite, perlite, glasses, and silica beads are also present. These minerals were used because of the hydrophilic surface's chemical stability and reaction, which was brought on by the presence of a silanol group. However, because of silica beads' lack of resistance towards the utilization of solutions that are alkaline, application as adsorbent was limited to environments with pH values less than 8 [20].

5.3 Zeolites

In nature, porous aluminosilicates with a range of cavity designs and common oxygen atoms bind together to produce zeolites, which are aluminosilicate porous materials. Zeolite comes in a wide variety of species [21]. The natural species include clinoptilolite and chabazite. On the other hand, clinoptilolite, a heulandite mineral, is the chemical that is most often studied because of its high selectivity for particular contaminants. A unique property of zeolite is its cage-like structure, which makes it ideal for the removal of trace contaminants like phenols and heavy metal ions [22].

Figure 3.
Agricultural waste-based adsorbents.

6. Agricultural waste-based adsorbents (e.g., coconut shell, rice husk, sugarcane bagasse)

Agricultural wastes are one of the expected possibilities for alternative absorption. There are many review articles on the elimination of hazardous chemicals using biomass resources. In **Figure 3**, various types of agricultural waste-based adsorbents are listed for absorbing water contaminants from industrial effluent [23]. Orange peel, coniferous Pinus bark powder, wheat husk, activated palm ash, cattail root, neem sawdust, peanut husk, Jujuba seeds, and coir pith activated carbon are among the agricultural wastes. To remove dye from industrial effluent, activated carbon can also be made from a variety of other agricultural wastes, including coconut tree sawdust, banana pith, silk cotton hull, maize cob, and sago waste [24].

7. Industrial waste-based adsorbents (e.g., fly ash, sludge, and sawdust)

7.1 Fly ash

A form of industrial waste called fly ash can be used to absorb colors shown in **Table 1**. Fly ash, which can contain some dangerous substances like heavy metals, is produced in vast amounts during combustion processes [25]. The sugar industry produces bagasse fly ash, which is free of dangerous metals and is frequently used for color adsorption. Its characteristics vary greatly and are influenced by its source. Adsorption tests on Congo red and MB textile dye solutions revealed exothermic monolayer formation and adsorption processes. Fly ash from combustion plants can effectively filter out colors from dyeing industry effluents [26]. Fly ash was used to remove methylene blue from a solution with an initial dye concentration of 65 mg/l at pH 6.75 and 900 mg/l absorbent. The removal rate ranged from 95 to 99%, with a Langmuir constant of 1.91 mg g^{-1}, Ka value of 48.94 L mg^{-1}, and a linear regression coefficient of 0.999 [27].

Table 1 contains information on several compounds and the colors they may absorb. Here's a quick rundown: Fly ash: This material absorbs no distinct color. It has nothing to do with color absorption. Carbon Black: Carbon Black can absorb all colors of light. It is thought to be a particularly effective black pigment. Chlorophyll absorbs

Substance	Color absorbed
Fly ash	N/A
Carbon black	All colors
Chlorophyll	Red & blue violet
Copper sulphate	Blue & green
Chromium oxide	Green
Iron oxide	Red, orange & yellow
Manganese dioxide	Black & brown
Indigo carmine	Blue
Sudan red	Red
Malachite green	Green

Table 1.
Explain the substance and colors adsorbed.

predominantly red and blue-violet light and is found in plants. These wavelengths are required for photosynthesis. Copper Sulphate absorbs light in the blue and green wavelengths. This chemical is frequently utilized in a variety of applications, including fungicides and electrolysis tests. Green light is absorbed by chromium oxide. It's a common pigment in paintings and ceramics. Iron Oxide is a light absorber that absorb red, orange, and yellow light. It is commonly used as an ingredient in a variety of applications, including paints, coatings, & pigments. Manganese Dioxide absorbs the black and brown colors of light. It's used in a variety of sectors, including the manufacture of batteries & ceramics. Indigo Carmine: It absorbs blue light specifically. It is commonly used as a coloring agent in food, textiles, & medicine. Sudan Red is a color that absorbs red light. It is an artificial color that is often used to color plastics, waxes, & oils. Malachite Green: This color absorbs green light. It is commonly used in textile manufacturing as a dye and as a fungicide in fish farming. The table gives a summary of these chemicals and the colors they absorb, helpful for learning their real-world applications and light absorption properties.

7.2 Metal hydroxide sludge

Metal hydroxide sludge, having maximal adsorption capacities of 45.87 and 61.73 mg/g at 30°C and pH 8–9, is used to remove azo colors from surfaces. pH controls the development of dye-metal complexes and adsorption. Metal hydroxide sludge exhibited a maximum adsorption capacity of 270.8 mg/g at 30°C and an initial pH of 10.4. The elimination of Remazol Brilliant Blue reactive dye from a solution with metal hydroxide has a monolayer adsorption capacity of 91.0 mg/g at 25°C and is inexpensive [28].

7.3 Red mud

Another industrial byproduct and waste material from the bauxite industry that is used to manufacture alumina is red mud. When the ability of discarded red mud to absorb dye from its solution was investigated, it was discovered to be successful. It was discovered that the Freundlich isotherm followed pH 2 as the pH where the most dye could be removed by adsorption [29]. Red mud was used as an adsorbent to obtain methylene blue, a basic dye, from water solutions. Its adsorption capacity was calculated at 7.8106 mol/g. Wasted red mud was used to separate Congo red from aqueous solutions and acid activated red mud was investigated for Congo red adsorption from sewage [30]. The experimental data closely matched the Langmuir isotherm, and red mud effectively removed Fast Green, Methylene Blue, and Rhodamine B from wastewater. The adsorption procedure followed the Langmuir and Freundlich isotherms and was exothermic [31, 32].

8. Removal of Cd (II) ions from water samples

The emission of effluent containing heavy metals, including Cd (II) ions, is extremely dangerous to both the environment and human health. Through a multistep process, a 2-aminopyridine group (2Ap) was chemically added to graphene oxide (GrO). The Gr2Ap adsorbent showed a high ability to adsorb Cd (II) at pH = 6, according to research on the effects of adsorbent quantity, pH, temperature, and equilibrium time on sorption. Additionally, empirical isotherm data were fitted to the Freundlich and Langmuir model to learn more about the adsorption isotherms of metal ions. Additionally, Roginsky-Zeldovich types and pseudo-first- and second-order kinetics were discovered to be comparable with adsorption kinetic data when the fundamental steps of the metal sorption mechanism were examined. A remarkable potential for eliminating the heavy metal ions from aqueous samples was shown by the Gr2Ap. The method is also effective, simple, affordable, and appropriate for determining the presence of Cd (II) ions in various water and wastewater samples, according to the most recent research [33].

9. Applied techniques of green adsorbents in water purification

9.1 Removal of heavy metals (e.g., Pb, Cd, As, Hg)

The use of green adsorbents to remove heavy metals from water is a crucial area of water purification research and development. Natural substances or related compounds that can remove heavy metal ions from water are referred to as "green adsorbents." The removal of heavy metals by some common green adsorbents is described below.

9.1.1 Biochar

Biochar is a substance rich in carbon that is created through the pyrolysis of biomass, such as wood or agricultural waste. Its very porous design provides a substantial surface area for adsorption. Through physical adsorption, in which metal ions are drawn to the surface of the biochar by electrostatic interactions or Van der Waals forces, biochar can absorb heavy metals [34]. Additionally, it can perform chemical reactions like ion exchange or complexation, in which metal ions connect to functional groups on the surface of the biochar.

9.1.2 Chitosan

Chitosan is a biopolymer made up of glucosamine units that are obtained from the shells of crustaceans. It possesses amino groups that can complex with or chelate with heavy metal ions. To increase its adsorption capability and selectivity for particular heavy metals, chitosan can be chemically altered. It is efficient at removing heavy metals from water, including lead (Pb), cadmium (Cd), arsenic (As), and mercury (Hg) [14].

9.1.3 Chelation

Chelating agents are organic substances that create stable complexes with heavy metal ions to speed up the elimination of those ions from water. Heavy metals can be removed by natural chelators including organic acids, amino acids, and plant extracts as green adsorbents. These chelating substances generate soluble or insoluble complexes with the metal ions they bind to through coordination bonds, which can then be removed from water by filtration or sedimentation [11].

9.1.4 Phytoremediation

Phytoremediation is a method for removing heavy metals from contaminated water and soil by using plants. Hyperaccumulation is a technique that some plant species can use to store heavy metals in their tissues. These plants can be grown in artificial wetlands or other treatment facilities, where they accumulate heavy metals in their leaves or stems after absorbing them through their roots. The heavy metals can then be efficiently removed from the environment by harvesting and properly discarding the plants [35].

9.1.5 Modified green adsorbents

Green adsorbents that have been changed can be used to increase their adsorption capacity and selectivity for particular heavy metals. Physical modifications like grinding, sieving, and palletization as well as chemical modifications like acid/base treatment, oxidation, and functionalization with certain groups or polymers are examples of modification procedures. These alterations may increase surface area, the addition of new functional groups, or an improvement in the adsorbent's ability to bind to the desired heavy metals.

9.1.6 Biosorption

Biological materials, such as plant biomass or microbial cells, are utilized as adsorbents in the biosorption process to remove heavy metals from water. Functional groups like carboxyl, amino, and hydroxyl groups can attach to heavy metal ions by a variety of mechanisms, including ion exchange, complexation, and electrostatic interactions, in the cell walls or biomass of these materials. Algae, bacteria, fungi, and agricultural waste including rice husks, coconut shells, and sawdust are a few examples of biosorbents [36].

10. Removal of organic pollutants (e.g., dyes, pesticides, pharmaceuticals)

Utilizing green adsorbents to remove organic pollutants from water, such as dyes, pesticides, and medicines, is a successful method of water filtration. Green adsorbents, made from natural resources, provide a long-term and sustainable option for the elimination of harmful organic pollutants [37]. The elimination of organic pollutants by various commonly used green adsorbents is described below.

10.1 Biomass-derived activated carbon

For the elimination of organic pollutants, activated carbon is frequently made from biomass sources such as agricultural waste, wood, and coconut shells. It has a huge surface area and a highly porous structure, which allows for the physical adsorption of organic molecules. Van der Waals forces or hydrophobic interactions can be used to bind organic molecules to the surface of activated carbon because of the porous structure [38].

10.2 Plant-based materials

A variety of plant-based materials, such as leaves, sawdust, or agricultural leftovers, can be used as green adsorbents to remove organic pollutants. These substances frequently have functional groups that can interact with organic molecules and porous structures. For example, lignocellulosic materials may attract organic contaminants through hydrophobic interactions and physical adsorption.

10.3 Minerals in clay

For the elimination of organic pollutants, clay minerals such as montmorillonite and bentonite have been employed as green adsorbents. The layered structure of these minerals has a large specific surface area. Cation exchange, Van der Waals forces, or hydrophobic interactions are a few of the processes by which organic contaminants might be absorbed onto the clay mineral surfaces or inter-layer spaces [39].

10.4 Membrane processes

Membrane filtration methods like reverse osmosis (RO), nanofiltration (NF), or ultrafiltration (UF) may effectively eliminate organic contaminants from water. These methods make use of semi-permeable membranes, which let water through while filtering out dissolved organic molecules based on their size and charge. Membrane procedures can be used with other treatment techniques to increase the removal efficiency of tiny organic molecules, which they are particularly good at eliminating.

10.5 Biological treatment

Techniques for biological treatment use microorganisms' capacity to break down organic contaminants. Microbial activity is used in processes like activated sludge, sequencing batch reactors (SBRs), and created wetlands to break down organic chemicals into simpler, less dangerous molecules. Biodegradable organic pollutants benefit most from biological treatment [40].

10.6 Chemical coagulation/flocculation

Chemical coagulation and flocculation is the process of adding coagulants or flocculants to water to aggregate and destabilize organic contaminants into bigger particles that may then be separated by sedimentation or filtering. Polymers are frequently employed as flocculants, while common coagulants include aluminum or iron salts. Certain organic contaminants, particularly those with larger molecular sizes, can be effectively removed with this technique [41].

10.7 Photocatalysis

To produce reactive species that can break down organic contaminants, photocatalysis uses photocatalysts like titanium dioxide (TiO2) which are activated by ultraviolet (UV) light. The photocatalyst, when exposed to radiation, generates electron-hole pairs that start oxidation processes, breaking down organic materials into inert byproducts.

10.8 Electrochemical techniques

The elimination of organic contaminants can be accomplished using electrochemical techniques such as electrooxidation and electrocoagulation [42]. These techniques use electrodes and electric currents to trigger chemical processes that decompose or remove organic pollutants. For a variety of organic contaminants, electrochemical approaches work well and can be combined with other treatment methods.

The characteristics of the organic contaminants, the water quality metrics, the treatment objectives, economic considerations, and regulatory requirements all play a role in determining which approach is most appropriate. In some instances, a combination of various procedures may be used to accomplish the best organic pollution removal and ensure the safety of the water [43].

11. Removal of inorganic contaminants (e.g., fluoride, nitrate, sulfate)

To ensure that water is safe to drink, inorganic pollutants like fluoride, nitrate, and sulfate must be removed. These inorganic pollutants can be effectively removed using a variety of techniques. Here are some methods for getting rid of them:

11.1 Ion exchange

Ion exchange is a common technique for clearing out inorganic impurities from water. A solid resin and the water exchange ions in this process. Anion exchange resins, which selectively absorb fluoride ions and release other ions (such as chloride) in exchange, are used to remove fluoride from water. Similarly, nitrate and sulfate ions can be eliminated using cation exchange resins.

11.2 Reverse osmosis (RO)

Reverse osmosis is a membrane-based filtering method that can successfully rid water of inorganic impurities. It functions by exerting pressure on the water and forcing it through a semi-permeable membrane that picks out specific ions, such as fluoride, nitrate, and sulfate, and removes them from the solution. Reverse osmosis is frequently employed in water treatment systems and is capable of removing a variety of inorganic impurities [44].

Adsorption is a procedure used to draw out and remove impurities from water using solid adsorbents. For the filtration of inorganic pollutants, a variety of adsorbents, including e.g., clay, zeolites, biochar, activated carbon, activated alumina, and zeolites, can be used. These adsorbents may selectively absorb and remove fluoride, nitrate, and sulfate due to their high surface area and affinity for ions.

11.3 Precipitation

Precipitation is the addition of chemicals that combine with the inorganic pollutants to create insoluble compounds that may then be removed using sedimentation or filtration. For instance, fluoride ions can be precipitated as calcium fluoride using calcium hydroxide (lime). Like how calcium or magnesium compounds can be used to precipitate sulfate ions, biological denitrification is one method that can be used to remove nitrate.

11.4 Ion-selective membranes

Ion-selective membranes are used in the electrochemical process of electrodialysis for dropping inorganic impurities from water. It entails applying an electric field across the membranes, which allows the flow of ions to be selectively allowed based on their charge. Fluoride, nitrate, and sulfate ions can all be successfully removed from water using electrodialysis [45].

12. Case studies of green adsorbents

The utilization of waste tea residue, a green biosorbent, for the removal of heavy metals from water, was investigated in a study by Wang et al. [39]. It was discovered that the discarded tea residue had significant potential for adsorbing elements like lead (Pb), cadmium (Cd), and copper (Cu). The performance of the biosorbent was on par with or even better than that of traditional adsorbents. The study showed the potential of using waste products as environmentally friendly adsorbents for heavy metal removal.

Chitosan-Based Adsorbent for Dye Removal: Chitosan, a green adsorbent derived from crustacean shells, was modified and used to remove dye pollution in a study by Zhang et al. [40]. For a variety of colors, the modified chitosan shows outstanding adsorption capacity and great removal efficiency. Chitosan may be used as an environmentally acceptable adsorbent for color removal in wastewater treatment, according to the study.

A Plant-Based Adsorbent for the Removal of Pharmaceuticals *Azadirachta indica* leaves, a waste plant product, was employed as a green adsorbent in a study by Verma and Dash [41] to remove pharmaceuticals from water. Ibuprofen, paracetamol, and naproxen were only a few of the pharmaceuticals that the plant-based adsorbent effectively absorbed. The study showed that using plant-based materials as green adsorbents for drug removal is feasible.

Ahmad et al.'s [42] study looked into the usage of biochar made from agricultural waste for the removal of organic contaminants, especially bisphenol A (BPA). The BPA levels in the water were significantly reduced by the biochar's strong BPA adsorption ability. The study demonstrated the potential of using biochar as an environmentally friendly adsorbent for the elimination of organic contaminants.

13. Challenges and future directions

13.1 Limitations of green adsorbents in water purification

Due to their potential as environmentally benign substitutes for traditional adsorbents, green adsorbents, which are obtained from natural sources, have attracted considerable attention in water purification procedures. However, they also have some restrictions that should be considered. Green adsorbents' drawbacks in the filtration of water include the following:

13.1.1 Limited availability and variability

Green adsorbents are made from organic materials including plants, agricultural byproducts, and biomaterials. Geographical location, seasonal fluctuations, and other factors can all affect the accessibility and caliber of these adsorbents. Consistent performance and supply may become difficult because of this [46].

13.1.2 Lower adsorption capacity

green adsorbents often have lower adsorption capacities as compared to synthetic adsorbents like e.g., clay, zeolites, biochar, and activated carbon. Due to this, more green adsorbents may be needed to accomplish the same level of purification, which would result in higher expenses and maybe more trash that would need to be disposed of.

13.1.3 Slow adsorption kinetics

Compared to manufactured adsorbents, green adsorbents frequently have slower adsorption kinetics, which means that the adsorption process may take longer to achieve equilibrium. In high-flow rate applications, this may have an impact on the effectiveness and efficiency of water purifying systems [47].

13.1.4 Limited selectivity

Green adsorbents may only have a little amount of selectivity for some water pollutants. They might eliminate a variety of pollutants, but they do not always successfully target dangerous chemicals. This may limit their usefulness in circumstances where exact removal of contaminants is necessary.

13.1.5 Regeneration challenges

Adsorption-based water purification procedures rely heavily on regeneration, which entails removing impurities that have been adsorbed to the adsorbent. Due to their organic makeup, green adsorbents may provide regeneration difficulties, making it challenging to efficiently recover and repurpose the adsorbent material.

13.1.6 Lack of standardized procedures

Green adsorbents frequently do not have standardized procedures for synthesis, characterization, and performance assessment. Due to this, it may be difficult to assess the effectiveness of various green adsorbents or develop uniform criteria for their use in water purification procedures.

13.1.7 Compatibility with current systems

To take advantage of green adsorbents' special features, existing water treatment systems may need to be modified or new procedures may need to be developed. It may be difficult and expensive to integrate green adsorbents into existing infrastructure for conventional water treatment.

While green adsorbents have a bright future in water filtration, overcoming their current drawbacks through research and development is essential to their widespread adoption and successful use in real-world applications [48].

14. Current scenario and challenges in adsorption for water treatment

14.1 Emerging trends in green adsorbent research

The future of the field of green adsorbents is being shaped by continuing research as well as several new trends. The following are some important developments in the study of green adsorbents:

14.1.1 Adaptation and nanotechnology

Researchers are looking into how nanotechnology might be used to increase the effectiveness and adsorption capacity of green adsorbents. The surface area, porosity, and reactivity of green adsorbents can be increased, resulting in improved adsorption performance. Nanoscale changes, such as the inclusion of nanoparticles or the functionalization of adsorbent surfaces, can achieve this [49].

14.1.2 Hybrid materials

Activated carbon, polymers, and nanoparticles are being combined with green adsorbents to create hybrid materials. To achieve better overall performance, these hybrid materials combine the benefits of many components, such as the environmental friendliness of green adsorbents and the high adsorption capacity of synthetic materials.

14.1.3 Waste valorization

Scientists are investigating the utilization of waste items as potential green adsorbents, such as agricultural byproducts, food waste, and biomass residues. By utilizing waste materials to create value-added adsorbents, this movement aims to address both the demand for sustainable water filtration technology and environmental concerns over waste management.

14.1.4 Advanced characterization methods

To better understand the adsorption mechanisms and behavior of green adsorbents, researchers are using advanced characterization methods like spectroscopy, microscopy, and computational modeling. This makes it possible to build and optimize adsorbent materials with better performance and property characteristics.

14.1.5 Regeneration and re-usability

Strategies for the regeneration and reusability of green adsorbents are currently being developed. These methods seek to recover pollutants that have been absorbed and restore the adsorption capacity of the adsorbents, lowering waste production and improving the process's efficiency and sustainability.

Although most research has been done in laboratories, there is rising interest in scaling up green adsorbent technology for use in real-world settings. To create affordable, effective, and efficient water filtration systems employing green adsorbents, researchers are exploring pilot-scale investigations [50].

These new trends highlight the growing significance of environmentally friendly and sustainable water filtration techniques. Researchers are working to create effective, affordable, and scalable green adsorbent technology for water treatment by addressing the limitations of green adsorbents and investigating novel methodologies [51].

15. Conclusions

In conclusion, natural-based green adsorbents present a viable strategy for water filtration. Sustainable, abundant, and potential for efficient pollution removal have made these eco-friendly adsorbents popular. The following are important points about using green adsorbents for water purification:

Environmentally friendly and sustainable: Green adsorbents reduce reliance on synthetic or non-renewable adsorbents by using natural materials like biomass, agricultural waste, or plant fibers. They provide a long-term fix with little harm to the environment.

Versatile removal capabilities: Green adsorbents have demonstrated efficacy in the removal of a wide range of contaminants from water, such as heavy metals, organic pollutants, dyes, medicines, and other inorganic compounds. Through changes and treatments, they can be customized and optimized for the removal of particular contaminants.

Cost-effective: Because they are readily available and have minimal production costs, green adsorbents are frequently commercially viable. A cost-effective solution for water filtration, many of these materials are leftovers or results of industrial or agricultural processes.

Green adsorbents have surface characteristics such as a large surface area, porosity, and functional groups that allow for adsorption through both chemical and physical interactions. These methods make it easier to remove impurities from water.

Alternative water treatment methods: To improve overall treatment effectiveness and address numerous contaminants, green adsorbents can be used in conjunction with alternative water treatment methods such as filtration, membrane processes, or advanced oxidation.

Although green adsorbents have a lot of commitment, their effectiveness may vary based on the contaminant, the water's chemistry, the adsorbent's properties, and the operational circumstances. To maximize their effectiveness, scalability, and potential regeneration for continuous usage, more study is still required. Overall, the use of green adsorbents in water purification shows promise for offering long-term and practical solutions to problems with water pollution, helping to achieve the objective of providing communities with clean and safe water resources.

Conflict of interest

The authors declare no conflict of interest.

Author details

Hafsa Muzammal[1,2], Muhammad Danish Majeed[3,4], Muhammad Zaman[1], Muhammad Safdar[1,2]*, Muhammad Adnan Shahid[1,2], Zahid Maqbool[2] and Tayyaba Majeed[5]

1 Department of Irrigation and Drainage, University of Agriculture, Faisalabad, Punjab, Pakistan

2 Agricultural Remote Sensing Lab of National Center of GIS and Space Applications (NCGSA-ARSL), University of Agriculture, Faisalabad, Punjab, Pakistan

3 Soil and Water Conservation Research Institute (SAWCRI), Chakwal, Punjab, Pakistan

4 Barani Agriculture Research Institute (BARI), Chakwal, Punjab, Pakistan

5 Department of Physics, University of Agriculture, Faisalabad, Punjab, Pakistan

*Address all correspondence to: safdarsani4340@gmail.com

References

[1] Sivakumar B. Water crisis: From conflict to cooperation—An overview. Hydrological Sciences Journal. 2011;**56**:531-552

[2] Schmoll O, Howard G, Chilton J, Chorus I, editors. Protecting Groundwater for Health. UK: World Health Organization; 2006

[3] Ayoob S, Gupta AK. Fluoride in drinking water: A review on the status and stress effects. Critical Reviews in Environmental Science and Technology. 2006;**36**:433-487

[4] Zamora ML, Tracy BL, Zielinski JM, Meyerhof DP, Meyerhof MA. Moss chronic ingestion of uranium in drinking water: A study of kidney bioeffects in humans. Toxicological Sciences. 1998;**43**:68-77

[5] Schwarzenbach RP, Escher BI, Fenner K, Hofstetter TB, Johnson CA, von Gunten U. Wehrli the challenge of micropollutants in aquatic systems. Science. 2006;**313**:1072-1077

[6] Binnie C, Kimber M. Basic Water Treatment. 4th ed. UK: Thomas Telford Limited; 2009

[7] Richards BS, Capão DPS, Schäfer AI. Renewable energy powered membrane technology. 2. The effect of energy fluctuations on performance of a photovoltaic hybrid membrane system. Environmental Science & Technology. 2008;**42**:4563-4569

[8] Kyzas GZ, Mitropoulos AC. Nanomaterials and nanotechnology in wastewater treatment .Nanomaterials (Basel). Jun 2021;**11**(6):1539. DOI: 10.3390/nano11061539

[9] Zekić E, Vuković Ž, Halkijević I. Application of nanotechnology in wastewater treatment. Građevinar. 2018;**70**(04):315-323

[10] Radespiel-Tröger M, Meyer M. Association between drinking water uranium content and cancer risk in Bavaria, Germany. International Archives of Occupational and Environmental Health. 2013;**86**:767-776

[11] Shen J, Schäfer A. Removal of fluoride and uranium by nanofiltration and reverse osmosis: A review. Chemosphere. Dec 2014;117:679-691. DOI: 10.1016/j.chemosphere.2014.09.090. Available from: https://www.sciencedirect.com/journal/chemosphere

[12] Water purification [Internet]. Ecologix Systems. Available from: https://www.ecologixsystems.com/library-water-purification/

[13] Kyzas GZ, Kostoglou M. Green adsorbents for wastewaters: A critical review. Materials (Basel). 2014, 2014;7(1):333-364. DOI: 10.3390/ma7010333

[14] Thakur AK, Singh R, Pullela RT, Pundir V. Green adsorbents for the removal of heavy metals from wastewater: A review. Journal of Materials Today: Proceedings. 2022;**57**(4):1468-1472. DOI: 10.1016/j.matpr.2021.11.373

[15] Hussain A, Madan S, Madan R. Removal of Heavy Metals from Wastewater by Adsorption. Submitted: 04 July 22nd, 2020. Reviewed: January 05, 2021. Published: July 12th, 2021. DOI: 10.5772/intechopen.95841. Available from: https://www.intechopen.com/chapters/75491

[16] Thambiraja S, Sharmilab G, Shankarana DR. Green adsorbents from

solid wastes for water purification application. Journal of Materials Today: Proceedings. 2018;**5**(8, part 3):16675-16683. DOI: 10.1016/j.matpr.2018.06.029

[17] Raval NP, Shah PU, Shah NK, Shah D. Adsorption of malachite green dye from aqueous solution using fruit peel powder. Journal of Environmental Chemical Engineering. 2016;**4**(4):4090-4102

[18] Banerjee K, Ramesh, ST, Gandhimathi R, Nidheesh PV. Bharathi KS. A novel agricultural waste adsorbent, watermelon shell for the removal of copper from aqueous solutions. Iranica Journal of Energy and Environment. 2012;**3**:143-156

[19] El-Said AG, Gamal AM, Mansour HF. Potential application of orange peel (OP) as an ecofriendly adsorbent for textile dyeing effluents. Journal of Textile & Apparel Technology & Management (JTATM). 2012

[20] Doğan M, Abak H, Alkan M. Adsorption of methylene blue onto hazelnut shell: Kinetics, mechanism and activation parameters. Journal of Hazardous Materials. 2009;**164**:172-181. DOI: 10.1016/j.jhazmat.2008.07.155

[21] Ahmed MN, Ram RN. Removal of basic dye from waste-water using silica as adsorbent. Environmental Pollution. 1992;**77**:79-86. DOI: 10.1016/0269-7491(92)90161-3

[22] Ozdemir O, Armagan B, Turan M, Çelik MS. Comparison of the adsorption characteristics of azo-reactive dyes on mezoporous minerals. Dyes and Pigments. 2004;**62**:49-60. DOI: 10.1016/j.dyepig.2003.11.007

[23] Calzaferri G, Brühwiler D, Megelski S, Pfenniger M, Pauchard M, Hennessy B, et al. Playing with dye molecules at the inner and outer surface of zeolite L. Solid State Sciences. 2000;**2**:421-447. DOI: 10.1016/S1293-2558(00)00129-1

[24] Ahmad R. Studies on adsorption of crystal violet dye from aqueous solution onto coniferous pinus bark powder (CPBP). Journal of Hazardous Materials. 2009;**171**:767-773. DOI: 10.1016/j.jhazmat.2009.06.060

[25] Sardar M, Manna M, Maharana M, Sen S. Remediation of dyes from industrial wastewater using low-cost adsorbents. In: Green Adsorbents to Remove Metals, Dyes and Boron from Polluted Water. Chapter: 15 (Part of the Environmental Chemistry for a Sustainable World book series (ECSW, volume 49)). Springer Nature; 2021. DOI: 10.1007/978-3-030-47400-3_15

[26] Acemioğlu B. Adsorption of Congo red from aqueous solution onto calcium-rich fly ash. Journal of Colloid and Interface Science. 2004;**274**:371-379. DOI: 10.1016/j.jcis.2004.03.019

[27] Rastogi K, Sahu JN, Meikap BC, Biswas MN. Removal of methylene blue from wastewater using fly ash as an adsorbent by hydrocyclone. Journal of Hazardous Materials. 2008;**158**:531-540. DOI: 10.1016/j.jhazmat.2008.01.105

[28] Saha P, Datta S. Assessment on thermodynamics and kinetics parameters on reduction of methylene blue dye using flyash. Desalination and Water Treatment. 2009;**12**:219-228. DOI: 10.5004/dwt.2009.967

[29] Santos SCR, Vilar VJP, Boaventura RAR. Waste metal hydroxide sludge as adsorbent for a reactive dye. Journal of Hazardous Materials. 2008;**153**:999-1008. DOI: 10.1016/j.jhazmat.2007.09.050

[30] Wang S, Boyjoo Y, Choueib A, Zhu ZH. Removal of dyes from aqueous solution using fly ash and red mud. Water Research. 2005;**39**:129-138. DOI: 10.1016/j.watres.2004.09.011

[31] Tor A, Cengeloglu Y. Removal of Congo red from aqueous solution by adsorption onto acid activated red mud. Journal of Hazardous Materials. 2006;**138**:409-415. DOI: 10.1016/j.jhazmat.2006.04.063

[32] Gupta VK, Ali I, Saini VK. Removal of chlorophenols from wastewater using red mud: An aluminum industry waste. Environmental Science & Technology. 2004;**38**:4012-4018. DOI: 10.1021/es049539d

[33] Hamad HN, Idrus S. Recent developments in the application of bio-waste-derived adsorbents for the removal of methylene blue from wastewater: A review. Polymers. 2022;**14**(4):783. DOI: 10.3390/polym14040783

[34] Abniki M, Moghimi A. Removal of Cd (II) ions from water solutions using dispersive solid-phase extraction method with 2-aminopyridine/graphene oxide Nano-plates. Current Analytical Chemistry. 2022;**18**(10):1070-1085. DOI: 10.2174/15734110186662205050 00009

[35] Al-Rashdi BAM, Johnson DJ, Hilal N. Removal of heavy metal ions by nanofiltration. Desalination. 2013;**315**:2-17

[36] Fu F, Wang Q. Removal of heavy metal ions from wastewaters: A review. Journal of Environmental Management. 2011;**92**(3):407-418

[37] Dalida MLP, Mariano AFV, Futalan CM, Kan CC, Tsai WC, Wan MW. Adsorptive removal of Cu (II) from aqueous solutions using non-crosslinked and crosslinked chitosan-coated bentonite beads. Desalination. 2011;**275**(1-3):154-159

[38] Jeevanantham S, Saravanan A, Hemavathy RV, Kumar PS, Yaashikaa PR, Yuvaraj D. Removal of toxic pollutants from water environment by phytoremediation: A survey on application and future prospects. Environmental Technology and Innovation. 2019;**13**:264-276

[39] Wang X, Liu R, Zhang Z. Waste tea residue as a green biosorbent for the removal of heavy metals from water. Journal of Environmental Management. 2019;**241**:314-321

[40] Zhang L, Luo Z, Liu Y, Xie F, Liu X, Luo S. Chitosan-based adsorbent for removal of dyes from aqueous solutions: A mini-review of recent progress. International Journal of Biological Macromolecules. 2018;**106**:697-705

[41] Verma AK, Dash RR. Adsorption of pharmaceutical compounds onto plant-based adsorbents: A review. Journal of Environmental Chemical Engineering. 2017;**5**(3):2782-2797

[42] Ahmad MA, Rafatullah M, Ghazali A, Sulaiman O, Ibrahim MH, Hashim R. Adsorption of bisphenol A onto biochar derived from agricultural waste: Equilibrium, kinetics, and thermodynamic studies. Journal of Environmental Management. 2020;**264**:110444

[43] Ullah H, Noreen S, Khan A, Nisar J. Recent advances in the synthesis and applications of green adsorbents for heavy metals removal: A review. Journal of Cleaner Production. 2019;**229**:1292-1318

[44] Tang Q, Zhang S. A review: Fundamental aspects and current

progress in the adsorption of heavy metals by zeolite. Journal of Hazardous Materials. 2013;**244-245**:98-109

[45] Rodrigues DF, Mohan D. Environmental sustainability of emerging technologies for water treatment. Chemical Society Reviews. 2016;**45**(23):6429-6442

[46] Poonam S, Kumar S, Kumar R. Removal of sulfate from water and wastewater using adsorbents: A review. Journal of Water Process Engineering. 2020;**33**:101064

[47] Tripathi M, Sahu JN. Adsorption of fluoride from aqueous solution using various adsorbents: A review. Journal of Environmental Chemical Engineering. 2019;**7**(2):102938

[48] Foo KY, Hameed BH. Insights into the modeling of adsorption isotherm systems. Chemical Engineering Journal. 2010;**156**(1):2-10

[49] Senthil Kumar P, Ramalingam S, Senthamarai C, Niranjanaa M, Vijayalakshmi P. Adsorption of heavy metals from water using banana peel as a low-cost adsorbent. Journal of Environmental Chemical Engineering. 2018;**6**(1):660-672

[50] Kratochvil D, Volesky B. Advances in the biosorption of heavy metals. Trends in Biotechnology. 1 July 1998;**16**(7):291-300

[51] Ng C, Foo KY, Hameed BH. Insights into the modeling of adsorption isotherm systems (II). Chemical Engineering Journal. 2013;**226**:336-347

Chapter 6

Silica Based Mesoporous Materials as Sorbents for Solid-Phase Extraction of Organic Pollutants in Water

María Dolores Gil García, María Martínez Galera and María Jesús Martínez Bueno

Abstract

Since the first applications of solids with adsorbent properties in water treatment, Solid-Phase Extraction (SPE) has been the most suitable sample preparation technique for the determination of organic pollutants in water. In the last decades, numerous articles have appeared in the literature describing a wide variety of SPE sorbents. Attention in this field has focused on the search for new SPE materials capable of extracting a wide variety of pollutants from water. However, each sorbent has advantages for certain compounds, but not when it is used with others. Thus, conventional sorbents showed poor performance for most high polarity analytes. Recently, mesoporous materials are attracting increasing interest in sample preparation due to their desirable characteristics. They present unique advantages as highly ordered and size-controlled mesoporous structures, high surface areas and large pore volumes, good performance in thermal and chemical stability, compositional controllability and flexibility for functionalization.

Keywords: solid-phase extraction, mesoporous materials, organic pollutants, water, mesoporous functionalized silica-based mesoporous

1. Introduction

The rate of growth of the world's population [1] has led to an increase in demand for products and, consequently, to an increase in the production of chemicals in the framework of sustainable development of industry, agriculture and the welfare state. Thus, the pursuit of short-term high productivity over ecological sustainability has led to high environmental pollution. Unfortunately, water and pollution are two words that are closely related, mainly due to human activity. The main sources of water pollution [2] are agricultural, industrial and domestic and hospital wastewater discharges (**Figure 1**). As a result, water quality and quantity have deteriorated, affecting not only the aquatic ecosystem, but also the availability of drinking water for human consumption.

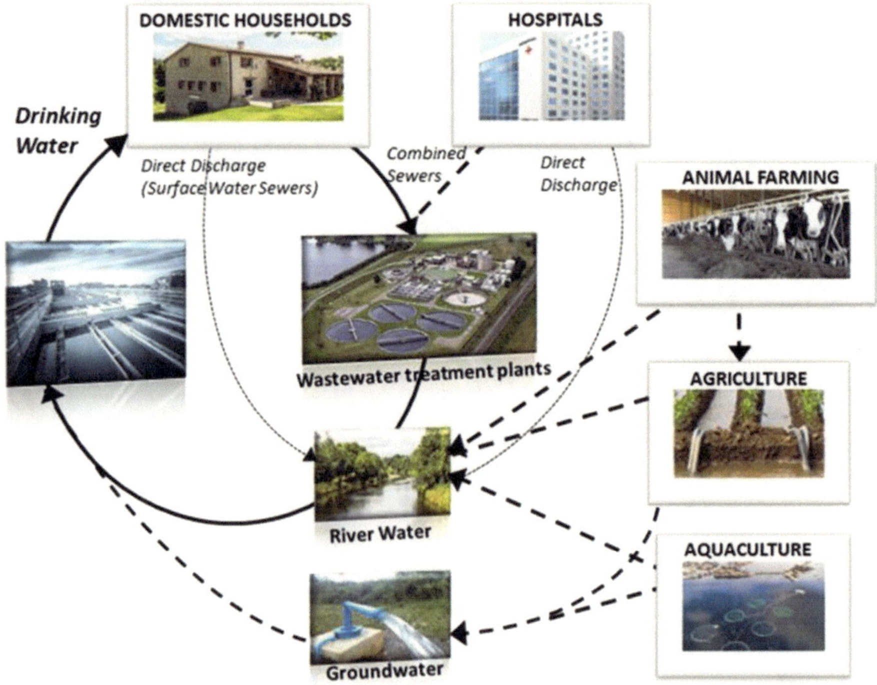

Figure 1.
Sources and pathways of PPCPs in the urban water cycle [3].

The occurrence of pesticides [2, 4] and fertilizers [5–7] in environmental waters is mainly due to their use in agriculture. According to FAO, the agricultural land has more than doubled in recent decades, which has increased the use of pesticides and eventually affected ground/underground water quality.

Currently, more than 1200 pesticides are used worldwide, posing a potential risk to human health and the environment [4]. Most of them are highly persistent and have been associated with mutagenic, teratogenic and carcinogenic effects [8]. International organizations have established strict standards for the control of "priority" contaminants in waters such as pesticides, metals, Polycyclic Aromatic Hydrocarbons (PAHs) and organochlorine compounds among other organic substances.

In recent years, research on pollutants has focused on a new generation of organic compounds, called "Emerging Contaminants" (CECs), which include pharmaceuticals and Personal Care Products (PCPs), among others. CECs are not commonly monitored in the environment but have the potential to enter the environment and cause known or suspected adverse ecological and (or) human health effect at very low environmental concentrations [9]. CECs have been applied for years, but only recently have been found in domestic and hospital wastewater, in surface and groundwater and even in drinking water [10].

For this reason, the priority objectives are to control water pollution by conventional "priority" (pesticides and metals) and "emerging" (pharmaceuticals and personal hygiene products) contaminants in environmental water. Some of these anthropogenic organic chemical pollutants are highly persistent in the environment and, therefore, they are called Persistent Organic Pollutants (POPs).

To control water quality, the European Union (EU) has implemented controls on specific substances through a series of directives and regulatory laws. The first directive (Directive 75/440/EEC) on the quality of surface water intended for human consumption was adopted in 1975 [11], establishing a list of physical, chemical and microbiological parameters that are characteristics of drinking water. In 1998, EU Directive 98/83/EC [12] set the legal framework to protect human health from the adverse effects of any contamination of water intended for human consumption by ensuring that it is wholesome and clean. At the beginning of the twenty-first century, the EU adopted the Water Framework Directive (WFD) 2000/60/EC [13], which aims to improve, protect and prevent further deterioration of water quality by controlling 33 priority substances in surface waters, and the list of contaminants is updated with new hazardous priority substances, every 2 years. In this sense, the Stockholm Convention recognizes that POPs possess toxic properties, resist degradation, bioaccumulate and are transported, through air, water and migratory species, across international boundaries and deposited far from their site of release, where they accumulate in terrestrial and aquatic ecosystems, establishing the need to drastically reduce or eliminate the release of these pollutants into the environment [14].

Furthermore, Directive 2015/1787/EC recommended the monitoring study of three CECs (diclofenac and two hormones) and, furthermore, other CECs were considered [15] in a watch list (WL). In 2018, Directive 2018/8402/EU concluded that the substances diclofenac, oxadiazon, 2,6-di-tert-butyl-4-methylphenol, triallate and 2-ethylhexyl-4-methoxycinnamate should be removed from the WL, while the insecticide metaflumizone and the antibiotics amoxicillin and ciprofloxacin were identified as suitable candidates and then included [16]. Currently, in the last directive approved by the EU (Directive 2020/2184/EC) the third WL has been established [17].

2. Occurrence of organic pollutants in water

In the last years, some reviews have focused on the study of the presence of organic pollutants in water such as pesticides [10, 18–20] and pharmaceuticals, personal care products (PCPs) and Endocrine Disrupting Compounds (EDCs) [20]. River and groundwater are the main sources for human consumption, hence in recent years there has been an increase in the number of articles to detect the presence of organic pollutants in them although the number of studies on surface water is particularly predominant compared to groundwater. For this reason, this section focuses mainly on the presence of pesticides in surface water.

The pesticide pollution of surface waters has been extensively investigated. Thus, RM de Souza et al. [18] have reviewed the literature published in the period 2012–2019 related to the occurrence of pesticides in surface water, with the USA, China and Spain being the countries with the largest number of studies. Atrazine and its metabolites, metalochlor, chlorpyrifos, tebuconazole, diuron, dimethoate and carbendazim were the most detected pesticides in surface water. The highest concentrations were found for dimethoate (61,000 ng/L) and diuron (22,770 ng/L) in Guanacaste (Cosa Rica) and for metalochlor (10,500 ng/L) and diazinon (9000 ng/L) in Georgia (USA). **Table 1** summarizes the surface waters with the highest concentrations of pesticides.

In the same sense, E. Pereira de Araujo et al. [19] have reviewed articles investigating the presence of pesticides in surface waters. A total of 46 articles published between 1976 and 2021 were included of which 127 studies used grab sampling, 10

Country	Site	Pesticide	Concentration (ng/L)	Ref.
Costa Rica	Guacanaste	Dimethoathe	61,200	[21]
		Diuron	22,700	
		Terbutryin	4760	
		Carbendazim	2239	
USA	Tifton	Metolachlor	10,500	[22]
		Metalaxyl	5760	
		Alachlor	1400	
Bangladesh	Jamalpur	Diazinon	9000	[23]
		Chlorpyrifos	3100	
		Malathion	1800	
India	Ganga	Heptachlor	2138	[24]
China	Yangtze	Imidacloprid	1887	[25]
		Atrazina	1726	
Canada	Ontario	2,4-D	1678	[26]
Portugal	Alqueva	Bentazone	1769	[27]

Table 1.
Sites in the world with the highest levels of pesticides in surface waters.

passive sampling and 9 both sampling techniques, the most detected pesticides were atrazine (43%), simazine (29%), metolachlor (28%) and imidacloprid (28%) using grab sampling and for passive sampling (68%), simazine (42%), chlorpyrifos (42%), p,p′-DDDD (42%) and p,p′-DDE (42%) were the most detected.

In the last decade, the concern of researchers has focused on the study of other pollutants in water, mainly pharmaceuticals and personal care products due to their high consumption. Wastewater from sources like hospitals and households often contains a significant amount of pharmaceutical and personal care products (PPCP) and they are not entirely degraded in the wastewater treatment plants (WWTPs) and, therefore, they enter into the aquatic environment (particularly surface water).

Recently, a review of the occurrence of Emerging Organic Contaminants (EOCs) in drinking water has been published [20], including studies of the presence of pesticides, pharmaceuticals, personal care products and other organic contaminants in surface and groundwater. Several antihypertensives, analgesics, antibiotics, psychoactives, among other pharmaceuticals, have been detected in natural waters around the world, with carbamazepine being the most frequently detected antibiotic in surface waters, with the highest concentrations recorded in Germany (11,581 ng/L). Detailed information on the number of ECOs detected and their concentrations can be consulted in this review [20].

In order to assess the ultra-trace levels of organic contaminants in drinking water set by EU Directives, reliable and relevant data are needed and, for this purpose, rapid, sensitive and selective analytical methods need to be developed. In this respect, chromatographic techniques coupled to mass spectrometry have enabled the detection of a large number of organic contaminants in water [28–30]. Generally, this methodology involves a pre-concentration stage of the sample in order to detect the target analytes at the required ultra-trace levels, with solid-phase extraction (SPE) being the most widely used pre-concentration technique for the analysis of organic pollutants in water.

3. Solid-phase extraction for organic pollutants in water

SPE is a sample treatment technique used to isolate and pre-concentrate analytes from a flowing sample stream of gas, fluid or liquid by transferring and retaining them onto a disposable solid-phase cartridge (disk, membrane, etc.), followed by the elution of the analytes with a suitable solvent [31, 32].

The first SPE work was carried out 70 years ago [33] using an iron cylinder filled with 1200–1500 g of Granular Activated Carbon (GAC) for the concentration of organic compounds from surface water by a group in the US Public Health Service [34]. After a volume of surface water pass through, GAC filter was removed, air-dried and the retained compounds were extracted with diethyl ether in a Soxhlet extractor. Next, the organic residue obtained was separated into five groups according to a modification of a method previously published by Shriner and Fuson [35]. A qualitative test for the elements nitrogen, sulfur and chlorine was made for each fraction. Since this pioneering work, thousands of articles applying SPE methodology with different solid sorbents have been published. In this sense, SPE has become the most suitable sample preparation technique in the analysis of organic contaminants in water [31].

The SPE technique can be performed by offline or online procedures. In recent years, advances in the automation of the sample SPE treatment have provided significant advantages over offline SPE, including fast analysis, reducing solvent consumption and avoiding human errors that they entail analyst chemical exposure [36]. Currently, the main limitation of online SPE methodology is the availability of commercial sorbent devices. Therefore, most of the published literature refers to the development of offline SPE methods in which a wide variety of sorbents in commercial formats can be purchased or it can be hand-made in the laboratory.

3.1 Solid-phase extraction steps

The most commonly used SPE device (**Figure 2a**) for off line pre-concentration of pollutants in water is a polypropylene syringe (1–25 mL volume capacity) that contains an amount of solid sorbent (50–200 mg or more) between two polypropylene frits, ensuring material compaction and preventing any loss of solid material during the SPE procedure.

Regardless of the SPE device used for pre-concentration and the type of methodology (offline or online), SPE procedures include the following four basic steps (**Figure 2b**):

1. *Sorbent conditioning.* Generally, a volume (2 to 4 times the sorbent volume) of suitable solvent or solvents mixture is passed through the cartridge to wet the SPE sorbent and to activate their functional groups. Usually, methanol is used as conditioning solvent for hydrophobic sorbents, and next, a similar volume of ultrapure water is passed through the cartridge to remove methanol [37]. In this step, care must be taken that the sorbent does not dry, as the retention capacity of the sorbent decreases and the analytes are poorly retained. Moreover, it has

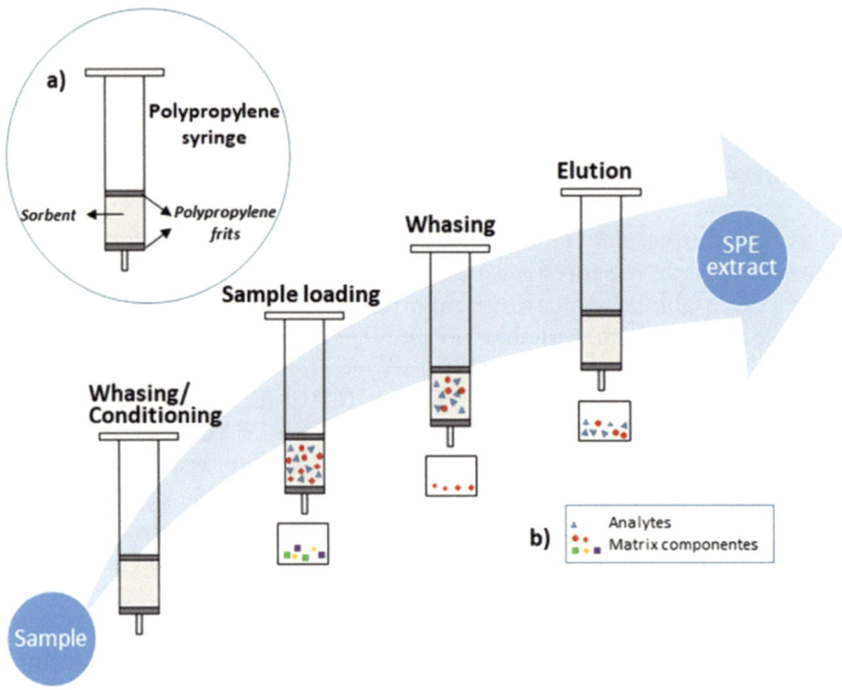

Figure 2.
(a) Components of a SPE syringe and (b) SPE operational steps.

been observed that this fact makes difficult the next step of the sample passing. Therefore, if the sorbent dries it must be reconditioned again. In some cases, the solid sorbent is washed with water or elution solvent before the conditioning step to remove impurities.

2. *Sample loading into the cartridge*. Usually, a volume of sample is passed through the cartridge using a semi-automated extraction manifold connected to a vacuum pump or using an automatic SPE extractor coupled on line with the chromatographic system [38].

To avoid losses of poorly retained analytes during the pre-concentration process, the sample volume to be pre-concentrated is a crucial parameter that must be optimized according to the retention capacity of the sorbent selected and the amount sorbent packed in the cartridge in order to avoid losses of poorly retained analytes (breakthrough volume). In addition, the sample flow rate selected should be low enough to allow efficient retention of the analytes, and high enough to avoid excessive duration of this step [37], with a range of 5 to 10 mL/min for pre-concentration of contaminants in water samples. In this step, both target analytes and some matrix compounds are retained on the sorbent, while other matrix interferents are eluted, allowing a partial purification of the sample.

3. *Sorbent washing*. To reduce Matrix Effect (ME) the sorbent can be washed (optional step) with a suitable solvent to remove most of the matrix components from the sorbent without loss of the target analytes [38, 39]. Normally, the sorbent is washed with a small volume of ultrapure water after the sample loading

step to remove the residual water sample remaining in the sorbent [40–43]. In order to reduce matrix effects, the washing solvent can contain a small percentage of organic solvent to remove matrix components [42, 44]. For aqueous matrices, the sorbent is then dried with air and/or N_2 stream to remove the washing aqueous solvent, which, in some cases, may hinder the subsequent concentration of the final extract [37].

4. *Elution of the analytes from the sorbent*. To quantify the target analytes in the chromatographic system, the analytes are eluted from the sorbent with a suitable solvent or several solvents, without eluting matrix interferences. In offline SPE procedures, the elution solvent is added to the cartridge and passed through by applying pressure. It is recommended to fractionate the solvent volume into several aliquots and allow it to soak the solid sorbent before elution [37].

To ensure their complete elution from the cartridge, the selection of the elution solvent depends on the nature of the analytes and the interactions between analytes and sorbent. For example, a polar solvent such as methanol, acetonitrile or ethyl acetate allows the elution of highly polar analytes while the best solvent for the elution of less polar analytes is hexane.

Figure 3 shows the experimental data obtained for the selection of the elution solvent for the SPE pre-concentration of 218 multiclass pesticides in water [45]. It can be observed that the most polar pesticides (pKow <4) were completely eluted with acetonitrile, whereas for the less polar pesticides (pKow >4) hexane was the best elution solvent.

Another parameter to take into account is the elution volume, which must be sufficient to completely elute the analytes, so elution volumes between 1 and 10 mL are usually applied for organic contaminants in water. Thus, larger volumes are not recommended to avoid dilution of the analytes for online SPE of direct injection into the chromatographic system and for offline SPE the solvent should be removed to dryness with a stream of nitrogen for subsequent redissolution in the injection solvent. Moreover, the elution flow rate of the solvent must be correctly adjusted to ensure interaction between the solvent and analytes and, therefore, their efficient elution.

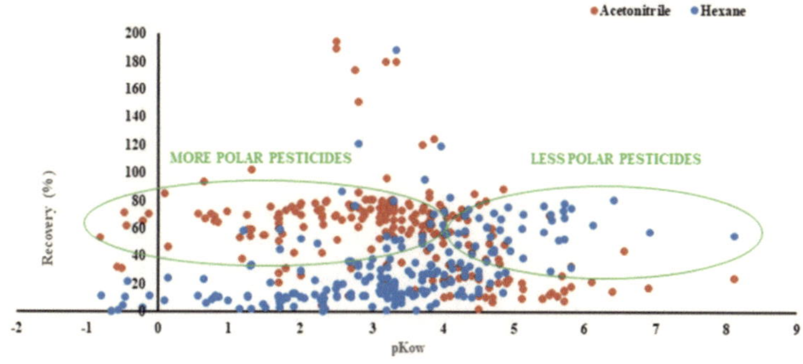

Figure 3.
Relationship between the recoveries and the polarity of pesticides using acetonitrile and hexane solvents for elution [45].

3.2 Solid-phase extraction optimization

Water pollution encompasses multiclass analytes with different chemical properties, including solubility, polarity and acid-basic characteristics. The main goal for the researchers is to develop a simple, efficient and fast method, which requires small amounts of chemicals as SPE sorbent and organic solvents, according to the requirements of green chemistry. In addition, the analytical method can achieve the pre-concentration of a large number of analytes with one sorbent in a short treatment time.

Usually, the development of a SPE procedure for the pre-concentration of analytes in water samples requires the study of the effect of several parameters on the retention/elution of these analytes on the SPE sorbent. Optimization of these parameters can be performed using step-by-step and/or multifactorial experimental design methodologies, being the last approach particularly useful for the pre-concentration of multiclass analytes [45–48]. In general, step-by-step optimization approach is employed to determine the optimal conditions for independent parameters (e.g. sample pH, elution solvent and salt content). On the other hand, for variables that are interdependent (sorbent amount, sample volume and solvent elution volume), experimental design is the best option [45, 48].

The main parameters affecting the effectiveness of a SPE procedure are described:

1. *Elution solvent.* The selection of the solvent for elution of the retained analytes on the SPE sorbent is the first parameter to be optimized. There are two options, a small volume of ultrapure water spiked with the analytes is passed through the cartridge and then the retained analytes are eluted with the organic solvent [49] or a small volume of the organic solvent for elution spiked with the analytes is passed through the cartridge [32].

2. *Sample pH, salinity and matrix components.* The pH of the sample is one of the most important parameters that need to be adjusted to achieve efficient retention of the analytes on the SPE sorbent. Its influence depends on the nature of both the solid sorbent and the analytes, mainly for those containing acidic and basic functional groups in their structures. The pH of the sample defines the molecular state of the target compounds (ionic or neutral state) and therefore, their adsorption onto the SPE sorbent. Hao et al. [50] founded a greater extraction of acidic drugs using Oasis HLB when the water sample was adjusted to a pH value of 2, prior to extraction (recoveries higher than 100%). However, under basic and neutral pH conditions the SPE extraction still provided some degree of selectivity for acidic compounds and better recoveries were obtained for all analytes (near 100%). For that, the pH of the water sample was adjusted to pH 7 in agreement with similar studies [44, 51].

In a recent study [49] using Multi-Walled Carbon Nanotubes (MWCNTs) as SPE sorbent for the pre-concentration of drugs with acid-base properties in river and wastewater, it was found that the pH of the sample affected the recovery and the best recoveries for most analytes were obtained at a pH value of 8 (**Figure 4**). In other similar work [52] we also used MWCNTs as SPE sorbent for the pre-concentration of pesticides in surface water, and similar recoveries were obtained in the pH range 3 to 7, except for methomyl which showed better recoveries at pH 5. Therefore, pH 5 was selected as the optimum pH to adjust the water sample.

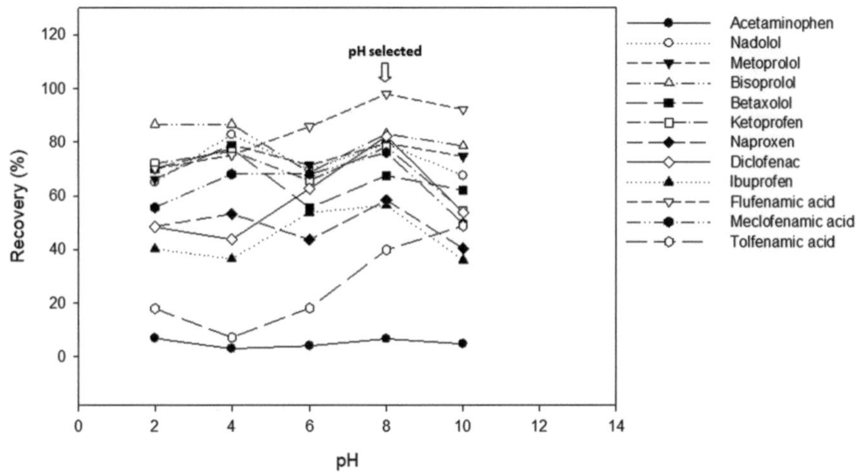

Figure 4.
Influence of the sample pH on the retention of the pharmaceuticals in the MWCNT sorbent [49].

On the other hand, environmental waters contain different amounts of salts depending on their origin. In this sense, the salt content can modify the retention of the target analytes in the SPE sorbent and it can also affect its analytical signal (matrix effect). Generally, the water solubility of hydrophobic organic compounds decreases with increasing salt concentration, which is likely to increase their retention on the SPE sorbent [53]. This effect is called "salting out" and it can be explained by the engagement of water molecules in the hydration spheres around the ionic salt and hence in the reduction of the water concentration available to dissolve the analytes [52, 54].

For this reason, the effect of salt content should be assessed using water samples with different amounts of salt, NaCl and KCl being the most commonly used for this purpose. Zhang and Zhou [53] studied the effect of salinity on the retention of 11 pharmaceuticals in environmental water using several conventional SPE sorbents and they found that the recovery enhanced for six of them with increasing salt content in the water. However, the presence of salt can also negatively affect the recovery of some analytes, and therefore, strategies to reduce or compensate for this effect are required, such as the standard addition calibration [45]. On the other hand, addition of Na_2.EDTA to water samples is recommended to improve extraction efficiency of the analytes, e.g. for the SPE extraction of pharmaceuticals [55, 56], because the analytes can interact with metals present in natural water and/or glass containers. Thus, soluble metals bind to the chelating agent, increasing the extraction efficiency of these pollutants.

3. *Sample volume (breakthrough volume), sorbent amount and elution volume.* In general, environmental waters (tap water, surface water and groundwater) may contain organic contaminants at concentrations below the detection limits achievable with current analytical instruments. Therefore, a trace enrichment step (SPE) is necessary prior to instrumental analysis. Thus, the amount of water needed for this task depends on the instrumental sensibility of the analytical technique and the legal requirements for the target compounds (e.g. the maximum residue level for one pesticide in ground water by EU Directive is 0.1 μ/L),

the recommended water volume being between 100 mL and 1 L. Therefore, one of the most important parameters that requires optimization for the pre-concentration of organic contaminants in water is the sample volume that can be passed through the SPE sorbent, known as the *breakthrough* volume.

The *breakthrough* volume represents the maximum sample volume, which can be pre-concentrated with a theoretical 100% recovery of the analytes. This means that no loss of analytes by elution occurs during the loading step.

The estimation of breakthrough can be achieved using different approaches [38]:

1. The water sample spiked with the analytes is continuously passed through the sorbent and the analyte signals in the water at the outlet of the SPE cartridge are monitored. This approach is particularly convenient for online SPE-LC procedure [57].

2. Different volumes of water sample spiked with the same amount of analytes are pre-concentrated and their concentration in the SPE extracts is analyzed after elution.

3. Various sample volumes spiked with different amounts of analytes are pre-concentrated and the analytes are determined in the SPE extracts, and finally curves representing peak areas versus sample volume are constructed.

Although the experimental determination of the breakthrough volume is time-consuming, it is necessary to optimize it, especially for the analysis of organic contaminants in environmental water requiring large sample volumes to achieve adequate sensitivity. However, losses of highly polar pesticides and pharmaceuticals have been observed with increasing sample volume resulting in very low recoveries [58, 59].

One strategy to avoid the elution of highly polar analytes when using the required sample water volumes is to increase the amount of sorbent [39, 59]. However, increasing the amount of sorbent can have drawbacks such as large volumes of organic solvent being required for the elution of less polar analytes that are strongly retained on the sorbent.

In the light of these evidences, the three parameters (sorbent amount, sample volume and elution volume) are related and experimental design is the best option for their optimization. Recently, we have applied step-by-step (elution solvent, sample pH and salt content) and multi-response surface (sorbent amount, water volume and elution volume) methodologies to optimize a SPE method for the pre-concentration of 218 multiclass pesticides in environmental waters [45]. The statistical significance of the three SPE variables was evaluated by ANOVA test. According to the obtained results, it can be concluded that the most polar pesticides were more retained as the amount of sorbent increased, but eluted during the pre-concentration step if the sample volume was larger. In contrast, the most significant factor affecting negatively the recovery of less polar pesticides was the amount of sorbent because they are strongly retained, making the subsequent elution step more difficult and larger volumes of elution solvent are needed.

Once the parameters described above have been optimized, if the recoveries or sensibilities are not satisfactory for some of them, the most appropriate alternative is to change the SPE sorbent and replace it with another one that has a higher affinity and more selective for these analytes.

In this sense, one of the main advantages of offline SPE is the wide variety of commercially available SPE sorbents and devices that are recommended for the pre-concentration of organic contaminants in water (C18, Oasis HLB, Oasis MAX, among others). In addition, there are both commercially available (Oasis HLB, Oasis MCX,..) and new synthesized sorbents (MCM-41) that can be packed in different SPE devices in the laboratory and they can be used for this purpose. A variety of literature has been published about the types of sorbents and devices used for SPE, including conventional and new sorbents [37, 60].

This chapter focuses on the recent application of silica-based mesoporous materials as sorbents for the pre-concentration of organic pollutants in water.

4. Silica-based mesoporous materials as sorbents for SPE of organic pollutants in water

The development of new mesoporous materials with large surface area is a field of extensive research, particularly with a view to potential applications in sectors such as adsorption of inorganic and organic pollutants, separation, catalysis, sensors and gas storage [61–66].

In the late of 1980s, a team of researchers of the Mobil Research and Development Corporation [67] discovered a family of nanostructured mesoporous materials known as M41S by hydrothermal reactions of aluminosilicate gels in the presence of quaternary ammonium surfactants as template, MCM-41 with hexagonal form being the most widely studied [68]. Other mesoporous silica materials of the MS41 family are MCM-48 with cubic form and MCM-50 with lamellar form [65, 68–70]. The discovery of the M41S family attracted the interest of researchers and other mesoporous silica materials were synthesized following similar procedures to the MS41 series such as SBA (University of California), MSU (Michigan State University material) and KIT (Korea Adv. Inst. of Science and Technology).

There are two kinds of templates to obtain mesoporous materials: (1) supramolecular aggregates such as surfactant micelle assemblies (soft-templating route) and (2) preformed rigid mesoporous solids such as ordered mesoporous silica, carbon and colloidal crystal (hard-templating route) [71]. Generally, mesoporous silica materials are prepared using soft-templating strategies [65] under hydrothermal conditions using basic or acidic media (sol-gel process).

The sol-gel process for the syntheses of mesoporous silica materials consists of the following steps [72]:

1. The surfactant is dissolved in an aqueous medium, resulting in the formation of the template via self-assembly also named Cooperative Template (CTM) or via "True" Liquid Crystal Template (TLCT) mechanisms.

2. The silica precursor is added to the surfactant solution, which is hydrolyzed under acidic or basic conditions, resulting in the formation of a silicate oligomer.

3. The oligomers are then condensed with the surfactant micelles by cooperative assembly and aggregated to form an inorganic-organic hybrid, which finally precipitates as a gel.

4. The gel is hydrothermally treated over a period of time.

5. The product obtained is cooled, filtered, washed and dried and, finally, the surfactant is removed by calcination or solvent extraction, resulting in a solid powder with an array of regular pores.

Figure 5 shows schematically the different steps of the MCM-41 synthesis procedure published by Grün et al. [73].

The morphological characteristics of mesoporous silica-based materials are influenced by different factors, such as the type of surfactant, silica source, concentration of both reagents, and synthesis conditions (medium pH, reaction time, temperature, among them) [66, 72].

Three kinds of surfactants can be used to obtain the template: cationic surfactants (hexadecyltrimethylammonium bromide, CTAB); anionic surfactants (sodium dodecyl sulphate, SDS) and non-ionic surfactants (Pluronic P123) and TEOS (tetraethyl orthosilicate), TMOS (tetramethyl orthosilicate) or inorganic sodium silicate are frequently used as silica sources.

Most mesoporous silica materials were developed in laboratory using six different synthesis pathways depending on the nature of the surfactant, the pH and the temperature [74] (**Table 2**). These mesoporous materials possess fascinating properties including regular, uniform and interpenetrating mesopores, tunable pore sizes and good chemical stability and biocompatibility. In addition, the properties of mesoporous materials can be easily modified by incorporation of different elements or organic groups to these silica structures, in order to improve their selectivity. For that, the number of published articles applying silica-based mesoporous materials as SPE sorbents for the pre-concentration of organic pollutants has increased in the recent years [71].

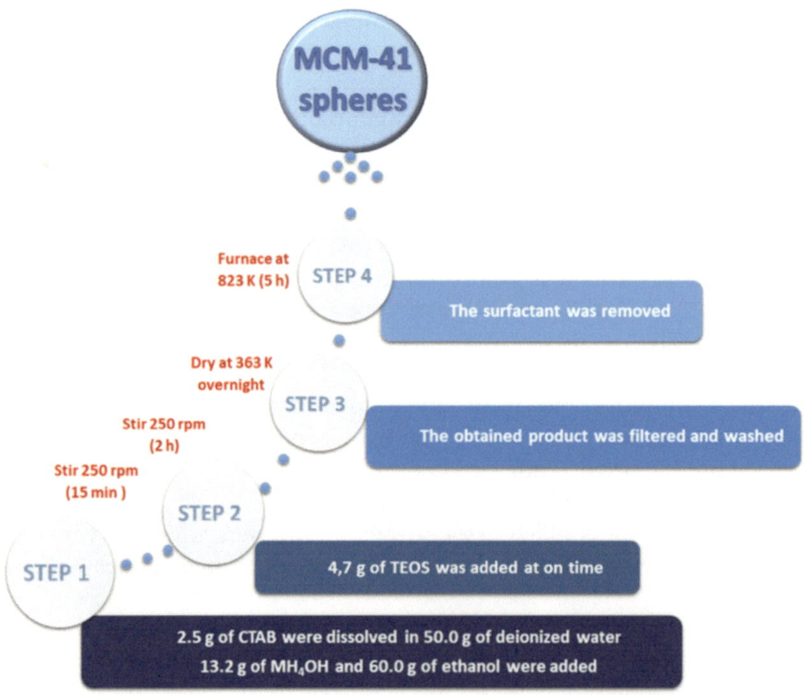

Figure 5.
Synthesis procedure of MCM-41.

Surfactant type	Interaction pathway	Interaction type	Examples
Cationic S^+	Electrostatic interaction	S^+I^-	Tungsten oxide, MCM41
		$S^+X^-I^+$	Lamellar zinc phosphate, cubic Pm3n
Anionic S^-	Electrostatic interaction	S^-I^+	Lamellar iron oxide
		S^-M^+I	Lamellar aluminum oxide, zinc oxide
Neutral S^0/N^0	H-bonding interaction	S^0I^0	Hexagonal silica HMS
		N^0I^0	MSU-X

S is the surfactant, I is the precursor, X and M are the intermediate species (anion and cation), S^0 is a neutral amine, I^0 is a hydrated inorganic oligomer and N^0 is the non-ionic template.

Table 2.
Types of interactions between the surfactants and the inorganic precursor.

Mesoporous materials present unique advantages for the adsorption of both small and large molecules because they can interact with them not only at the external surface, but also throughout the whole internal pore system [67]. These materials have special characteristics such as: (1) mesoporous structures with controlled pore sizes (2–50 nm) and highly ordered pores, which allow size-selective extraction of molecules, retaining small molecules and excluding larger ones, (2) extremely large surface areas, (3) a high degree of stability and (4) they can be easily functionalized with organic groups or imprinted molecular polymers. Therefore, the number of published articles applying silica-based mesoporous materials as SPE sorbents for the pre-concentration of organic contaminants in environmental samples has increased considerably in recent years [71].

In the last decades, the main applications of silica-based mesoporous materials in the sample treatment of inorganic and organic pollutants in different matrices have been summarized in different reviews [71, 75–77].

This chapter focuses on the optimization and validation of analytical methods for the pre-concentration of organic pollutants in water by SPE using different silica-based mesoporous materials as SPE sorbents.

4.1 Application of pure silica-based mesoporous materials as SPE sorbents in water

Due to the high capacity of adsorption of mesoporous materials, MCM-41 and SB-15 were firstly used to remove organic pollutants from environmental water [78–80] and air [81], while the first application as SPE sorbents (SBA-15) occurred in 2015 for the enrichment of flavanones from citrus fruits [82]. The results obtained demonstrated the potential of these mesoporous materials to be used as sorbents in SPE.

Table 3 shows some of the applications of pure silica-based mesoporous materials as sorbent for the SPE of organic pollutants in water.

In 2016, Dahane et al. [58] published the first application of MCM-41 as a SPE sorbent to pre-concentrate pharmaceuticals of very different polarity (atenolol, nadolol, pindolol, timolol, bisoprolol, metoprolol, betaxolol, ketoprofen, naproxen, ibuprofen, diclofenac, tolfenamic acid, flufenamic acid and meclofenamic acid) in surface water, obtaining recoveries close to 100% for all analytes. The MCM-41 was synthesized in the laboratory and 100 mg of this sorbent were hand-packed in 3 mL syringe bodies of polypropylene, obtaining the SPE cartridge. The optimization of

SPE sorbent	Analytes	Analytical method	Amount (mg)	Samples	Ref.
MCM-41	Drugs	UPLC-MS/MS	100	Surface water	[58]
	Pesticides	UPLC-MS/MS	100	Surface water	[59]
MSU-1	Pesticides	UPLC-MS/MS	50	Surface and Groundwater	[39]
	Pesticides	UPLC-MS/MS	100	Groundwater	[45]
TiUVM-7	OPPs	GC-MS	75	Irrigation and WWTP water	[83]
	FRs	GC-MS	200	Irrigation and WWTP water	[84]
Au/TiUVM-7	OPPs	GC-ECD	300	Environmental water	[85]
MTMOS-TEOS	OPPs	GC-MS	100	Tap, drinking and river water	[86]

Table 3.
Application of mesoporous silica-based as SPE sorbents in water.

the experimental conditions of both retention and elution of the target analytes into the MCM-41 cartridge was performed. The proposed method was validated and applied to determination of drugs in surface waters, atenolol, timolol, betaxolol, nadolol and diclofenac being found in some of them, at levels higher than their LOQs (4.5–10 ng/L). Subsequently, the same authors [59] investigated the use of MCM-41 sorbent for the pre-concentration of eight different polar pesticides (organophosphates, carbamates and triazoles) in water samples. The precision study showed that most pesticides pre-concentrate satisfactorily in MCM-41, with recoveries between 65 and 126%, except methomyl (the most polar) with recoveries <30%, demonstrating the usefulness of MCM-41 as a SPE sorbent for low and intermediate polarity pesticides, but it is not suitable for the pre-concentration of high polarity pesticides.

To overcome the low adsorption capacity of MCM-41 for highly polar compounds, Karbouche et al. [39] synthesized a silica-based MSU-1 mesoporous solid for the pre-concentration of 13 pesticides in water, including methomyl. The optimized SPE methodology was validated in river water and satisfactory recoveries were obtained for all pesticides at LOQ levels (70.1–113.5%). Recently, increasing the amount of sorbent MSU-1 (100 mg) enabled the pre-concentration of 218 multiclass pesticides in water [45]. The effect of the matrix compounds in the analyte signals and the low adsorption of some pesticides in the MSU-1 were corrected using standard addition methodology by adding pesticide standards before SPE-pre-concentration. A monitoring study of groundwater samples was performed applying the proposed methodology and 38 different pesticides were found but most of them in concentrations below 0.1 μg/L, with imidacloprid, hexaflumuron and oxadixyl being the most frequently detected. However, in five groundwater samples, high level of methiocarb (three samples), pyrethrin I and pytethrin II (one sample) and hezaflumuron (one sample) were found.

Researchers at the University of Valencia [87] developed a nanoparticulate bimodal silica material (UVM-7) to extract phospholipids from human milk. This mesoporous material exhibited selective extraction of target analytes and restricted access for biomacromolecules. Next, UVM-7 was doped with titanium (Ti) to pre-concentrate Organophosphorus Pesticides (OPPs) [83] and Organophosphorus Flame Retardants (FRs) [84] in water samples. The matrix compounds of the WWTPs samples affected negatively the recoveries obtained for OPPs and FRs, in the range 48–87% and 65–78%, respectively. Both analytical methods were applied to the determination of the target analytes in irrigation water and in effluent and influent WWTP water, with only three FRs found in concentrations in the range of 0.2–1.5 ng/mL in three water samples of the WWTPs. Recently, TiUVM-7 mesoporous silica material doped with gold (Au) nanoparticles was tested to improve OPP recoveries in real water samples [85]. The TiUVM-7 modified with Au was proved to increase the retention of the OPPs pesticides, with analytical parameters comparable to or even better than other alternative sorbents (C18). In addition, the authors argue that this new sorbent (Au/TiUVM-7) is cheap and environmentally friendly.

In the same way, Ibrahim et al. [86] synthesized a new mesoporous organic-inorganic material based on Methyltrimethoxysilane-Tetraethoxysilane (MTMOS-TEOS) and 100 mg of this sorbent were used for the pre-concentration of 5 OPPs using only 10 mL of sample (water and fruit samples). The incorporation of MTMOS helped to increase the hydrophobic character of the material and promoted the interaction between the pesticides and the sorbent. LOQs ranged from 1 to 3 pg/L, which were 500–600 times lower than those obtained using commercial C18. The reuse of the mesoporous sorbent was reported for the pre-concentration of at least 25 samples by washing the cartridge with 5 mL MeOH followed by 10 mL ultrapure water after each extraction.

4.2 Application of functionalized silica-based mesoporous materials as SPE sorbents in water

The functionalization of mesoporous materials with organic functional groups (-OH, $-CH_2-$, -COOH, $-NH_2$, -CN, -SH, $-SO_3H$) by inserting them into the channel walls is another approach to increase extraction, mainly for the most polar analytes. Thus, SBA-15 functionalized with different organic groups has been one of the most widely used for the pre-concentration of inorganic and organic species in various matrices such as water.

Table 4 summarizes some application for the pre-concentration of organic pollutants in water using functionalized mesoporous silica-based SPE sorbents.

Gañán et al. [88] synthesized two ordered mesoporous silicas with different morphology and structure and they were modified with octadecyl (C18) groups to obtain SBA-15-C18 and SM-C18 materials. The functionalized and non-functionalized silicas (200 mg) were studied as sorbents for pre-concentration of 17 β-estradiol in drinking water, and SBA-15-C18 was found to have the highest adsorption capacity against the extremely hydrophobic target analyte, which was attributed to its very high degree of functionalization. Moreover, these authors applied SBA-15-C18 sorbent for the pre-concentration of 12 EDCs [89] and four β-blockers [90, 91] in several environmental water matrices. Recoveries were obtained between 70 and 112% and between 58 and 105% for EDCs and β-blockers in real samples, respectively. To improve the recoveries obtained for EDCs, a bi-functionalized SBA-15 with octadecyl and amino groups (SBA-15-C18-NH_2) sorbent was prepared [92] and the recoveries were compared with those obtained using C18-functionalized amorphous silica sorbent (ExtraBondR). Recoveries close to 100% were obtained for all analytes with this new bi-functionalized SBA-15, except for estriol and prednisolone (66 and 82%, respectively), being in all cases better than those obtained with ExtraBond C18. The bi-functionalized SBA-15 provided mixed retention mechanisms for the analytes: reversed-phase sorption by hydrophobic interactions to the C18 and hydrogen bonding through the NH_2, being suitable for the extraction of a wide variety of moderately polar to non-polar EDCs.

SPE sorbent	Analytes	Analytical method	Amount (mg)	Samples	Ref.
SBA-15-C18	β-estradiol	HPLC-UV	200	Drinking water	[88]
	EDCs	HPLC-UV	100	Environmental water	[89]
	β-blockers	CE	100	Environmental water	[90]
	β-blockers	CE	200	River and wastewater	[91]
SBA-15-C18-NH_2	EDCs	HPLC-UV	100	Ultrapure water	[92]
SBA-15-multifunctionalized	NSAIDs	HPLC-UV	500	River water and wastewater	[93]
MCM-41-SO_3H	PCPs	UPLC-MS/MS	100	Swimming pool water and wastewater	[94]
MCM-41-CN	PCPs	UPLC-MS/MS	100	Swimming pool and river water	[95]
SBA-15-MIP	Bisphenol A	HPLC	50	Ultrapure water	[96]

Table 4.
Application of functionalized mesoporous silica-based as SPE sorbents in water.

Li et al. [93] prepared a functionalized SBA-15 with 3-[2-[2-[2-(2-(2-(2-aminoethylamino)ethylamino)ethylamino]propyl-trimethoxysilane material and after reacting with an excess of finyl glycidyl ether obtained a reversed-phase/anion exchange mixed-mode sorbent. 500 mg of the material were used for the pre-concentration of 4 Non-Steroidal Anti-Inflammatory drugs (NSAIDs) in river water and wastewater. The enrichment factors in the range 806–1109 were achieved, as well as LODs between 0.006 and 0.070 μg/L for tap water and between 0.014 and 0.16 μg/L for river water and wastewater and good recoveries ranging from 80.6 to 110.9% were obtained in all cases.

On the other hand, two organic-inorganic hybrid silica-based mesoporous MCM-41 functionalized with sulfonic [94] and cyanopropyl [95] groups were synthesized in the laboratory and they were tested as SPE sorbents for the pre-concentration of personal care products (PCPs) of very different polarity in environmental water. Similar recoveries were obtained for PCPs with both functionalized MCM-41 sorbents, but MCM-41-SO_3H showed better adsorption capacity for the more polar PCPs (75–100%). Four parabens and two UV filters were detected in the wastewater sample at ng/L and, in addition, the same four parabens were detected in the swimming pool water, but at lower concentrations than those found in wastewater [94]. In the case of the MCM-41-CN sorbent [95], all PCPs were found in the swimming pool water sample, whereas only methylparaben was detected in the river water.

Finally, W Cheng et al. [96] prepared silica-based Mesoporous Molecular Imprinted Polymers (M-MIP), using MCM-41 and SBA-15 mesostructures to extract bisphenol A from aqueous samples. MIP-SAB-15 maintained the mesoporous structure of SBA-15 and also exhibited excellent selectivity toward the bisphenol A, with recoveries of over 87% in fortified water samples.

4.3 Other applications of mesoporous silica-based materials

In addition to the SPE extraction of organic pollutants from water using silica-based mesoporous sorbents, these materials have been applied as sorbents in

Solid-Phase Microextraction (SPME) [97, 98], Micro-Solid-Phase Extraction (μ-SPE) [99], Dispersive Solid-Phase Extraction (DSPE) [100] and Stir-Bar Micro-Solid-Phase Extraction (SB- μ-SPE) [101].

Jiang et al. [97] have prepared a SPME fiber coated with Ordered Silica-based Mesoporous (OMS) film for the pre-concentration of chlorophenols in aqueous samples. The extraction capacity of the OMS fiber was 1.5–1.8 times higher than those of a commercial polyacrylate (PA) fiber. In addition, the selectivity of the OMS fiber for further application was evaluated using three types of organic compounds with different physico-chemical properties (non-polar compounds with benzene rings, non-polar compounds without benzene rings and polar compounds), and it was observed that the extraction efficiency of the OMS coating increased as the polarity of the compounds increased. Likewise, a fiber coated with uncalcined MCM-41 (MCM-41/CTAB) was used for SPME pre-concentration of four Polycyclic Aromatic Hydrocarbons (PAHs) in well water and surface water [98]. The durability of the MCM-41/CTAB fiber was investigated and found to maintain more than 95% of the initial efficacy over five SPME extractions. The proposed method was applied to the analysis of surface and well water and phenanthrene and anthracene were founded in concentrations between 8.8 and 99 ng/L in surface water.

The μ-SPE technique using calcined and uncalcined MCM-41 sorbents was applied to the pre-concentration of Perfluorocarboxylic Acids (PFCA) in water samples [99]. The μ-SPE device consisted of a porous polypropylene membrane bag containing 5 mg of mesoporous sorbents, which did not allow the passage of some matrix components. The extraction efficiency of both mesoporous sorbents was compared with that of other conventional sorbents, obtaining the best results with the uncalcined MCM-41. The proposed method offered good sensitivity and reproducibility (recoveries between 64 and 127%) and was successfully applied to PFCA extraction in river and rainwater samples.

MCM-48 sorbents functionalized with different organic groups, including amine-functionalized MCM-48, phenyl-functionalized MCM-48, octadecyl-functionalized MCM-48 and quaternary ammonium-functionalized MCM-48 [100] were applied to extract EDCs in water by DSPE technique. The best results for the determination of EDCs in water were obtained using MCM-48 functionalized with quaternary ammonium as sorbent, with recoveries close to 95%. However, with the other synthesized materials, recoveries of around 25–30% were obtained.

Finally, 4-phenyl-1,2,3-triazole-functionalized SBA-15 sorbent was synthesized from the reaction of azide-functionalized SBA-15 with phenylacetylene [101]. The mesoporous sorbent was inserted inside a polypropene membrane along with a tiny stir-bar and sealed with a heater. The stir-bar extraction increased the interaction between the polypropene membrane and the water sample due to continuous stirring and rotation, thus increasing the effective surface area of the sorbent. The method was applied to the determination of phenols in wastewater. The recovery studies were performed with wastewater samples spiked at three concentration levels (1, 100 and 400 ng/mL), showing recoveries in the range of 88.5–99.2% for all selected phenols.

5. Conclusions

The synthesis of new silica-based mesoporous materials offers a new field of research in the sample treatment by their application as sorbents for inorganic and organic pollutants in water. Silica-based mesoporous materials have demonstrated to show several advantages over conventional sorbents such as:

1. High adsorption capacity for both organic and inorganic compounds because they possess high surface areas and large pore volumes.

2. Good chemical stability, allowing its use with water samples with a wide range of pH values.

3. Selectivity to pre-concentrate a specific type of compounds, as they offer the possibility to modify their properties by functionalization with different organic groups (hydrophilic, hydrophobic, polar or charged functional moieties) or by doping with some metals depending on the nature of the analytes. But, they can also be functionalized with two different organic groups (bi-functionalized sorbents), increasing their potential use for the extraction of analytes with different nature and properties. Thus, silica-based mesoporous materials exhibit different types of interactions with target analytes, including chelation, hydrophobic, hydrophilic, hydrogen bonding and electrostatic interactions or mixed-mode mechanisms.

4. Flexibility linked to the synthesis of mesoporous materials in the laboratory allows a tailor-made design to solve specific problems in the treatment of samples.

5. Versatile to be applied in a wide variety of pre-concentration techniques such as MSPE, SPE, SPME and SBSE because the mesoporous material can be added directly to the water sample or used inside different devices such as cartridges, metallic fibers and stir-bars.

6. Easy to prepare in the laboratory and inexpensive as most synthesis procedures for mesoporous materials require both inexpensive instrumentation and reagents. Regarding the cost, the price of one Oasis HLB cartridge was 7.5 euros, whereas an MSU-1 cartridge (100 mg) cost 1.5 euros, including the polypropylene syringe and frits.

7. Reuse of the sorbent for several samples (recycling), which is in line with the Circular Economy, and also means a reduction of the waste generated and the cost of analysis, in agreement with the Green Analytical Chemistry.

Therefore, it can be concluded that the possibility of synthesizing tailor-made mesoporous materials in the laboratory improves and expands the possibilities in the sample processing stage.

Acknowledgements

The authors are grateful to the Spanish Ministry of Science, Innovation and Universities (MICINN) and the Andalusian Government (FEDER 2014-202) for financial support. Projects "CERTAIN" (PID2020-116230RB-I00) and "ARCO" (UAL2020-FQM-B2087).

Author details

María Dolores Gil García*, María Martínez Galera and María Jesús Martínez Bueno
Department of Chemistry and Physics, University of Almería, Agrifood Campus of International Excellence (ceiA3), Almería, Spain

*Address all correspondence to: mdgil@ual.es

References

[1] FAO. The State of Food and Agriculture. 2019. Available from: https://www.tandfonline.com/doi/full/10.1080/23308249.2019

[2] Schweitzer L, Noblet J. Water contamination and pollution. In: Török B, Dransfield T, editors. Green Chemistry. An Inclusive Approach. Amsterdam, Netherlands: Elsevier; 2018. pp. 261-290. DOI: 10.1016/B978-0-12-809270-5.00011-X

[3] Ellis JB. Pharmaceutical and personal care products (PPCPs) in urban receiving waters. Environmental Pollution. 2006;**144**:184-118. DOI: 10.1016/j.envpol.2005.12.018

[4] Rad SM, Ray AK, Barghi S. Water pollution and agriculture pesticide. Clean Technologies. 2022;**4**:1088-1102. DOI: 10.3390/cleantechnol4040066

[5] Scanlon BR, Jolly I, Sophocleous M, Zhang L. Global impacts of conversions from natural to agricultural ecosystems on water resources: Quantity versus quality. Water Resources Research. 2007;**43**:1-18. DOI: 10.1029/2006WR005486

[6] Bijay-Singh CE. Fertilizers and nitrate pollution of surface and ground water: An increasingly pervasive global problem. SN Applied Sciences. 2021;**3**:518. DOI: 10.1007/s42452-021-04521-8

[7] Mancuso G, Bencresciuto GF, Lavrnić S, Toscano A. Diffuse water pollution from agriculture: A review of nature-based solutions for nitrogen removal and recovery. Water. 2021;**13**:1893. DOI: 10.3390/w13141893

[8] Obinna IB, Ebere EC. A review: Water pollution by heavy metal and organic pollutants: Brief review of sources, effects and progress on remediation with aquatic plants. Analytical Methods in Environmental Chemistry Journal. 2019;**2**:5-38. DOI: 10.24200/amecj.v2.i03.66

[9] Poynton HC, Robinson WE. Contaminants of emerging concern, with an emphasis on nanomaterials and pharmaceuticals. In: Török B, Dransfield T, editors. Green Chemistry. An Inclusive Approach. Amsterdam, Netherlands: Elsevier; 2018. pp. 291-315. DOI: 10.1016/B978-0-12-809270-5.00012-1

[10] Morin-Crini N, Lichtfouse E, Liu G, Balaram V, Lado Ribeiro AR, Lu Z, et al. Worldwide cases of water pollution by emerging contaminants: A review. Environ Chem Letters. 2022;**20**:2311-2338. DOI: 10.1007/s10311-022-01447-4

[11] Directive 75/440/EEC of 16 June 1975 concerning the quality required of surface water intended for the abstraction of drinking water in the member states. Off J EC. 1975;**L194**:26-31

[12] Directive 98/83/EC of 3 November 1998 on the quality of water intended for human consumption. Off J EC. 2013;**L330**:32-54

[13] Directive 2000/60/EC of the European Parliament and of the council of 23 October 2000 establishing a framework for community action in the field of water policy. Off J Eur Commun. 2000;**L327**:1-73

[14] Stockholm convention on persistent organic pollutants. Off J EU. 2006;**L209**:3-29 Available from: http://data.europa.eu/eli/convention/2006/507/oj

[15] Directive 2015/1787/EC of the European Parliament and of the

council of 6 October 2015 amending annexes II and III to council directive 98/83/EC on the quality of water intended for human consumption. Off J EU. 2015;**L206**:6-17

[16] Directive 2018/840 of the commission implementing decision of 5 June 2018 establishing a watch list of substances for union-wide monitoring in the field of water policy pursuant to directive 2008/105/EC of the European Parliament and of the council and repealing commission implementing decision (EU) 2015/495. Off J EU. 2018;**L141**:9-12

[17] Directive 2020/2184/EC of the European Parliament and of the council of 16 December 2020 on the quality of water intended for human consumption. Off. J EU. 2020;**L435**:1-61

[18] de Souza RM, Seibert D, Quesada HB, Bassetti F, Fagundes-Klen MR, Bergamasco R. Occurrence, impacts and general aspects of pesticides in surface water: A review. Process Safety and Environment Protection. 2020;**135**:22-37. DOI: 10.1016/j.psep.2019.12.035

[19] Pereira de Araújo E, Dutra Caldas E, Cyrino O-FE. Pesticides in surface freshwater: A critical review. Environmental Monitoring and Assessment. 2022;**194**(52):1-25. DOI: 10.1007/s10661-022-10005-y

[20] Mukhopadhyay A, Duttagupta S, Mukherjee A. Emerging organic contaminants in global community drinking water sources and supply: A review of occurrence, processes and remediation. Journal of Environmental Chemical Engineering. 2022;**10**:107560. DOI: 10.1016/j.jece.2022.107560

[21] Carazo-Rojas E, Pérez-Rojas G, Pérez-Villanueva M, Chinchilla-Soto C, Chin-Pampillo JS, Aguilar-Mora P, et al. Pesticide monitoring and ecotoxicological risk assessment in surface water bodies and sediments of a tropical agro-ecosystem. Environmental Pollution. 2018;**241**:800-809. DOI: 10.1016/j.envpol.2018.06.020

[22] Glinski DA, Purucker ST, Van Meter RJ, Black MC, Henderson WM. Analysis of pesticides in surface water, stemflow, and throughfall in an agricultural area in South Georgia, USA. Chemosphere. 2018;**209**:496-507. DOI: 10.1016/j.chemosphere.2018.06.116

[23] Sumon KA, Rashid H, Peeters ETHM, Bosma RH, Van den Brink PJ. Environmental monitoring and risk assessment of organophosphate pesticides in aquatic ecosystems of north-West Bangladesh. Chemosphere. 2018;**206**:92-100. DOI: 10.1016/j.chemosphere.2018.04.167

[24] Khuman SN, Chakraborty P. Air-water exchange of pesticidal persistent organic pollutants in the lower stretch of the transboundary river ganga India. Chemosphere. 2019;**233**:966-974. DOI: 10.1016/j.chemosphere.2019.05.223

[25] Peng Y, Fang W, Krauss M, Brack W, Wang Z, Li F, et al. Screening hundreds of emerging organic pollutants (EOPs) in surface water from the Yangtze River Delta (YRD): Occurrence, distribution, ecological risk. Environmental Pollution. 2018;**241**:484-493. DOI: 10.1016/j.envpol.2018.05.061

[26] Metcalfe CD, Helm P, Paterson G, Kaltenecker G, Murray C, Nowierski M, et al. Pesticides related to land use in watersheds of the Great Lakes basin. Science of the Total Environment. 2019;**648**:681-692. DOI: 10.1016/j.scitotenv.2018.08.169

[27] Palma P, Köck-Schulmeyer M, Alvarenga P, Ledo L, Barbosa IR, López

de Alda M, et al. Risk assessment of pesticides detected in surface water of the Alqueva reservoir (Guadiana basin, southern of Portugal). Science of the Total Environment. 2014;**488-489**:208-219. DOI: 10.1016/j.scitotenv.2014.04.088

[28] Kuster M, López de Alda M, Barceló D. Liquid chromatography–tandem mass spectrometric analysis and regulatory issues of polar pesticides in natural and treated waters. Journal of Chromatography. A. 2009;**1216**:520-529. DOI: 10.1016/j.chroma.2008.08.031

[29] Kampioti AA, Borba da Cunha AC, de AldaM L, Barceló D. Fully automated multianalyte determination of different classes of pesticides, at picogram per litre levels in water, by on-line solid-phase extraction–liquid chromatography–electrospray–tandem mass spectrometry. Analytical and Bioanalytical Chemistry. 2005;**382**:1815-1825. DOI: 10.1007/s00216-005-3332-z

[30] Tóth E, Tölgyesi A, Simon A, Bálint M, Ma X, Sharma VK. An alternative strategy for screening and confirmation of 330 pesticides in ground- and surface water using liquid chromatography tandem mass spectrometry. Molecules. 2022;**27**:1872. DOI: 10.3390/molecules27061872

[31] Poole CF. Principles and practice of solid-phase extraction. In: Pawliszyn J, editor. Sampling and Sample Preparation for Field and Laboratory. Comprehensive Anal Chem. Amsterdam, Netherlands: Elsevier; Vol. 37. 2002. pp. 341-387. DOI: 10.1016/S0166-526X(02)80049-6

[32] Thurman FM, Mills MS. Solid-Phase Extraction: Principles and Practice. Vol. 147. Hoboke, New Yersey, US: Wiley & Sons; 1998

[33] Liška I. Fifty years of solid-phase extraction in water analysis – Historical development and overview. Journal of Chromatography. A. 2000;**885**:3-16. DOI: PII:S0021-9673 (99)01144-9

[34] Braus H, Middleton FM, Walton G. Organic chemical compounds in raw and filtered surface waters. Analytical Chemistry. 1951;**23**:1160-1164. DOI: 10.1021/ac60056a031

[35] Shriner RL, Fuson RC. Systematic Identification of Organic Compounds. 2nd ed. Hoboken, New Jersey, US: Wiley & Sons; 1940. pp. 252-253

[36] de la Serna Calleja MA, Bolado S, Jiménez JJ, López-Serna R. Performance critical comparison of offline SPE, online SPE, and direct injection for the determination of CECs in complex liquid environmental matrices. Microchemical Journal. 2023;**187**:108395. DOI: 10.1016/j.microc.2023.108395

[37] Camel V. Solid phase extraction of trace elements. Spectrochimica Acta B. 2003;**58**:1177-1233. DOI: 10.1016/S0584-8547-03.00072-7

[38] Hennion MC. Solid-phase extraction: Method development, sorbents, and coupling with liquid chromatography. Journal of Chromatography. A. 1999;**856**:3-54. DOI: PII-S0021-9673(99)00832-8

[39] Kharbouche L, Gil García MD, Lozano A, Hamaizi H, Martínez GM. Solid phase extraction of pesticides from environmental waters using an MSU-1 mesoporous material and determination by UPLC-MS/MS. Talanta. 2019;**199**:612-619. DOI: 10.1016/j.talanta.2019.02.092

[40] Kadokami K, Ueno D. Comprehensive target analysis for 484 organic micropollutants in environmental waters by the combination of tandem solid-phase extraction and quadrupole time-of-flight

mass spectrometry with sequential window acquisition of all theoretical fragment-ion spectra acquisition. Analytical Chemistry. 2019;**91**:7749-7755. DOI: 10.1021/acs.analchem.9b01141

[41] Portolés T, Pitarch E, López FJ, Hernández F. Development and validation of a rapid and wide-scope qualitative screening method for detection and identification of organic pollutants in natural water and wastewater by gas chromatography time-of-flight mass spectrometry. Journal of Chromatography. A. 2011;**1218**:303-315. DOI: 10.1016/j.chroma.2010.11.010

[42] Huidobro-López B, López-Heras I, Alonso-Alonso C, Martínez-Hernández V, Nozal L, de Bustamante I. Analytical method to monitor contaminants of emerging concern in water and soil samples from a non-conventional wastewater treatment system. Journal of Chromatography. A. 2022;**1671**:463006. DOI: 10.1016/j.chroma.2022.463006

[43] Hoi BV, Vu CT, Phung-Thi LA, Nguyen TT, Nguyen PT, Mai H, et al. Determination of pharmaceutical residues by UPLC-MS/MS method: Validation and application on surface water and hospital wastewater. Journal of Analytical Methods in Chemistry. 2021;**6628285**:1-12. DOI: 10.1155/2021/6628285

[44] Vieno NM, Tuhkanen T, Kronberg L. Analysis of neutral and basic pharmaceuticals in sewage treatment plants and in recipient rivers using solid phase extraction and liquid chromatography–tandem mass spectrometry detection. Journal of Chromatography. A. 2006;**1134**:101-111. DOI: 10.1016/j.chroma.2006.08.077

[45] Kharbouche L, Martínez Galera M, Díaz Galiano FJ, Gil García MD. Pre-concentration of 218 multiclass pesticide in groundwater samples using MSU-1 mesoporous sorbent. Microchemical Journal. 2023;**184**:108168. DOI: 10.1016/j.microc.2022.108168

[46] Nurerk P, Llompart M, Donkhampa P, Bunkoed O, Dagna T. Solid-phase extraction based on MIL-101 adsorbent followed by gas chromatography tandem mass spectrometry for the analysis of multiclass organic UV filters in water. Journal of Chromatography. A. 2020;**1610**:460564. DOI: 10.1016/j.chroma.2019.460564

[47] Chávez-Moreno CA, Hinojosa-Reyes L, Ruiz-Ruiz EJ, Hernández-Ramírez A, Guzmán-Mar JL. Optimization of solid-phase extraction of parabens and benzophenones in water samples using a combination of Plakett-Burman and box-Behnken designs. Journal of Separation Science. 2018;**41**:4488-4497. DOI: 10.1002/jssc.201800796

[48] Vaudreuil MA, Duy SV, Munoz G, Furtos A, Sauvé S. A framework for the analysis of polar anticancer drugs in wastewater: On-line extraction coupled to HILIC or reverse phase LC-MS/MS. Talanta. 2020;**220**:121407. DOI: 10.1016/j.talanta.2020.121407

[49] Dahane S, Gil García M, Martínez Bueno MJ, Uclés Moreno A, Martínez Galera M, Derdour A. Determination of drugs in river and wastewaters using solid-phase extraction by packed multi-walled carbon nanotubes and liquid chromatography–quadrupole-linear ion trap-mass spectrometry. Journal of Chromatography. A. 2013;**1297**:17-28. DOI: 10.1016/j.chroma.2013.05.002

[50] Hao C, Lissemore L, Nguyen B, Kleywegt S, Yang P, Solomon K. Determination of pharmaceuticals in environmental waters by liquid

chromatography/electrospray ionization/tandem mass spectrometry. Analytical and Bioanalytical Chemistry. 2006;**384**:505-513. DOI: 10.1007/s00216-005-0199-y

[51] Pedrouzo M, Borrull F, Marc RM, Pocurull E. Simultaneous determination of macrolides, sulfonamides, and other pharmaceuticals in water samples by solid-phase extraction and LC-(ESI) MS. Journal of Separation Science. 2008;**31**:2182-2188. DOI: 10.1002/jssc.200700571

[52] Dahane S, Gil García MD, Uclés Moreno A, Martínez Galera M, Socías Viciana MM, Derdour A. Determination of eight pesticides of varying polarity in surface waters using solid phase extraction with multiwalled carbon nanotubes and liquid chromatography-linear ion trap mass spectrometry. Microchimica Acta. 2014;**182**(1-2):95-103. DOI: 10.1007/s00604-014-1290-x

[53] Zhang ZL, Zhou JL. Simultaneous determination of various pharmaceutical compounds in water by solid-phase extraction–liquid chromatography–tandem mass spectrometry. Journal of Chromatography. A. 2007;**1154**:205-213. DOI: 10.1016/j.chroma.2007.03.105

[54] Carabias-Martínez R, Rodríguez-GonzaloE,Herrero-HernándezE, Sánchez-San Román FJ, Prado Flores MG. Determination of herbicides and metabolites by solid-phase extraction and liquid chromatography: Evaluation of pollution due to herbicides in surface and groundwaters. Journal of Chromatography. A. 2002;**950**:157-166. DOI: 10.1016/S0021-9673(01)01613-2

[55] Ferrando-Climent L, Rodriguez-Mozaz S, Barceló D. Development of a UPLC-MS/MS method for the determination of ten anticancer drugs in hospital and urban wastewaters, and its application for the screening of human metabolites assisted by information-dependent acquisition tool (IDA) in sewage samples. Analytical and Bioanalytical Chemistry. 2013;**405**:5937-5952. DOI: 10.1007/s00216-013-6794-4

[56] Gros M, Rodríguez-Mozaz S, Barceló D. Fast and comprehensive multi-residue analysis of a broad range of human and veterinary pharmaceuticals and some of their metabolites in surface and treated waters by ultra-high-performance liquid chromatography coupled to quadrupole-linear ion trap tandem mass spectrometry. Journal of Chromatography. A. 2012;**1248**:104-121. DOI: 10.1016/j.chroma.2012.05.084

[57] Poole CF, Gunatilleka AD, Sethuraman R. Contributions of theory to method development in solid-phase extraction. Journal of Chromatography. A. 2000;**885**:17-39. DOI: PIIS0021- 9673(00)00224-7

[58] Dahane S, Martínez Galera M, Marchionni ME, Socías Viciana MM, Derdour A, Gil García MD. Mesoporous silica based MCM-41 as solid-phase extraction sorbent combined with micro-liquid chromatography–quadrupole-mass spectrometry for the analysis of pharmaceuticals in waters. Talanta. 2016;**152**:378-391. DOI: 10.1016/j.talanta.2016.02.013

[59] Gil García MD, Dahane S, Arrabal-Campos FM, Socías Viciana MM, García MA, Fernández I, et al. MCM-41 as novel solid phase sorbent for the pre-concentration of pesticides in environmental waters and determination by microflow liquid chromatography-quadrupole linear ion trap mass spectrometry. Microchemical Journal. 2017;**134**:181-190. DOI: 10.1016/j.microc.2017.06.008

[60] Płotka-Wasylka J, Szczepańska N, de la Guardia M, Namieśnik J. Modern

trends in solid phase extraction: New sorbent media. Trends in Analytical Chemistry. 2016;77:23-43. DOI: 10.1016/j.trac.2015.10.010

[61] Jenekhe SA, Chen XL. Self-assembly of ordered microporous materials from rod-coil block copolymers. Science. 1999;**283**:372-375. DOI: 10.1126/science.283.5400.372

[62] Hartmann M. Ordered mesoporous materials for bioadsorption and biocatalysis. Chemistry of Materials. 2005;**17**:4577-4593. DOI: 10.1021/cm0485658

[63] Wu Z, Zhao D. Ordered mesoporous materials as adsorbents. Chemical Communications. 2011;**47**:3332-3338. DOI: 10.1039/c0cc04909c

[64] Coasne B, Galarneau A, Pellenq RJM, Di Renzo F. Adsorption, intrusion and freezing in porous silica: The view from the nanoscale. Chemical Society Reviews. 2013;**42**:4141-4171. DOI: 10.1039/c2cs35384a

[65] Pal N, Bhaumik A. Soft templating strategies for the synthesis of mesoporous materials: Inorganic, organic-inorganic hybrid and purely organic solids. Advances in Colloid and Interface Science. 2013;**189-190**:21-41. DOI: 10.1016/j.cis.2012.12.002

[66] Li W, Yue Q, Deng Y, Zhao D. Ordered mesoporous materials based on interfacial assembly and engineering. Advanced Materials. 2013;**25**:5129-5152. DOI: 10.1002/adma.201302184

[67] Kresge CT, Leonowicz ME, Roth WJ, Vartuli JC, Beck JS. Ordered mesoporous molecular sieves synthesized by a liquid-crystal template mechanism. Nature. 1992;**359**:710-712. DOI: 10.1038/359710a0

[68] Kresge CT, Vartuli JC, Roth WJ, Leonowicz ME. The discovery of ExxonMobil's M41S family of mesoporous molecular sieves. Studies in Surface Science and Catalysis. 2004;**148**:53-72. DOI: 10.1016/S0167-2991(04)80193-9

[69] Beck JS, Vartuli JC, Roth WJ, Leonowicz ME, Kresge CT, Schmitt KD, et al. A new family of mesoporous molecular sieves prepared with liquid crystal templates. Journal of the American Chemical Society. 1992;**114**:10834-10843. DOI: 10.1021/ja00053a020

[70] Suib SL. A review of recent developments of mesoporous materials. Chemical Record. 2017;**17**:1169-1183. DOI: 10.1002/tcr.201700025

[71] Wang D, Chen X, Feng J, Sun M. Recent advances of ordered mesoporous silica materials for solid-phase extraction. Journal of Chromatography. A. 2022;**1675**:463157. DOI: 10.1016/j.chroma.2022.463157

[72] Zimny K, Carteret C, Stébé MJ, Blin JL. Multitechnique investigation of mesoporous titanosilicate materials prepared from both the self-sssembly and the liquid crystal mechanisms. Journal of Physical Chemistry C. 2011;**115**:8684-8692. DOI: 10.1021/jp111876b

[73] Grün M, Unger KK, Matsumoto A, Tsutsumi K. Novel pathways for the preparation of mesoporous MCM-41 materials: Control of porosity and morphology. Microporous and Mesoporous Materials. 1999;**27**:207-216. DOI: 10.1016/S1387-811(98)00255-8

[74] Beck JS, Vartuli JC, Kennedy GJ, Kresge CT, Roth WJ, Schramm SE. Molecular or supramolecular templating: Defining the role of surfactant chemistry in the formation of microporous and mesoporous molecular sieves. Chemistry of Materials. 1994;**6**:1816-1821. DOI: 10.1021/cm00046a040

[75] Zhao L, Qin H, Wu R, Zou H. Recent advances of mesoporous materials in sample preparation. Journal of Chromatography. A. 2012;**1228**:193-204. DOI: 10.1016/j.chroma.2011.09.051

[76] Sierra I, Pérez-Quintanilla D. Heavy metal complexation on hybrid mesoporous silicas: An approach to analytical applications. Chemical Society Reviews. 2013;**42**:3792-3807. DOI: 10.1039/C2CS35221D

[77] Casado N, Pérez-Quintanilla D, Morante-Zarcero S, Sierra I. Current development and applications of ordered mesoporous silicas and other sol-gel silica-based materials in food sample preparation for xenobiotics analysis. TrAC Trends in Analytical Chemistry. 2017;**88**:167-184. DOI: 10.1016/j.trac.2017.01.001

[78] Diagboya PNE, Dikio ED. Silica-based mesoporous materials; emerging designer adsorbents for aqueous pollutants removal and water treatment. Microporous and Mesoporous Materials. 2018;**266**:252-267. DOI: 10.1016/j.micromeso.2018.03.008

[79] Walcarius A, Mercier L. Mesoporous organosilica adsorbents: Nanoengineered materials for removal of organic and inorganic pollutants. Journal of Materials Chemistry. 2010;**20**:4478-4511. DOI: 10.1039/B924316J

[80] Gibson LT. Mesosilica materials and organic pollutant adsorption: Part B: Removal from aqueous solution. Chemical Society Reviews. 2014;**43**:5173-5182. DOI: 10.1039/c3cs60095e

[81] Gibson LT. Mesosilica materials and organic pollutant adsorption: Part a: Removal from air. Chemical Society Reviews. 2014;**43**:5163-5172. DOI: 10.1039/c3cs60096c

[82] Cao W, Ye LH, Cao J, Xu JJ, Peng LQ, Zhu QY, et al. Quantitative analysis of flavanones from citrus fruits by using mesoporous molecular sieve-based miniaturized solid phase extraction coupled to ultrahigh-performance liquid chromatography and quadrupole time-of-flight mass spectrometry. Journal of Chromatography. A. 2015;**1406**:68-77. DOI: 10.1016/j.chroma.2015.06.035

[83] Pellicer-Castell E, Belenguer-Sapiña C, Amorós P, El Haskouri J, Herrero-Martínez JM, Mauri-Aucejo AR. Study of silica-structured materials as sorbents for organophosphorus pesticides determination in environmental water samples. Talanta. 2018;**189**:560-567. DOI: 10.1016/j.talanta.2018.07.044

[84] Pellicer-Castell E, Belenguer-Sapiña C, Amorós P, El Haskouri J, Herrero-Martínez JM, Mauri-Aucejo AR. Enhancing extraction performance of organophosphorus flame retardants in water samples using titanium hierarchical porous silica materials as sorbents. Journal of Chromatography. A. 2021;**1639**:461938. DOI: 10.1016/j.chroma.2021.461938

[85] Pellicer-Castell E, Belenguer-Sapiña C, Amorós P, El Haskouri J, Herrero-Martínez JM, Mauri-Aucejo AR. Mesoporous silica sorbent with gold nanoparticles for solid-phase extraction of organochlorine pesticides in water samples. Journal of Chromatography. A. 2022;**1662**:462729. DOI: 10.1016/j.chroma.2021.462729

[86] Ibrahim WAW, Veloo KV, Sanagi MM. Novel sol-gel hybrid methyltrimethoxysilane-tetraethoxysilane as solid phase extraction sorbent for organophosphorus pesticides. Journal of Chromatography. A. 2012;**1229**:55-62. DOI: 10.1016/j.chroma.2012.01.022

[87] Martínez Pérez-Cejuela H, Ten-Doménech I, El Haskouri J, Amorós P, Simó-Alfonso EF, Herrero-Martínez JM. Solid-phase extraction of phospholipids using mesoporous silica nanoparticles: Application to human milk samples. Analytical and Bioanalytical Chemistry. 2018;**410**:4847-4854. DOI: 10.1007/s00216-018-1121-8

[88] Gañán J, Pérez-Quintanilla D, Morante-Zarcero S, Sierra I. Comparison of different mesoporous silicas for off-line solid phase extraction of 17β-estradiol from waters and its determination by HPLC-DAD. Journal of Hazardous Materials. 2013;**260**:609-617. DOI: 10.1016/j.jhazmat.2013.06.016

[89] Gañán J, Morante-Zarcero S, Pérez-Quintanilla D, Sierra I. A novel hybrid mesostructured silica for the solid-phase extraction of estrogenic hormones from waters. Analytical Methods. 2015;7:4740-4749. DOI: 10.1039/C5AY00983A

[90] Silva M, Morante-Zarcero S, Pérez-Quintanilla D, Marina ML, Sierra I. Preconcentration of β-blockers using functionalized ordered mesoporous silica as sorbent for SPE and their determination in waters by chiral CE. Electrophoresis. 2017;**38**:1905-1912. DOI: 10.1002/elps.201600510

[91] Silva M, Morante-Zarcero S, Pérez-Quintanilla D, Marina ML, Sierra I. Environmental chiral analysis of β-blockers: Evaluation of different n-alkyl-modified SBA-15 mesoporous silicas as sorbents in solid-phase extraction. Environment and Chemistry. 2018;**15**:362-371. DOI: 10.1071/EN18030

[92] Gañán J, Morante-Zarcero S, Pérez-Quintanilla D, Sierra I. Evaluation of bi-functionalized mesoporous silica for solid-phase extraction of twelve endocrine disrupting compounds from water. Materials Letters. 2014;**132**:19-22. DOI: 10.1016/j.matlet.2014.06.042

[93] Li Y, Huang C, Yang J, Peng J, Jin J, Ma H, et al. Multifunctionalized mesoporous silica as an efficient reversed-phase/anion exchange mixed-mode sorbent for solid-phase extraction of four acidic nonsteroidal anti-inflammatory drugs in environmental water samples. Journal of Chromatography. A. 2017;**1527**:10-17. DOI: 10.1016/j.chroma.2017.10.051

[94] Mokhtari M, Hamaizi H, Gil García MD, Martínez GM. Synthesis and characterization of a sulfonic species-based mesoporous sorbent for the pre-concentration of nine personal care products in wastewater and swimming pool water. Microchemical Journal. 2020;**153**:104515. DOI: 10.1016/j.microc.2019.104515

[95] Kharbouche L, Gil García MD, Lozano A, Hamaizi H, Martínez GM. Determination of personal care products in water using UHPLC–MS after solid phase extraction with mesoporous silica-based MCM-41 functionalized with cyanopropyl groups. Journal of Separation Science. 2020;**43**:2142-2153. DOI: 10.1002/jssc.201901148

[96] Cheng W, Ma H, Zhang L, Wang Y. Hierarchically imprinted mesoporous silica polymer: An efficient solid-phase extractant for bisphenol a. Talanta. 2014;**120**:255-261. DOI: 10.1016/j.talanta.2013.12.001

[97] Jiang H, Lia J, Hua X, Shen J, Sun X, Han W, et al. Ordered mesoporous silica film as a novel fiber coating for solid-phase microextraction. Talanta. 2017;**174**:307-313. DOI: 10.1016/j.talanta.2017.06.026

[98] Razmi H, Khosrowshahi EM, Farrokhzadeh S. Introduction of coiled solid phase microextraction fiber coated by mesoporous silica/cetyltrimethylammonium bromide for ultra-trace environmental analysis. Journal of Chromatography. A. 2017;**1506**:1-8. DOI: 10.1016/j.chroma.2017.04.010

[99] Lashgari M, Basheer C, Kee H. Application of surfactant-templated ordered mesoporous material as sorbent in micro-solid phase extraction followed by liquid chromatography-triple quadrupole mass spectrometry for determination of per fluorinated carboxylic acids in aqueous media. Talanta. 2015;**141**:200-206. DOI: 10.1016/j.talanta.2015.03.049

[100] Zhang S, Lu F, Ma X, Yue M, Li Y, Liu J, et al. Quaternary ammonium–functionalized MCM-48 mesoporous silica as a sorbent for the dispersive solid-phase extraction of endocrine disrupting compounds in water. Journal of Chromatography. A. 2018;**1557**:1-8. DOI: 10.1016/j.chroma.2018.05.011

[101] Tanimu A, Muhammad S, Jillani S, Alluhaidan A, Ganiyu S, Alhooshani K. 4-phenyl-1,2,3-triazole functionalized mesoporous silica SBA-15 as sorbent in an efficient stir bar-supported micro-solid-phase extraction strategy for highly to moderately polar phenols. Talanta. 2019;**194**:377-384. DOI: 10.1016/j.talanta.2018.10.012

Section 3

Colors, Dyes, and Antibiotics

Chapter 7

A Review on Elimination of Colour and Dye Products from Industrial Effluent

Saranyadevi Subburaj, Manikandan Paramasivam, Methaq H. Lafta and Lekshmi Gangadhar

Abstract

Every industry now takes clean technology very seriously. Particularly in textile dying facilities, a lot of water, energy, dyeing colours, and chemicals are used. Significant issues with air and water pollution may result from this. The wastewater is quite colourful and smells poisonous. It also has high chemical oxygen demand (COD) and biochemical oxygen demand (BOD) levels. Reactive dyes found in wastewater include orange OGR, red DB-8, black DN, and turquoise DG. It has been demonstrated that the type of dye, the dosage of coagulant, and the pH of the sample all affect how effectively dye is removed. Also, the effectiveness of COD and colour elimination in single-stage and multi-stage ozonation-biological process methods were studied. The functional groups of azo dye are changed by ozonation to produce more readily removable biodegradable byproducts. Ozonation changes the functional groups in azo dye, resulting in more biologically degradable byproducts that are simply eliminated through biological treatment. Activated sludge treatment as well as the coagulation-flocculation technique with 80% waste water removal efficiency. Sequencing batch reactor is a fill-and- draw activated sludge-based system for wastewater treatment and has been proposed as an alternative technique for treating industrial effluents. We briefed a promising novel technology for textile effluent de-colorisation in this chapter.

Keywords: dye, industrial waste water, methods, characteristics of wastewater, colour elimination

1. Introduction

There is a growing freshwater issue worldwide as a result of the fast growth in the world's populace, climate change, and industrial expansion. As a result, numerous freshwater users and polluters make a major contribution to its depletion [1, 2]. Among them, the majority of significant contributors of industrial pollutants come from different industries including fabric, leather, pharmacy, varnish, and pulp-based sectors. These dyes contain methylene blue (MB), methyl orange (MO), Disperse Violet 26, crystal violet and so on [3, 4]. A recent estimate places the yearly global production of dyes at over 7,000,000 tonnes [5]. Humans as well as the environment are put at risk by the discharge of this industrial waste dye into the water.

Nowadays, one of the main problems disturbing the whole world was water pollution brought on through textile manufacturers' incapacity to suitably dispose its waste water. In many nations, particularly China and the estuaries of South Africa, the textile industry has an important negative effect on the environment and the world economy [6]. Because of its textile industry's large-scale production of extremely coloured effluent. This effluent comprising a diversity of persistent pollutants, effluent like dyes were an important ecological polluter that also negatively influences public health [7]. Around 107 tonnes of synthetic dyes are developed per year; the fabric industry employs over 10,000 tonnes of these colours [8]. As per their origin, assembly, and intended usage, dyes are often divided to several groups [9, 10].

The textile industry frequently employs azo, mordant, acid-basic, dispersion, and sulphide dyes among these synthetic pigments (**Figure 1**). In addition, a lot of toxic chemicals were utilised in the fabric industry at diverse stages of manufacturing method like agents for sizing, softening, desizing, brightening, and finishing. Textile dyes, are released as effluent in addition to wastewater into aquatic ecosystems including lakes, streams, and ponds without being treated. This will create main ecotoxicological risks with detrimental impacts on living things. Textile colourants does not adhere closely to fabric; therefore, they are harmful to living things [11, 12].

Heavy elements like mercury, cadmium, arsenic, etc. that are necessary in the synthesis of fabric dye colour pigments, and aromatic complexes have been detected in textile effluent. The creation of textile dye colour pigments requires an abundance of heavy metals like mercury, cadmium, arsenic, etc. [13]. Along with the wastewater,

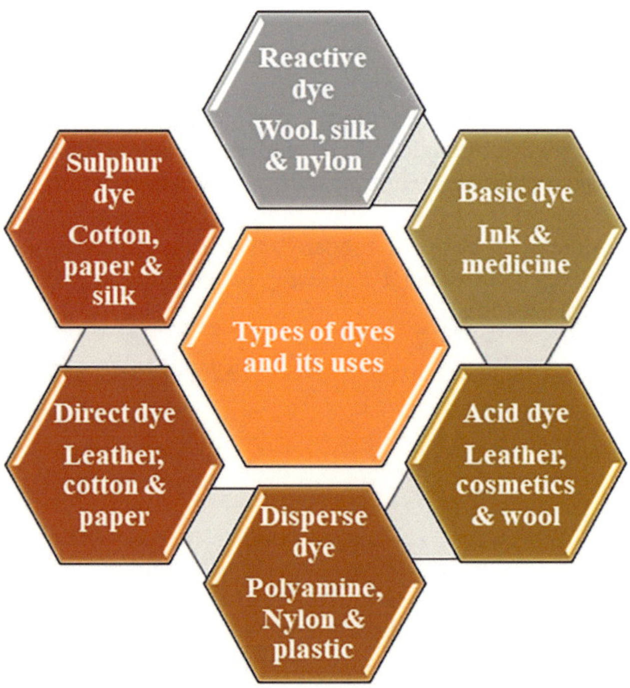

Figure 1.
Different types of dyes and its probable industrial uses.

these hazardous substances are transported over great distances. They persist in the water as well as soil for an extended period, causing major health concerns to living things, depleting soil fertility, impairing aquatic plants' ability to photosynthesize, and ultimately creating anoxic circumstances for aquatic fauna and flora [14]. The aesthetic value of water bodies is also diminished by textile dyes, which increase the biochemical and chemical oxygen demand (COD). It will hinder photosynthesis, prevents the development of plants, enters the food chain, causes resistance and biological accumulation, and might even enhance toxicity, muta-genicity, and carcino-genicity [15]. Due to the exceptionally long lifespan and solubility of artificial colours in water, the substantial quantities of water utilised in fabric production. This leads to wastewater that is similarly large and contains a significant quantity of dissolved solids, metals, and refractory dyes [16]. Since traditional alternatives to treatment are frequently ineffective, it is necessary to use advanced methods due to secondary pollution and incompetent elimination of natural load upon discoloration. Consequently, it was urgently necessitated to grow management process which are both economical and environmentally friendly for properly cleaning effluent that comprises dye before it is finally disposed of [17]. According to this premise, this assessment will give readers a thorough understanding of the damaging effects. Moreover, dye-comprising textile effluent has on living things and natural ecological systems, as well as various cutting-edge approaches for managing textile wastewater in a way that promotes ecological safety.

2. Toxicity and its effect of fabric dyes

Most of the organic contaminants found in the unprocessed effluents discharged by the textile sector are textile colours. Azo dyes were quite common dyes in fabric industry and comprise the highest class among the major classes of textile dyes. Azo dyes chemically include one or more azo groups. Nearly 15–50 percent of azo dyes which were not attached to fibres and fabrics are discharged to produced effluent as a result of ineffective textile dyeing procedures [18]. Though few textile industries cleanse their effluent to break down the free azo dyes discharged to the atmosphere, others dump industrial waste straight to water bodies, posing main ecotoxicological risks and toxic impacts on living things (**Figure 2**). The soil quality and crop germination rates suffered when farmers in underdeveloped nations used wastewater containing untreated industrial effluents to irrigate their agricultural holdings. Azo dyes released into water diminish light penetration, has an impact on aquatic plant and algae development [13]. Furthermore, the colours that fish as well as living creatures perceive may be converted in its bodies to dangerous intermediate chemicals that could be hazardous to the fish's welfare and to the well-being of their predators [19].

Azo dyes included in industrial effluent can be ingested through humans as well as other species. Several bodily tissues are adversely affected by the transformation of azo dyes into toxic amino acids by the microbes that live in the human gut [20]. Furthermore, human skin-derived microbes have the capacity of converting azo dyes into hazardous amines. Because of its growing number of textile firms as well as the massive quantities of effluent comprising dyes, proper and efficient treatment measures are needed to avert ecosystem infection and to advance sustainability [11]. Nevertheless, it is vital to consider the existence of other artificial and organic compounds, their harmful effects, and the relevant environmental outflow levels in order to choose the best option [21].

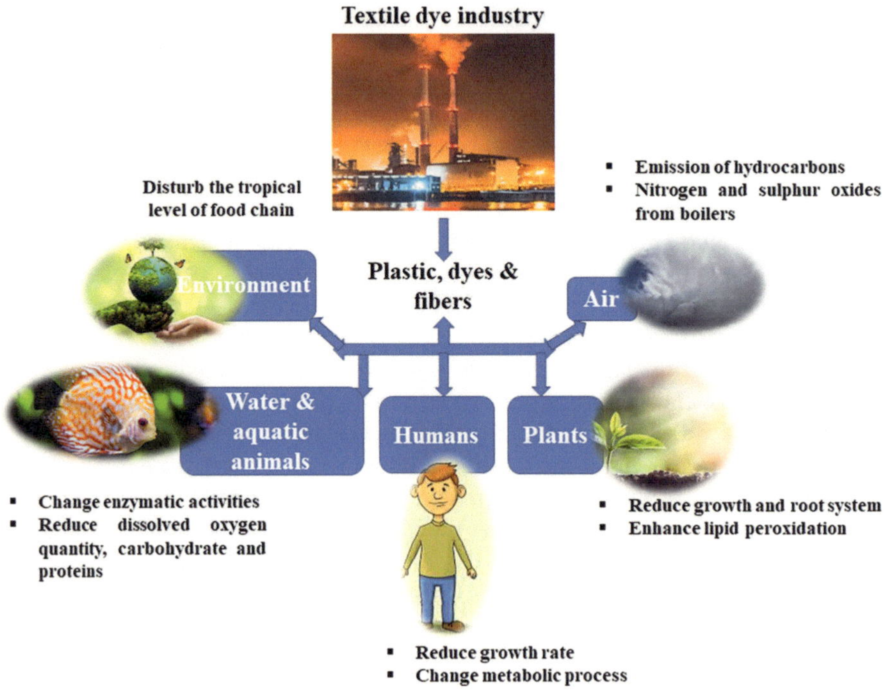

Figure 2.
Effects of dye fabric sewage on its surroundings and live beings from an ecotoxicological perspective.

3. Physicochemical characteristics of textile wastewater

The wastewater from the textile factory was thought to be one of the main causes of non-biodegradable chemicals entering natural water sources. During the dyeing process, a considerable amount of chemicals and colours do not adhere to the fibre. Moreover, surplus was released to atmosphere as effluent. Eighty percent of dyes and chemicals can be absorbed by textile fabrics [22]. **Figure 3** depicts the many steps in the manufacturing of textiles as well as the primary contaminants found in textile wastewater. Starch, which is produced from the tissue, is one of the substances that are employed to produce biological oxygen demand (BOD) during the desizing stage [23]. The process of scouring involves removing the fabric's other non-cellulosic components (fats, oils, and herbicides) through alkaline washing having detergents, producing effluent with a pH level that is elevated. Caustic soda is used to wash the fabrics during the mercerizing step in order to enhance fabric strength and colour retention, which raises the pH of the wastewater [24]. The bleaching stage creates white cloth by using hypochlorite, peracetic acid, and enzymes to create light and bright colours. Since several heavy metals are employed as auxiliaries in the method of dyeing to enhance the adsorption of the dye on the fibres through complexation processes. They might be introduced to the wastewater during the dyeing stage [25]. Since a significant portion of the dye is not absorbed by the fabric during this stage (20–30%), colour, BOD, and COD are significantly elevated. This phase is thought to provide the highest degree of contaminants. Urea, additional colours, or metals are added during printing. Adsorbable organic halogens, enzymes, surfactants, detergents, sodium or disodium thiosulfate, and BOD are primarily added during the

Figure 3.
Green technology necessary properties for wastewater treatment.

washing processes of prewash bleaching, stone or acid or enzyme wash, as well as softener. Utilising compounds made for crosslinking, softening, and water resistance, the final stage softens the materials [25].

4. Numerous techniques for wastewater treatment

Obtaining safe water to consume is the most important world concerns impacting people nowadays due to human activity and dangerous access. By 2030, the availability of clean water might be a problematic for nearly 47 percent of the world's populace [26]. The circumstance for removing microbes and natural impurities like textile dyes are attaining support globally to attain the aim of having clean and secure water to drink. But the materials needed to purify water are costly, making them inaccessible for a great deal of middle- and low-income nations. China is becoming a significant manufacturer of textiles for roughly 55 percent of global textile consumption [27]. China have occurred as a big textile producer, nearly for more than 55 percent of world fabric usage. Nearly 10,000 distinct synthetic colours and dyes are employed in the fabric and paper industries that is harmful to both the natural world and human wellness [28]. Cellulosic fibres, protein fabrics like silk, and synthetic fibres like nylon, and polyester are formed in textile industries by wet and dry techniques [7].

The final stage of textile colouring is the washing cycle. It eliminates excess colour and dyes which are poured to water and create water pollution since they have dyes with different chromophoric families which are extensively hazardous and possibly cancer-causing. In addition, colours from textiles discharged into the atmosphere and groundwater resources persist there for a very long period [28]. The development of a procedure that can be kept up with is necessary for colour eradication technology to evolve. Regarding regulation, there was still no worldwide agreement on the discharge of wastewater comprising dyes and related textile wastes, particularly azo dyes. Usually, azo dye limits are included with restrictions on the physico-chemical characteristics of treated effluent. The treatment of wastewater can be physico-chemical, biological/amalgamation of this methods, depending on the characteristics of the effluent. The following are a few of the most popular strategies [29].

4.1 Physical methods for fabric dye effluent

According to the mass transfer system, different physical methods, including adsorption, and membrane filtration, were employed to treat dye-comprising effluent with extremely higher elimination efficiencies that range from 85 to 99 percent [17]. Physical approaches give several advantages, such as uncomplicated designs, ease of use, cost, fewer chemical needs. It has no repressive effects because of its lack of toxic substances. These technologies are, however, largely unpopular owing to its variety of problems like the production of toxic by-products and sludge, followed by their inadequate applicability. Higher temperatures, COD, BOD, colour, and heavy metals are often obstacles to their usage in the process of textile effluent [30].

4.1.1 Adsorption and its mechanism

In the surface-based method of adsorption, adsorbed ions or molecules are drawn to a solid adsorbent surface. Physisorption and chemisorption were the two main kinds of adsorption. Depends on how the dye compound was adsorbed to the adsorbent surface, this classification is made. Several forces, including van der wall, and electrostatic connections might be present during the adsorption of dye molecules. Adsorbents are recycled, the procedure is extremely efficient, and colour elimination from effluent only takes a shorter time [31]. Adsorbents that frequently have porous shapes that enhance the total exposed surface area. It is needed for the quick and effective adsorption of dye substances from effluent, are the foundation of the adsorption method. For the elimination of dye from wastewater, several adsorbents like zeolites, silica gel, and activated carbon, etc. are extensively employed. On the other side, activated carbon is a frequently utilised adsorbent in industry [32, 33].

Numerous adsorption techniques are utilised to achieve the adsorption of dye from polluted water on the surface of an adsorbent. It must be denoted that electrostatic attraction, pi-pi interactions, van der Waals forces, hydrogen bonds, acid-base reactions, and hydrophobic interaction plays a main part in the adsorption of water contaminants on adsorbents [1]. According to Shen and Gondal's findings, the adsorption of Rhodamine dye on the surface of the adsorbent was controlled through electrostatic and intermolecular connections [34]. According to Zheng et al., ion exchange and electrostatic attraction are both involved in the adsorption of anionic dyes [35]. **Figure 4** exemplifies the mechanism of dye adsorption in wastewater effluent.

According to Ebrahimian study, the elimination of MB dye from wheat straw preserved with sodium hydroxide and permeated with Fe_3O_4 involves two main mechanisms: the development of an exterior complex and ion exchange among the dye substance and adsorption surfaces [36]. The adsorption method was characterised

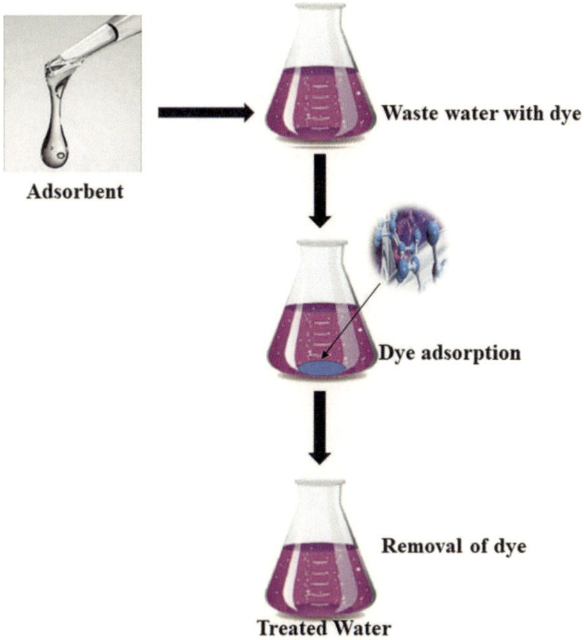

Figure 4.
Dye adsorption method for dye wastewater effluent.

through ion binding to different surface functional groups exist on the surface of adsorbent and electrostatic interaction. Among the adsorbent-adsorbate surfaces, includes the mechanism of the development of a surface complex. According to Cojocaru et al., the adsorption process was elucidated through the creation of hydrogen bonds among the adsorbents and Acid Orange 7 dye [37]. Siddiqui et al. claim that, interaction among the -OH groups in MnO_2/BC as well as the acceptor in MB substances are what causes H-bonds to form between MB and MnO_2/BC [38]. Like electrostatic interactions, pi-effects involve pi systems, in which positively charged substances connect with negatively charged surfaces. Furthermore, many mechanisms may operate concurrently throughout the adsorption process [39].

4.1.2 Membrane filtration (MF)

Membrane filtration (MF) are the most cutting-edge physical techniques for treating of effluent comprising colours. This method employs membranes with tiny pores, and a dye-free solution is created as a result of solutes greater than these pores getting trapped between them. Although simple and effective, this procedure occasionally necessitates replacing the membranes [17]. **Figure 5** represents the MF method for dye wastewater effluent.

In order to distinguish between particles that are suspended and colours in wastewater, the microfiltration (MF) method employs a conventional membrane with pore sizes that vary from 0.1 to 10 m. The purification of wastewater comprising dyes has recently been carried out using nanofiltration (NF), a cutting-edge membrane technique with average membranes with a diameter ranging from 0.5 to 0.2 nm. Consequently, nanofiltration technologies can remove colour components from effluent solutions by utilising size and electrostatic repulsion concepts. Ultrafiltration (UF) membranes

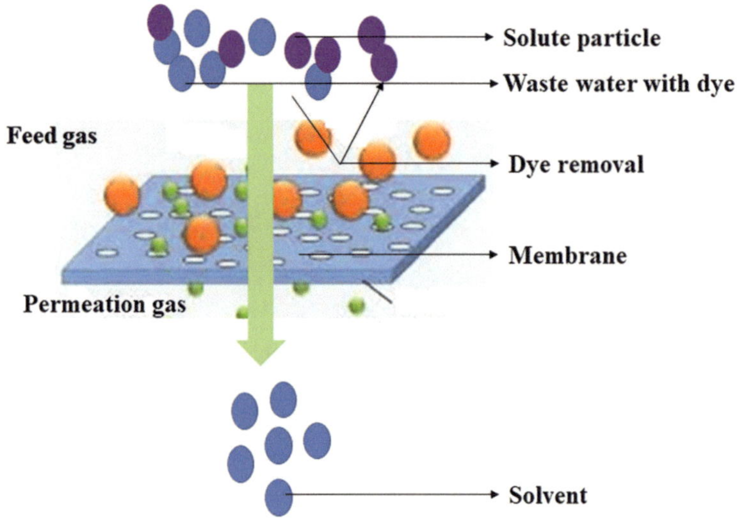

Figure 5.
Membrane filtration method for dye wastewater effluent.

are additionally suitable to eliminate natural colourants from textile effluent, with membrane widths that vary from 0.1 to 0.001 m [30]. Despite being fewer costly and requiring lesser pressure than nanofiltration, ultrafiltration still has a lower separation rate due to the enormous membrane pore size. Reverse osmosis (RO), a membrane filter technique was commonly employed in industry to remove dye-containing effluent and create higher-quality water. The advantage of RO technology is that it makes concentration and separation possible without using heat or state-changing energy [40]. **Table 1** summarises the benefits and effects various methods in dye removal.

S. No.	Techniques used	Pros	Cons
1	Physical method		
	(i) Adsorption process	Complete elimination of an extensive range of colours	Expensive
	(ii) Membrane filtration	All the dyes were removed	Development of concentrated sludge
2	Chemical method		
	(i) Advanced Oxidation	Easy to use	Needs activation agent
3	Biological method		
	(i) Microbial process	Elimination in 24 to 30 hr.	Azo-based dyes are not simply metabolised below aerobic circumstances
	(ii) Enzyme assisted method	A combination of two or more enzymes used to convert toxic intercedes to insoluble end-products	Higher cost of pure enzymes
	(iii) Yeast	Viable, cost-effective	Time consuming

Table 1.
An outline of the pros and cons of various dye removal techniques.

4.2 Chemical methods for fabric dye effluent

Dye-containing wastewater is processed chemically utilising methods like electrochemistry, and sophisticated processes for oxidation. With the notable exception of electrochemical technological advances, these techniques are frequently more costly than physical and biological methods [11]. The major difficulties for marketable usage of chemical methods for eliminating dyes from textile effluent include the higher electrical energy needs, massive volumes of utilised chemicals as well as the requirement for the appropriate equipment. Chemical approaches for colour elimination are more difficult since they produce potentially dangerous metabolites and residues [41, 42].

4.2.1 Advanced oxidation approaches

Diverse methods for cleaning up dye-containing effluent are being studied. Advanced oxidation constitutes a few of these techniques, and it is based on the idea that dye effluent can be treated by creating hydroxyl radicals (OH•), which are strong oxidising agents. Photocatalysis, ozonation, and electrochemical procedures are a few examples of advanced oxidation methods. With these methods, dye may be quickly and efficiently eliminated in challenging situations without creating sludge. They have their own disadvantages such higher cost, pH dependence, and dangerous by-products. The development of OH• as well as the deterioration of textile effluent have both been extensively investigated via the photocatalysis approach [43]. The process through which dangerous coloured molecules are destroyed through photocatalysis. Nanoparticles like zinc oxide and titanium peroxide are employed as photocatalytic materials to develop the free radicals required for the breakdown of dye through photocatalysis. Superoxide was created by electrons acting as reducing agents, and OH• and mineralisation byproducts were created by holes oxidising the organic dye components. The Fenton and photo-Fenton methods are two extensively employed progressive oxidation procedures for the disintegration of natural contaminants in effluent, like pigments. This procedure depends on an existence of H_2O_2 and soluble iron [44, 45].

4.3 Biological methods for fabric-based dye effluent

Biological methods for the cleansing of dye-comprising sewage more ensuring than both chemical and physical methods were readily necessary to develop fewer sludge. It will use smaller number of chemical reagents that are cheaper, has energy-saving characteristics are ecologically safer, as well as the by-products produced during bacterial metabolic processes are non-toxic [7]. Furthermore, applying biological techniques results in total dye mineralisation and is cost-effective for use in underdeveloped countries. The aim of biological therapy was to turn refractory organic dyes to non-toxic products. The major benefit of a biological approach is the usage of microbes by an excellent biodegradable capability, either alone or in consortia, to treat effluent dyes. Adsorption and deterioration are the two main techniques used to treat textile effluent for dye decolorization [46]. These activities can take place in either aerobic or anaerobic circumstances. Since the primary by-products of the aerobic process are biomass, carbon dioxide, and water, while the major by-product of anaerobic processing is methane. Appropriate biological alternatives for treating textile dye effluent and changing dye substances to non-toxic products including microbes, algae, fungi, and enzyme-based methods [47].

4.3.1 Microbial-assisted deprivation of dye-comprising effluent

It has been demonstrated that many bacterial species are more effective than other microbes at breaking down wastewater that contains dye. The adaptability and capacity of the bacteria to digest dye under the existing environmental conditions determines the efficiency of dye deprivation by microbes [48]. It has been demonstrated that a wide variety of bacterial species, like *Klebsiella*, *Bacillus*, *Shigella*, etc. are capable of degrading azo dyes. The major benefits of utilising microbe in dye deprivation include their ease of cultivation, higher specific growth rates, and varied catalytic abilities for mineralizing azo dyes contained in effluent [49]. Numerous microbial species have utilised azo reductase enzymes to biologically degradable azo dyes via azo reductase underneath aerobic or anaerobic circumstances as the initial phase of the microbial deterioration mechanism. Under anaerobic conditions, bacterial azo dye decolorization frequently takes place via azo bond reduction, resulting in the synthesis of colourless amino acids. However, because these derived chemicals are mutagenic and carcinogenic, they must undergo an additional aerobic step of microbial deprivation in order to lessen their toxicity and change to compounds that are harmless to the environment [50]. Reactive Orange M2R's decolorization by *Lysinibacillus* sp. KMK-A was predominantly attributable to azo bond decrease and the creation of metabolites; as a result, the rate of dye decolorization can be gauged in terms of BOD and COD decrease. The evaluation of natural load and level of mineralisation depend on these two variables [51].

Microbial consortia usually outclass a single strain in terms of dye elimination efficiency, and microbes have shown nearly 100 percent efficacy in the biodegradation of dye-comprising textile effluent. *Bacillus megaterium*, *B. cereus*, and *B. subtilis* are only a few of the five strains of distinct Bacillus species that make up the bacterial consortium known as SKBII found to be more efficient at dye decolorization compared with discrete strains [52]. Additionally, degradation of Red HE3B by a consortium of bacteria made up of *Pseudomonas aeuroginosa* BCH and *Providencia* sp. SDS that were separated from dye-contaminated soil. It was shown that the consortium's intense metabolic action caused the decolorization and degradation of 50 mg/L of Red HE3B dye to occur at a pace that was 100 percent faster than that of individual bacterial strains within 1 hour. By employing a microbial consortium BP made up of *Bacillus flexus* TS8, and *Pseudomonas aeruginosa* NCH separated from textile sewage and dye-contaminated soil, Mohanty and Kumar examined the decolorization of Indanthrene Blue RS. The BP consortium outperformed the individual strains in terms of dye decolorization. The performance of the BP consortium was also increased by an introduction of residual farming wastes. Enhanced intracellular enzyme concentrations and non-toxic produced metabolites were a result of oxido reductive enzymes' involvement in the complete deprivation mechanism. Thus, it was inferred that a wide range of specific microbial strains as well as consortia can degrade a wide range of colours employed in fabric production. This green method holds promise for the processing of effluent that comprises dyes, and it might be a cutting-edge and ground-breaking method for large-scale dye decolorization [53].

4.3.2 Enzyme-based deprivation of dye-containing effluent

Pure enzymes were not usually the first option for processing of wastewater comprising dyes due to their expensive cost. On the other hand, industrial enzymes stand out due to their affordability, effectiveness, and availability in liquid form. As an illustration, laccases and azo reductases were efficient at breaking down wastewater containing azo dyes. These dyes convert such complex natural pollutants to simpler compounds, and eliminating them from effluent containing textiles through flocculation. Enzymes are therefore commonly utilised in the chemical and biotechnology fields. Biocatalyst deactivation as a result of denaturation, however, is one of the most challenging elements of enzymatic degradation [17]. The action of numerous enzymes during the breakdown of azo dyes has recently been the subject of several investigations, the oxidoreductase enzyme class being one of the extensively studied. The most common enzymes against a variety of industrial dyes are peroxidase enzymes, which also have a wider pH range and higher temperature tolerance [54]. With immobilisation technologies making substantial development and enhancing enzyme performance in a variety of applications, enzyme immobilisation emerges as a capable element in textile dye biological degradation. Almost every type of enzyme can be utilised with cross-linking, making it a widely used immobilisation technique. Several items, such as charcoal and calcium alginate gel capsules are employed to immobilise dye-degrading enzymes. To enhance azo dye (such as Acid Red) decolorization, Malani et al. employed immobilised horse radish peroxidase (HRP) and a sono-enzymatic combination processing technique, obtaining up to 61 percent. For the breakdown of azo dyes with 85 percent azoreductase action has also been created [55, 56].

Additionally, Bilal et al. examined the dye decolorization effectiveness of chitosan-encapsulating HRP in a packed-bed reactor and discovered that Remazol Brilliant Blue R, Crystal Violet, Congo Red, and Reactive Black-5 were decolored by 82.17, 87.43, 94.35, and 97.82 percent. This immobilisation technique further enhanced the heavy metal inhibition resistance of the enzyme [57]. Additionally, Bilal et al. looked into how HRP immobilised on an environmentally friendly agarose-chitosan hydrogel support material degraded the Reactive Blue 19 dye. At 50 and 70°C, the immobilised enzyme system had more catalytic action than the free enzyme under alkaline conditions. Finally, Indigo Carmine through laccase revealed 82 percent better dye deprivation when ionic solutions were utilised [58].

4.3.3 Yeast-based deterioration of dye-comprising effluent

When it comes to bacterial and filamentous fungus degradation of textile colours, yeasts have several advantages. Yeasts has the possible to be employed as an alternative for the treatment of wastewater comprising dyes because of its quick development rates and capacity to withstand challenging ecological circumstances like lower pH. Only a few accounts, nevertheless, specifically mention yeasts' destruction and removal of colours. The ascomycetous yeast species *Debaryomyces*, *Candida*, *Kluyveromyces*, and *Saccharomyces* have all been observed to digest wastewater containing colour. On the other side, the most capable basidiomycetous yeast include Trichosporon and Rhodotorula. Numerous yeasts, like *Scheffersomyces spartinae*, and *Sterigmatomyces halophilus*, has revealed the capacity to destroy textile colours, like azo dyes, and also to endure difficult circumstances, including higher salt concentrations in textile effluent [59]. The two main mechanisms through which yeast strains can eliminate higher amounts of various colours from effluent are biosorption as well as reductive cleavage. *Debaryomyces hansenii* F39A, an Antarctic yeast, was examined by Ruscasso et al. for its potential to be used as a biological sorbent for the reactive textile dyes. The collected data demonstrated that no poisonous or dangerous compounds were released which results in adsorption during the first 15 minutes of the operation [17, 60].

5. Conclusion

Scientists made an effort to reconsider the accessibility of clean and safe water due to the environmental impact of coloured dye wastewater disposal. It was documented that an occurrence of poisonous and coloured substances in dye-containing effluent has consequences on living organisms such as cancer, mutagenesis, allergies, and dermatitis. Textile effluent comprises a variety of detrimental xenobiotics which were found in higher concentrations and are hazardous to atmosphere as well as common people well-being. Since textile industries around the world releasing millions of gallons of contaminated sewage, toxicity dye-comprising effluent was an important problem. Contrarily, treating effluent that comprises dye is a substantial difficulty because no effective treatment that is also economically feasible currently exists. There are many tried-and-true and cutting-edge techniques for handling dye-containing wastewater. Wastewater containing dye can be effectively removed or degraded by employing physico-chemical process. Nevertheless, these techniques come with higher operational costs and unfavourable effects. In comparison to physico-chemical process, the bacteria-based technique to dye-comprising effluent treatment is more practical, ecologically safe, and socially and acceptable. Nevertheless, a drawback of biological techniques is that they are ineffective and necessitate a long period of administration. More study is needed to provide an advanced, waste-free technique and to lower ecological and human health threats when scaling up from lab use to the pilot level.

Acknowledgements

This work was supported by the Nanodot Research Pvt. Ltd. Neyyattinkara, Trivandrum.

Conflict of interest

The authors declare that they have no conflict of interests for this chapter.

Author details

Saranyadevi Subburaj[1], Manikandan Paramasivam[2], Methaq H. Lafta[3] and Lekshmi Gangadhar[4]*

1 Paavai Engineering College (Autonomous), Namakkal, India

2 Dhanalakshmi Srinivasan Engineering College (Autonomous), Perambalur, India

3 Urmia University, Iran

4 Nanodot Research Pvt. Ltd. Neyyattinkara, Trivandrum, India

*Address all correspondence to: lekshmigangadhar.nano@gmail.com

References

[1] Lu F, Astruc D. Nanocatalysts and other nanomaterials for water remediation from organic pollutants. Coordination Chemistry Reviews. 2020;**408**:213180

[2] Wong JK, Tan HK, Lau SY, Yap PS, Danquah MK. Potential and challenges of enzyme incorporated nanotechnology in dye wastewater treatment: A review. Journal of Environmental Chemical Engineering. 2019;**7**(4):103261

[3] Benkhaya S, M'rabet S, El Harfi A. A review on classifications, recent synthesis and applications of textile dyes. Inorganic Chemistry Communications. 2020;**115**:107891

[4] Chen J, Xiong Y, Duan M, Li X, Li J, Fang S, et al. Insight into the synergistic effect of adsorption–photocatalysis for the removal of organic dye pollutants by Cr-doped ZnO. Langmuir. 2019;**36**(2):520-533

[5] Zhou Y, Lu J, Zhou Y, Liu Y. Recent advances for dyes removal using novel adsorbents: A review. Environmental Pollution. 2019;**252**:352-365

[6] Almroth BC, Cartine J, Jönander C, Karlsson M, Langlois J, Lindström M, et al. Assessing the effects of textile leachates in fish using multiple testing methods: From gene expression to behavior. Ecotoxicology and Environmental Safety. 2021;**207**:111523

[7] Ali SS, Al-Tohamy R, Sun J. Performance of Meyerozyma caribbica as a novel manganese peroxidase-producing yeast inhabiting wood-feeding termite gut symbionts for azo dye decolorization and detoxification. Science of the Total Environment. 2022;**806**:150665

[8] Chandanshive V, Kadam S, Rane N, Jeon BH, Jadhav J, Govindwar S. In situ textile wastewater treatment in high rate transpiration system furrows planted with aquatic macrophytes and floating phytobeds. Chemosphere. 2020;**252**:126513

[9] Akpomie KG, Conradie J. Advances in application of cotton-based adsorbents for heavy metals trapping, surface modifications and future perspectives. Ecotoxicology and Environmental Safety. 2020;**201**:110825

[10] Silva AC, Silvestre AJ, Freire CS, Vilela C. Modification of textiles for functional applications. In: Fundamentals of Natural Fibres and Textiles. Duxford: Woodhead Publishing; 2021. pp. 303-365

[11] Kishor R, Purchase D, Saratale GD, Saratale RG, Ferreira LF, Bilal M, et al. Ecotoxicological and health concerns of persistent coloring pollutants of textile industry wastewater and treatment approaches for environmental safety. Journal of Environmental Chemical Engineering. 2021;**9**(2):105012

[12] Parmar S, Daki S, Bhattacharya S, Shrivastav A. Microorganism: An ecofriendly tool for waste management and environmental safety. In: Development in Wastewater Treatment Research and Processes. Netherland: Elsevier; 2022. pp. 175-193

[13] Singha K, Pandit P, Maity S, Sharma SR. Harmful environmental effects for textile chemical dyeing practice. In: Green Chemistry for Sustainable Textiles. Netherland: Elsevier, Woodhead Publishing; 2021. pp. 153-164

[14] Dutta S, Bhattacharjee J. A comparative study between physicochemical and

biological methods for effective removal of textile dye from wastewater. In: Development in Wastewater Treatment Research and Processes. Netherland: Elsevier; 2022. pp. 1-21

[15] Patil R, Zahid M, Govindwar S, Khandare R, Vyavahare G, Gurav R, et al. Constructed wetland: A promising technology for the treatment of hazardous textile dyes and effluent. In: Development in Wastewater Treatment Research and Processes. Netherland: Elsevier; 2022. pp. 173-198

[16] Shindhal T, Rakholiya P, Varjani S, Pandey A, Ngo HH, Guo W, et al. A critical review on advances in the practices and perspectives for the treatment of dye industry wastewater. Bioengineered. 2021;**12**(1):70-87

[17] Samsami S, Mohamadizaniani M, Sarrafzadeh MH, Rene ER, Firoozbahr M. Recent advances in the treatment of dye-containing wastewater from textile industries: Overview and perspectives. Process Safety and Environmental Protection. 2020;**143**:138-163

[18] Oyeniran DO, Sogbanmu TO, Adesalu TA. Antibiotics, algal evaluations and subacute effects of abattoir wastewater on liver function enzymes, genetic and haematologic biomarkers in the freshwater fish, Clarias gariepinus. Ecotoxicology and Environmental Safety. 2021;**212**:111982

[19] Sojobi AO, Zayed T. Impact of sewer overflow on public health: A comprehensive scientometric analysis and systematic review. Environmental Research. 2022;**203**:111609

[20] Elgarahy AM, Elwakeel KZ, Mohammad SH, Elshoubaky GA. A critical review of biosorption of dyes, heavy metals and metalloids from wastewater as an efficient and green process. Cleaner Engineering and Technology. 2021;**4**:100209

[21] Li F, Zhao K, Ng TS, Dai Y, Wang CH. Sustainable production of bio-oil and carbonaceous materials from biowaste co-pyrolysis. Chemical Engineering Journal. 2022;**427**:131821

[22] Babu AN, Reddy DS, Sharma P, Kumar GS, Ravindhranath K, Mohan GK. Removal of hazardous indigo carmine dye from waste water using treated red mud. Materials Today: Proceedings. 2019;**17**:198-208

[23] Chowdhury MF, Khandaker S, Sarker F, Islam A, Rahman MT, Awual MR. Current treatment technologies and mechanisms for removal of indigo carmine dyes from wastewater: A review. Journal of Molecular Liquids. 2020;**318**:114061

[24] Periyasamy AP, Militky J. Denim processing and health hazards. In: Sustainability in Denim. Netherland: Woodhead Publishing; 2017. pp. 161-196

[25] Madhav S, Ahamad A, Singh P, Mishra PK. A review of textile industry: Wet processing, environmental impacts, and effluent treatment methods. Environmental Quality Management. 2018;**27**(3):31-41

[26] Islam A, Teo SH, Taufiq-Yap YH, Ng CH, Vo DV, Ibrahim ML, et al. Step towards the sustainable toxic dyes removal and recycling from aqueous solution-A comprehensive review. Resources, Conservation and Recycling. 2021;**175**:105849

[27] Darwesh OM, Ali SS, Matter IA, Elsamahy T. Nanotextiles waste management: Controlling of release and remediation of wastes. In: Nanosensors and Nanodevices for Smart

Multifunctional Textiles. Netherland: Elsevier; 2021. pp. 267-286

[28] Tounsadi H, Metarfi Y, Taleb M, El Rhazi K, Rais Z. Impact of chemical substances used in textile industry on the employee's health: Epidemiological study. Ecotoxicology and Environmental Safety. 2020;**197**:110594

[29] Al-Tohamy R, Kenawy ER, Sun J, Ali SS. Performance of a newly isolated salt-tolerant yeast strain Sterigmatomyces halophilus SSA-1575 for azo dye decolorization and detoxification. Frontiers in Microbiology. 2020;**11**:1163

[30] Behera M, Nayak J, Banerjee S, Chakrabortty S, Tripathy SK. A review on the treatment of textile industry waste effluents towards the development of efficient mitigation strategy: An integrated system design approach. Journal of Environmental Chemical Engineering. 2021;**9**(4):105277

[31] Mudhoo A, Ramasamy DL, Bhatnagar A, Usman M, Sillanpää M. An analysis of the versatility and effectiveness of composts for sequestering heavy metal ions, dyes and xenobiotics from soils and aqueous milieus. Ecotoxicology and Environmental Safety. 2020;**197**:110587

[32] Abu-Nada A, Abdala A, McKay G. Removal of phenols and dyes from aqueous solutions using graphene and graphene composite adsorption: A review. Journal of Environmental Chemical Engineering. 2021;**9**(5):105858

[33] Madan S, Shaw R, Tiwari S, Tiwari SK. Adsorption dynamics of Congo red dye removal using ZnO functionalized high silica zeolitic particles. Applied Surface Science. 2019;**487**:907-917

[34] Shen K, Gondal MA. Removal of hazardous Rhodamine dye from water by adsorption onto exhausted coffee ground. Journal of Saudi Chemical Society. 2017;**21**:S120-S127

[35] Zheng Y, Cheng B, You W, Yu J, Ho W. 3D hierarchical graphene oxide-NiFe LDH composite with enhanced adsorption affinity to Congo red, methyl orange and Cr (VI) ions. Journal of Hazardous Materials. 2019;**369**:214-225

[36] Pirbazari AE, Saberikhah E, Kozani SH. Fe3O4–wheat straw: Preparation, characterization and its application for methylene blue adsorption. Water Resources and Industry. 2014;**7**:23-37

[37] Cojocaru C, Samoila P, Pascariu P. Chitosan-based magnetic adsorbent for removal of water-soluble anionic dye: Artificial neural network modeling and molecular docking insights. International journal of biological macromolecules. 2019;**123**:587-599

[38] Siddiqui SI, Manzoor O, Mohsin M, Chaudhry SA. Nigella sativa seed based nanocomposite-MnO2/BC: An antibacterial material for photocatalytic degradation, and adsorptive removal of methylene blue from water. Environmental Research. 2019;**171**:328-340

[39] Thamer BM, Aldalbahi A, Moydeen M, El-Hamshary H, Al-Enizi AM, El-Newehy MH. Effective adsorption of Coomassie brilliant blue dye using poly (phenylene diamine) grafted electrospun carbon nanofibers as a novel adsorbent. Materials Chemistry and Physics. 2019;**234**:133-145

[40] Liang C, Wei D, Zhang S, Ren Q, Shi J, Liu L. Removal of antibiotic resistance genes from swine wastewater by membrane filtration treatment.

Ecotoxicology and Environmental Safety. 2021;**210**:111885

[41] Wang X, Xia J, Ding S, Zhang S, Li M, Shang Z, et al. Removing organic matters from reverse osmosis concentrate using advanced oxidation-biological activated carbon process combined with Fe3+/humus-reducing bacteria. Ecotoxicology and Environmental Safety. 2020;**203**:110945

[42] Rahmani AR, Mousavi-Tashar A, Masoumi Z, Azarian G. Integrated advanced oxidation process, sono-Fenton treatment, for mineralization and volume reduction of activated sludge. Ecotoxicology and Environmental Safety. 2019;**168**:120-126

[43] Zhao H, Cao P, Lu L, Li F, Pang CH, Wu T. Co-regulation of dispersion, exposure and defect sites on CeO2 (111) surface for catalytic oxidation of Hg^0. Journal of Hazardous Materials. 2022;**424**:126566

[44] Siwińska-Stefańska K, Kubiak A, Piasecki A, Goscianska J, Nowaczyk G, Jurga S, et al. TiO2-ZnO binary oxide systems: Comprehensive characterization and tests of photocatalytic activity. Materials. 2018;**11**(5):841

[45] Hassanshahi N, Karimi-Jashni A. Comparison of photo-Fenton, O3/H2O2/UV and photocatalytic processes for the treatment of gray water. Ecotoxicology and Environmental Safety. 2018;**161**:683-690

[46] Coria-Oriundo LL, Battaglini F, Wirth SA. Efficient decolorization of recalcitrant dyes at neutral/alkaline pH by a new bacterial laccase-mediator system. Ecotoxicology and Environmental Safety. 2021;**217**:112237

[47] Jafari N, Soudi MR, Kasra-Kermanshahi R. Biodegradation perspectives of azo dyes by yeasts. Microbiology. 2014;**83**:484-497

[48] Liu J, Chen J, Zuo K, Li H, Peng F, Ran Q, et al. Chemically induced oxidative stress improved bacterial laccase-mediated degradation and detoxification of the synthetic dyes. Ecotoxicology and Environmental Safety. 2021;**226**:112823

[49] Chen YG, Huang JH, Luo R, Ge HZ, Wołowicz A, Wawrzkiewicz M, et al. Impacts of heavy metals and medicinal crops on ecological systems, environmental pollution, cultivation, and production processes in China. Ecotoxicology and Environmental Safety. 2021;**219**:112336

[50] Brüschweiler BJ, Merlot C. Azo dyes in clothing textiles can be cleaved into a series of mutagenic aromatic amines which are not regulated yet. Regulatory Toxicology and Pharmacology. 2017;**88**:214-226

[51] Chaudhari AU, Tapase SR, Markad VL, Kodam KM. Simultaneous decolorization of reactive Orange M2R dye and reduction of chromate by Lysinibacillus sp. KMK-A. Journal of Hazardous Materials. 2013;**262**:580-588

[52] Guo G, Hao J, Tian F, Liu C, Ding K, Xu J, et al. Decolorization and detoxification of azo dye by halo-alkaliphilic bacterial consortium: Systematic investigations of performance, pathway and metagenome. Ecotoxicology and Environmental Safety. 2020;**204**:111073

[53] Mohanty SS, Kumar A. Enhanced degradation of anthraquinone dyes by microbial monoculture and developed consortium through the production of specific enzymes. Scientific Reports. 2021;**11**(1):7678

[54] Jun LY, Yon LS, Mubarak NM, Bing CH, Pan S, Danquah MK, et al. An overview of immobilized enzyme technologies for dye and phenolic removal from wastewater. Journal of Environmental Chemical Engineering. 2019;**7**(2):102961

[55] Malani RS, Khanna S, Moholkar VS. Sonoenzymatic decolourization of an azo dye employing immobilized horse radish peroxidase (HRP): A mechanistic study. Journal of Hazardous Materials. 2013;**256**:90-97

[56] Misra N, Kumar V, Goel NK, Varshney L. Laccase immobilization on radiation synthesized epoxy functionalized polyethersulfone beads and their application for degradation of acid dye. Polymer. 2014;**55**(23):6017-6024

[57] Bilal M, Asgher M, Parra-Saldivar R, Hu H, Wang W, Zhang X, et al. Immobilized ligninolytic enzymes: An innovative and environmental responsive technology to tackle dye-based industrial pollutants-A review. Science of the Total Environment. 2017;**576**:646-659

[58] Bilal M, Rasheed T, Zhao Y, Iqbal HM. Agarose-chitosan hydrogel-immobilized horseradish peroxidase with sustainable bio-catalytic and dye degradation properties. International Journal of Biological Macromolecules. 2019;**124**:742-749

[59] Jamee R, Siddique R. Biodegradation of synthetic dyes of textile effluent by microorganisms: An environmentally and economically sustainable approach. European Journal of Microbiology and Immunology. 2019;**9**(4):114-118

[60] Ruscasso F, Bezus B, Garmendia G, Vero S, Curutchet G, Cavello I, et al. Debaryomyces hansenii F39A as biosorbent for textile dye removal. Revista Argentina de Microbiología. 2021;**53**(3):257-265

Chapter 8

The Efficiency of Wastewater Treatment Plants for the Removal of Antibiotics

Raed S. Al-Wasify, Majid M. Alruwaili, Fahad S. Aljohani, Shimaa R. Hamed and Samar Ragab

Abstract

Undoubtedly domestic Wastewater Treatment Plants (WWTPs) are not designed for the removal of some pollutants such as antibiotics. This chapter summarizes the occurrence and fate of six groups of the most widely used antibiotics (β-lactams, sulfonamides, quinolones, tetracyclines, macrolides, and others) in domestic WWTPs. The literature showed that the six groups of antibiotics have been frequently detected during wastewater treatment train (influent, primary treatment, secondary treatment, tertiary treatment, effluent, and sludge treatment) of domestic WWTPs. Also, it was clear that the main removal routes of antibiotics during sewage treatment of domestic WWTPs were adsorption, biodegradation, membrane filtration, and disinfection. Domestic WWTPs cannot remove most of the antibiotics which finally enter the environment through treated effluent and sludge.

Keywords: antibiotics, adsorption, biodegradation, domestic wastewater, organic pollutants, wastewater treatment plants

1. Introduction

Nowadays, environmental researchers have extended their focus beyond classic environmental contaminants such as pesticides, Polychlorinated Biphenyls (PCBs), and dioxins [1, 2]. Antibiotics are one of the new serious environmental contaminants. Antibiotics, a group of pharmaceuticals used as a medicine and growth promoter for both humans and animals, are considered a new serious environmental contaminant due to their continuous input into the environment and persistent presence [3–5].

Antibiotics are chemicals classified depending on their nature into three main categories: natural, semi-synthetic, and synthetic. Moreover, antibiotics can be classified depending on their mode of action into two categories: bactericidal (kill microorganisms) and bacteriostatic (impede microbial growth).

In developing countries, there is an increase in the consumption of antibiotics, without any prescription from physicians, to cure the different kinds of diseases that originate from improper general hygiene and poor sanitation systems. Also, antibiotics are used widely in animal farming for the protection of animal health to maintain the high demand for animal products [6].

Boyles et al. [7] reported that the high consumption of antibiotics by humans (households, hospitals, and industry) and in veterinary results in the increasing release of unchanged active ingredients and partially metabolized antibiotics into the sewer system (directly or indirectly) which consequently reach into the domestic wastewater treatment plants [8–10].

Conventional wastewater treatment plants cannot completely remove antibiotics and these antibiotics will finally contaminate the environment through effluent or sludge [11, 12]. Therefore, local wastewater treatment plants act as one of the main pathways for antibiotics to transfer into the environment [13].

Although antibiotics residues in water are very low (ng/L to μg/L), they still draw the researchers' attention in the whole world since these antibiotics are the main source for the occurrence and transfer of Antibiotic-Resistant Genes (ARG) and Antibiotic-Resistant Bacteria (ARB) which have serious impacts on the environment [14, 15].

During last years, simultaneous detection of trace concentrations of antibiotics in wastewater and sludge samples is no longer difficult as a result of the invention of new detection methods such as liquid chromatography-mass spectrometry, solid-phase extraction, and ultra-performance liquid chromatography-mass spectrometry with the rapid development of analytical methods such as Solid Phase Extraction (SPE), High-Performance Liquid Chromatography Mass Spectrometry (HPLC-MS/MS) and Ultra-Performance Liquid Chromatography Mass Spectrometry (UPLC-MS/MS) [16, 17].

The occurrence of antibiotics in water environments such as groundwater and surface water was summarized by previous studies [18–22] as well as sediments, sludge, and soil [19]. Nevertheless, most of these studies focused only on the occurrence of antibiotics in different environments with little focus on the fate of antibiotics. For example, summarizing the removal of antibiotics in sediment, water, and soil environments, instead of the elimination of antibiotics in wastewater treatment plants [21]. Also, the elimination of Personal Care Products (PPCPs) and pharmaceuticals via biodegradation and other pathways in wastewater treatment plants, with limited content on antibiotics elimination [23].

Thus, in this review, we summarize the data and information on the occurrence and fate of antibiotics in wastewater treatment plants (WWTPs) to provide the overall profile of antibiotics concentrations in influent, treatment stages, sludge, and effluent of wastewater treatment plants, and to understand the elimination routes and fate of antibiotics in WWTPs.

2. Occurrence of antibiotics in the aquatic environment

Domestic wastewater treatment plants receive most of the used antibiotics through the sewer network, while the rest of the antibiotics are dumped directly into rivers and streams or escape as leachate from landfills. **Figure 1** summarizes the introduction pathways of antibiotics into the aquatic environment.

3. Occurrence of antibiotics in wastewater treatment plants (WWTPs)

Antibiotics can be classified using different ways such as their chemical structure. According to the chemical structure, there are 12 different classes of antibiotics (**Table 1**) such as β- lactams, aminoglycosides, macrolides, glycopeptides,

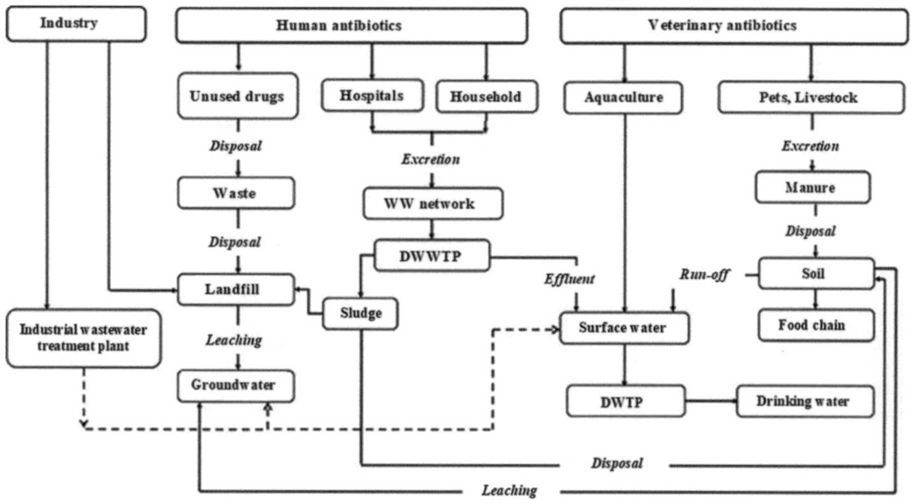

Figure 1.
Pathways of antibiotics in the aquatic environment, WWTP: Wastewater treatment plant, DWTP: Drinking water treatment plant.

oxazolidinones, sulfonamides, quinolones (fluoroquinolones), polymyxins, tetracyclines, streptogramins, and others such as chloramphenicol, thiamphenicol, lincomycin, trimethoprim, and clindamycin [5, 8]. **Table 1** summarizes the chemical structure of the different 12 classes of antibiotics and their mode of action, mechanism of action as well as discovery dates.

In this review, **Tables 2–7** summarize data about the occurrence of the most common antibiotics (6 classes) in influent and effluent samples of wastewater treatment plants. These antibiotics include β-lactams, quinolones, sulfonamides, macrolides, tetracyclines, and others, while **Table 8** summarizes the occurrence of these antibiotics in sludge generated from wastewater treatment plants.

Figure 2 shows the typical structure and the treatment train of a domestic wastewater treatment plant. The treatment train begins consists of three stages including primary, secondary, and tertiary. The primary treatment stage contains preliminary treatment units (screens and sand & grit removal), an oil and grease removal unit, and primary settling tanks (not common in most WWTPs). The secondary treatment stage contains a biological treatment unit (aeration tank) using different methods such as activated sludge (AS) and trickling filters (TF), followed by secondary settling tanks (secondary clarifiers). The tertiary treatment stage contains a sand filtration unit followed by a disinfection unit using different technologies such as chlorination, ozonation, and ultraviolet (UV), and finally nutrient removal unit (for treated effluent which contains high concentrations of phosphorus and nitrogen).

3.1 Presence of antibiotics in influent and effluent samples of WWTPs

3.1.1 β-Lactams

β-lactams are a group of antibiotics characterized by the presence of the β-lactam ring. The β-lactams ring is the main structure that gives the antibacterial activity

Antibiotic group	β-Lactams	Sulfonamides	Aminoglycosides	Chloramphenicol	Tetracyclines	Macrolides
Chemical structure	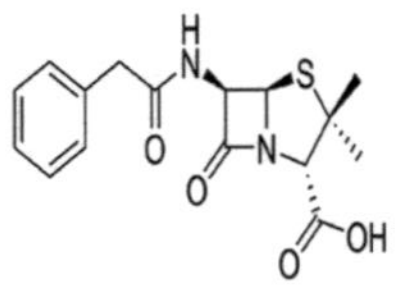	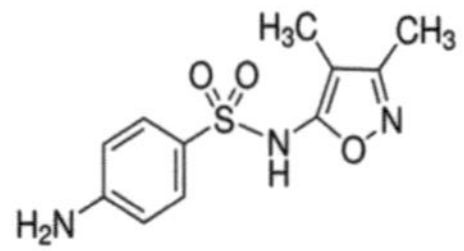	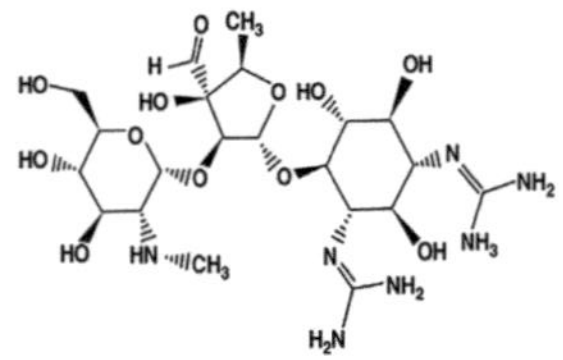	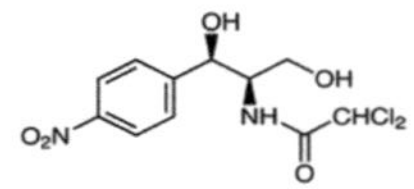	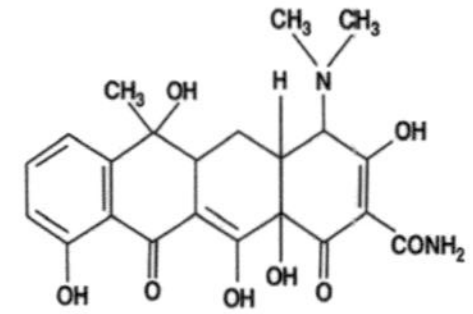	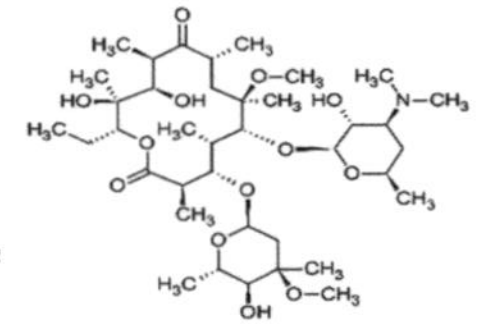
Main feature	All contain a beta-lactam ring.	All contain the sulfonamide group	All contain aminosugar substructures	Distinct individual compound (shown)	All contain 4 adjacent cyclic hydrocarbon rings	All contain a 14-, 15-, or 16-membered macrolide ring
Examples	• Penicillins (penicillin G shown) such as flucloxacillin and amoxicillin. • Cephalosporins such as cefalexin.	Sulfisoxazole (shown), sulfanilamide, prontosil	Streptomycin (shown), kanamycin, neomycin		Tetracycline (shown), oxytetracycline, limecycline	Clarithromycin (shown), erythromycin, azithromycin
Mode of action	Bactericidal	Bacteriostatic	Bactericidal	Bacteriostatic	Bacteriostatic	Bacteriostatic
Mechanism of action	Inhibit bacterial cell wall synthesis	Inhibit bacterial growth and multiplication	Inhibit bacterial protein synthesis	Inhibit bacterial protein synthesis	Inhibit bacterial protein synthesis	Inhibit bacterial protein synthesis
Discovery	1920s	1935	1943	1947	1948	1950s

Antibiotic group	Glycopeptides	Ansamycins	Quinolones	Streptogramins	Oxazolidinones	Lipopeptides
Chemical structure	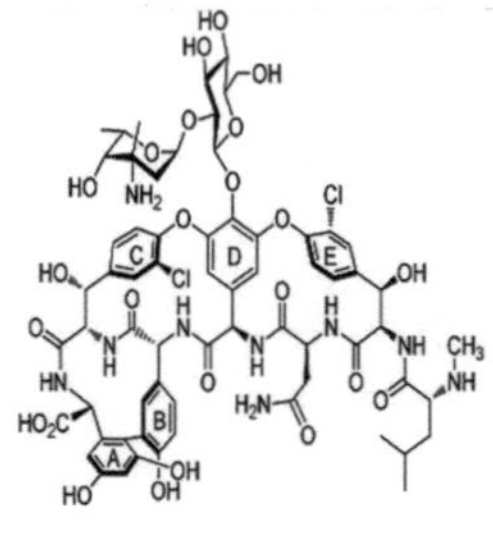	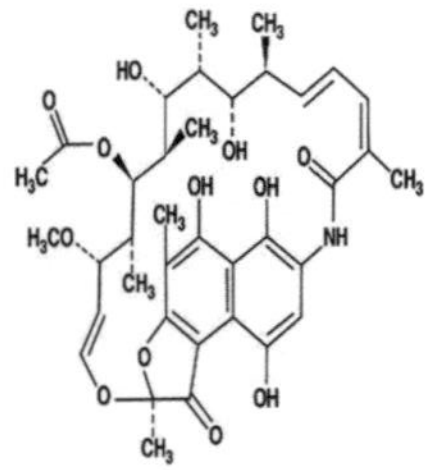	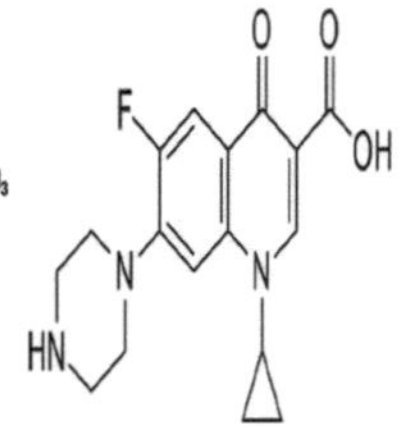	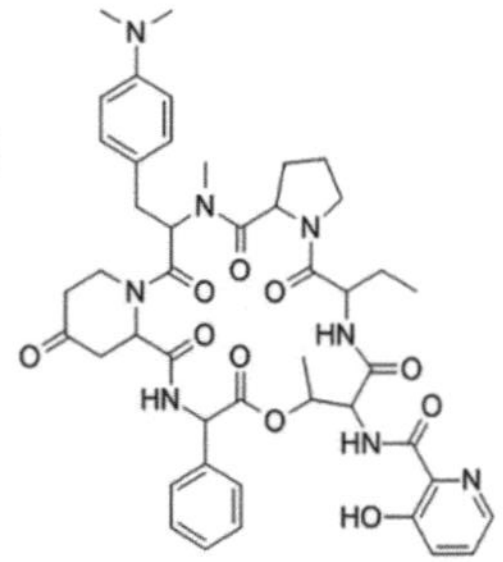	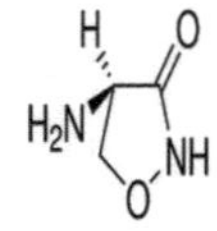	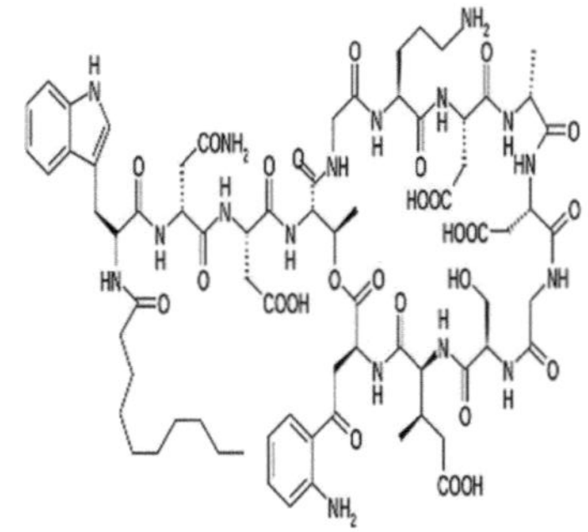
Main feature	Consist of carbohydrate linked to a peptide formed of amino acids	All contain an aromatic ring bridged by an aliphatic chain	All contain fused aromatic rings with an attached carboxylic acid group	Combination of two structurally different compounds	All contain 2-oxazolidone somewhere in their structure	All contain a lipid bonded to a peptide
Examples	Vancomycin (shown), teicoplanin	Rifamycin (shown), geldanamycin	Ciprofloxacin (shown), trovafloxacin	Pristinamycin IA (shown), Pristinamycin IIA	Cycloserine (shown), linezolid	Daptomycin (shown), surfactin
Mode of action	Bactericidal	Bactericidal	Bactericidal	Bactericidal	Bacteriostatic	Bactericidal
Mechanism of action	Inhibit bacterial cell wall synthesis	Inhibit bacterial RNA synthesis	Inhibit bacterial DNA replication and transcription	Inhibit bacterial protein synthesis.	Inhibit bacterial protein synthesis	Disturb multiple cell membrane functions
Discovery	1958	1959	1962	1962	1978	1986

Table 1.
Different classes of antibiotics and their mode of action and mechanism of action.

Antibiotic	Concentration (ng/L)		Sampling type	Treatment technology
	Influent	Effluent		
Penicillin G	29.0	NA	Grab	Primary treatment
	ND-10	ND-300	Composite	—
Penicillin V	20–13,800	ND–2000	Grab	Primary treatment
Amoxicillin	190–280	ND–30	24 h FPC	Membrane filtration/Reverse osmosis
	1400–6940	ND-50	Composite	—
Ampicillin	NA	7	Grab	Activated sludge
Oxacillin	NA	<20	24 h FPC	Activated sludge + chlorination
Cloxacillin	ND–320	ND	Grab	Activated sludge
	ND-4600	ND-700	Composite	—
Cefalexin	1200	980	Grab	Activated sludge + denitrification and nitrification
	NA	1110–1410	24 h FPC	Chemically enhanced primary treatment
	2800–64,000	ND-250	Composite	—
Cefotaxime	24	<12	Grab	Activated sludge + denitrification and nitrification
Cefaclor	500–980	ND–60	Grab	Activated sludge
	500–6150	ND-1800	Composite	—
Cefradine	NA	12	Grab	Activated sludge
Cefazolin	NA	27	Grab	Activated sludge

ND: Not detected, NA: Not analyzed, FPC: Flow proportional composite.

Table 2.
Concentrations of β-lactams in wastewater.*

Antibiotic	Concentration (ng/L)		Sampling type	Treatment technology
	Influent	Effluent		
Sulfamethoxazole	360–500	270–320	Grab	Activated sludge
	255–333	ND	Grab	Membrane filtration/Reverse osmosis
	<80–674	<80–304	1-week FPC	Activated sludge + denitrification
	390	310	48 h composite	—
	230–750	211–860	24 h FPC	Activated sludge + sand filtration
N^4-acetylsulfamethoxazole	850–1600	<20–180	72 h FPC	Activated sludge + sand filtration
	NA	690–2200	Grab	—
Sulfathiazole	2–40	ND–5	Composite	Activated sludge
Sulfasalazine	ND–60	ND–10	Grab	Activated sludge
	ND-100	4–150	Composite	—
Sulphapyridine	60–150	40–350	24 h FPC	Activated sludge + sand filtration
	NA	124	24 h composite	—
	NA	85–88	27 h FPC	Activated sludge + sand filtration

Antibiotic	Concentration (ng/L)		Sampling type	Treatment technology
	Influent	Effluent		
Sulfacetamide	NA	64–151	24 h FPC	Activated sludge + chlorination
	110–210	ND	24 h composite	Activated sludge
Sulfacholoropyridazine	<30–476	<30–149	Grab	Activated sludge
Sulfadiazine	3.7	1.9–3.8	Grab	Activated sludge
Sulfisoxazole	NA	0.13	Grab	Activated sludge
Sulfadimethoxine	<10–213	<10 – 70	Grab	Activated sludge
Sulfamethizole	ND–710	ND–10	Grab	—
Sulfamerazine	29–73	12–42	Grab	—
Sulfanilamide	NA	26	Grab	Activated sludge
Sulfaguanidine	NA	2	Grab	—

ND: Not detected, NA: Not analyzed, FPC: Flow proportional composite.

Table 3.
Concentrations of sulfonamides in wastewater.*

Antibiotic	Concentration (ng/L)		Sampling type	Treatment technology
	Influent	Effluent		
Norfloxacin	110	85	Grab	Activated sludge + denitrification and nitrification
	339	85	24 h composite	Activated sludge
	72–174	7–37	1-week FPC	Activated sludge + denitrification
	NA	250–546	24 h FPC	Chemically enhanced primary treatment
Ciprofloxacin	3800–4600	640–720	Grab	Activated sludge
	210	60	24 h composite	Activated sludge
	113–300	7–32	1-week FPC	Activated sludge + denitrification
Enrofloxacin	10–100	10	Grab	Activated sludge
	ND-40	2–50	Composite	—
Nalidixic acid	ND–200	55	Grab	Activated sludge
Ofloxacin	137	41	Grab	Activated sludge + chlorination
	ND-200	ND-450	Composite	—
	470	110	48 h composite	—
	NA	740–1220	24 h FPC	Chemically enhanced primary treatment
Lomefloxacin	NA	<45	Grab	Activated sludge + chlorination + sand filtration
Flumequine	NA	15	Grab	Activated sludge
Pipemidic acid	NA	70	Grab	—
	54	12	24 h composite	—
Flerofloxacin	28	5.8	24 h composite	Activated sludge

Antibiotic	Concentration (ng/L)		Sampling type	Treatment technology
	Influent	Effluent		
Lomefloxacin	98	17	24 h composite	—
Gatifloxacin	111	56	24 h composite	—
Moxifloxacin	44	17	24 h composite	—

ND: Not detected, NA: Not analyzed.

Table 4.
Concentrations of quinolones in wastewater.*

Antibiotic	Concentration (ng/L)		Sampling type	Treatment technology
	Influent	Effluent		
Tetracycline	240–790	50–160	24 h composite	Activated sludge
	96	180	24 h FPC	Activated sludge + denitrification and nitrification
	NA	683–1420	Grab	Primary treatment
Oxytetracycline	NA	<5.0–100	24 h FPC	Chemically enhanced primary treatment
	ND	ND–20	Grab	Activated sludge
Chlortetracycline	10–35	ND	Grab	Membrane filtration/Reverse osmosis
	270	60	24 h FPC	Activated sludge + chlorination
Demeclocycline	50	ND	Grab	—
Doxycycline	ND–65	ND–40	Grab	Activated sludge
	20–650	10–150	Composite	—
	<64–2210	<64–915	1-week FPC	Activated sludge + denitrification

ND: Not detected, NA: Not analyzed, FPC: Flow proportional composite.

Table 5.
Concentrations of tetracyclines in wastewater.*

Antibiotic	Concentration (ng/L)		Sampling type	Treatment technology
	Influent	Effluent		
Erythromycin-H_2O	810	850	Grab	Primary treatment
	470	520	Grab	Activated sludge +DN
	NA	105	24 h composite	—
	60–190	60–110	24 h FPC	Activated sludge + sand filtration
	NA	1310–4330	Grab	Primary treatment
Erythromycin	71–141	145–290	24 h FPC	Trickling filters + activated sludge + UV
	830	620	24 h FPC	Activated sludge + denitrification

Antibiotic	Concentration (ng/L)		Sampling type	Treatment technology
	Influent	Effluent		
Tylosin	ND–55	ND–65	Grab	Activated sludge
	20–40	1–5	Grab	Membrane filtration/Reverse osmosis
	ND-60	3–3400	Composite	—
	1150	60	24 h composite	—
Roxithromycin	ND–18	ND–100	Grab	Activated sludge
	140–175	10–15	Grab	Membrane filtration/Reverse osmosis
	10–40	10–30	24 h FPC	Activated sludge + sand filtration
	75	35	24 h composite	Activated sludge + denitrification
Oleandomycin	20–190	5–30	Grab	Membrane filtration/Reverse osmosis
	ND-5	ND-150	Composite	—
Azithromycin	6–53	9–28	Grab	Oxidation ditch + chlorination
	90–380	80–400	24 h FPC	Activated sludge + sand filtration
Clarithromycin	NA	172	Grab	Activated sludge
	NA	57–328	24 h FPC	Activated sludge + sand filtration
	NA	220–329	27 h FPC	Activated sludge + sand filtration

*ND: Not detected, NA: Not analyzed, FPC: Flow proportional composite.

Table 6.
Concentrations of macrolides in wastewater.*

Antibiotic	Concentration (ng/L)		Sampling type	Treatment technology
	Influent	Effluent		
Trimethoprim	213–300	218–322	Grab	Trickling filters + activated sludge + UV
	7900	2400	Grab	Activated sludge + sand filtration
	NA	68–81	72 h FPC	Activated sludge + sand filtration
	210–440	20–310	24 h FPC	Activated sludge + sand filtration
	140–1300	66–700	1-week FPC	Activated sludge + denitrification
Chloramphenicol	31	17	24 h FPC	Oxidation ditch + UV
	NA	92–1050	Grab	Primary treatment
	NA	50	24 h FPC	Activated sludge
Thiamphenicol	NA	138	Grab	Activated sludge
Clindamycin	2–5	5	Grab	Activated sludge
	20–60	5–70	Composite	—
	1–10	ND–5	Grab	Membrane filtration/Reverse osmosis
Lincomycin	60–80	50–60	Grab	Activated sludge
	20–500	3–30	Composite	—

*ND: Not detected, FPC: Flow proportional composite.

Table 7.
Concentrations of others in wastewater.*

Antibiotic	Concentration (ng/L)		Treatment technology	Country or region
	Influent	Effluent		
Sulfonamides				
Sulphapyridine	28	1	Activated sludge + sand filtration	Switzerland
Sulfamethoxazole	20	NA	Oxidation ditch + UV	China
Sulfadimidine	31	NA	Oxidation ditch + UV	China
Sulfadimethoxine	<2.0–8.1	NA	—	USA
Sulfisoxazole	<4.1–21.9	NA	—	USA
Macrolides				
Roxithromycin	61–131	NA	Activated sludge + denitrification and nitrification	Switzerland and Germany
Azithromycin	52–158	2.3	Activated sludge + denitrification and nitrification	Switzerland and Germany
Clarithromycin	27–63	0.8	Activated sludge + denitrification and nitrification	Switzerland and Germany
Erythromycin-H_2O	76	NA	Activated sludge + chlorination	China
Quinolones				
Ofloxacin	886	NA	Oxidation ditch + UV	China
Norfloxacin	301	NA	Activated sludge + chlorination	China
Ciprofloxacin	NA	1400–4800	Activated sludge + denitrification and nitrification	Sweden
Tetracyclines				
Tetracycline	<7.5–15.8	NA	—	USA
Chlortetracycline	<6.9–14.7	NA	—	USA
Doxycycline	NA	1300–1500	Activated sludge + denitrification and nitrification	Sweden
Others				
Clindamycin	3.7–15.4	NA	—	USA
Trimethoprim	21–133	0.1	Activated sludge + denitrification and nitrification	Switzerland and Germany

*NA: Not analyzed.

Table 8.
Concentrations of antibiotics in sludge.*

for β-lactams. The variation in pharmacological properties depends on the differentiation in the side chains. β-lactams include two subclasses which are cephalosporins and penicillins [24].

The β-lactam antibiotics industry has annual sales of about $15 billion [25]; this makes up 65% of the total antibiotics market [26]. 6-Aminopenicillanic acid is the precursor to produce β-lactam antibiotics, which can be produced using free and immobilized penicillin G acylase [27]. Immobilized enzymes are preferred over the free enzyme for many reasons including product-free enzymes as reported by Elnashar [28].

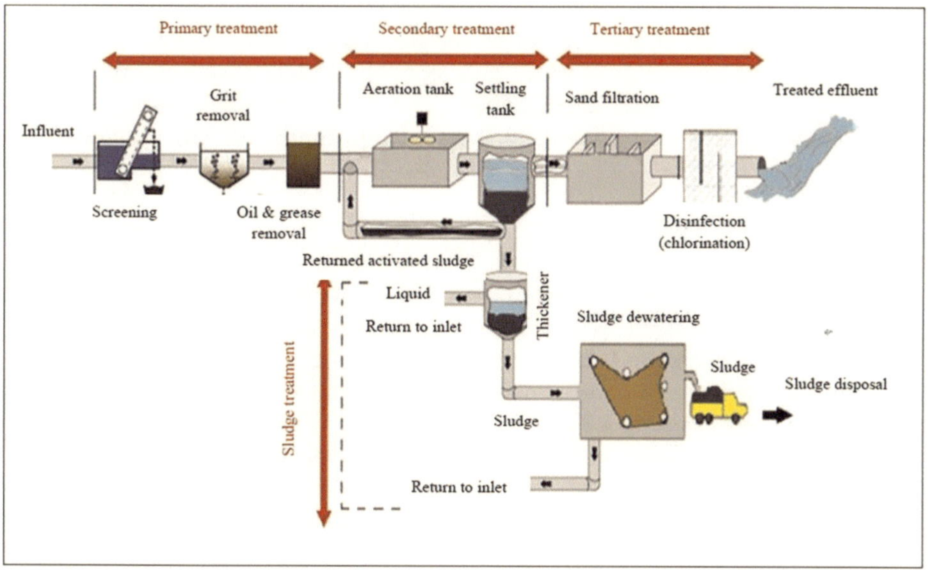

Figure 2.
Typical structure of domestic wastewater treatment plant.

In the samples collected from WWTPs (**Table 2**), some penicillins were detected such as ampicillin, penicillin G & V, amoxicillin, cloxacillin, and oxacillin. Penicillin V showed the highest concentrations (2000–2013,800 ng/L) in influent and effluent wastewater samples [29]. For cephalosporins, six types were detected including cefotaxime, cefalexin, cefaclor, cloxacillin, cefazolin, and cephradine [5, 30]. Generally, β-lactams, especially penicillins were rarely detected in domestic wastewater though they are used in great amounts [14, 31]. This may be attributed to the unstable nature of the β-lactam ring. The β-lactam ring can be broken by a popular bacterial enzyme group called β-lactamases [22] or can be broken through the chemical hydrolysis process [32].

3.1.2 Sulfonamides

Sulfonamides are used commonly since 1968 and consist of a large group of broad-spectrum antibiotics [5, 13, 33]. In WWTPs, 16 types of sulfonamides have been detected (**Table 3**). The most detected sulfonamides were sulfamethoxazole (5597–6000 ng/L) followed by sulfamethazine, sulphapyridine, and sulfadiazine, sequentially. N^4-acetylsulfamethoxazole and many other N^4-acetylated sulfonamides, dominant human metabolites of sulfonamides, were also detected in WWTPs and it was discovered that these metabolites can be transformed into their parent compounds [11, 34, 35].

3.1.3 Quinolones (fluoroquinolones)

Quinolones are a class of antibiotics characterized by the presence of quinolone branches as their basic structure. Fluoroquinolones are a subclass of quinolones that contains a fluorine-substituted central ring. Nalidixic acid was the first quinolone antibiotic discovered in the 1960s followed by newly developed four generations of quinolones [3, 35]. As a result of the universal extensive usage of quinolones, all generations of these antibiotics were detected worldwide in WWTPs as shown in **Table 4**. The most commonly detected quinolones in WWTPs were ciprofloxacin, norfloxacin, and ofloxacin [8, 12, 36]. The highest concentrations detected in influent and effluent samples were 4600 and 7870 ng/L for ciprofloxacin and ofloxacin, respectively [11, 37, 38].

3.1.4 Tetracyclines

Tetracyclines are composed of eight antibiotics (semisynthetic and natural) which inhibit the synthesis of bacterial proteins. Tetracyclines are widely used for human use, the poultry industry, and animal agriculture [39]. In WWTPs, five types of tetracyclines were detected in influent and effluent samples and tetracycline was the most detected one as shown in **Table 5** [5, 14]. Doxycycline showed the highest concentration (2210 ng/L) in influent samples [40], while tetracycline showed the highest concentration (1420 ng/L) in effluent samples [37]. Tetracyclines were discovered in sludge samples with no or little biodegradation. Generally, as shown in **Table 8**, tetracyclines have a relatively rare presence in WWTPs because tetracyclines are used rarely by humans [41, 42].

3.1.5 Macrolides

Macrolides are a group of antibiotics characterized by the presence of a lactone ring that is substituted with alkyl, hydroxyl, and ketone groups, which inhibit the synthesis of bacterial proteins and are usually used as substitutes for penicillin [1, 34]. In WWTPs, six types of macrolides and one metabolite of erythromycin (erythromycin-H_2O) were detected in influent and effluent samples. The most frequently detected macrolide was erythromycin-H_2O followed by roxithromycin, clarithromycin, azithromycin, and tylosin in sequence as shown in **Table 6**. The lowest detected macrolides in sequence were erythromycin-H_2O, roxithromycin, clarithromycin, azithromycin, and tylosin. Erythromycin-H_2O showed the highest concentrations in the influent (10,025 mg/L) and in the effluent (4330 ng/L) wastewater samples [37].

3.1.6 Others

In this review, the other category of antibiotics consists of five different types including chloramphenicol, trimethoprim, lincomycin, clindamycin, and thiamphenicol. All these other antibiotics were detected in WWTPs, and trimethoprim was the most abundant and widely distributed one as shown in **Table 7**. Trimethoprim showed the highest concentrations in influent (7900 ng/L) and in effluent (3052 ng/L) samples, while Chloramphenicol and thiamphenicol showed the lowest concentrations (<4 to 1050 ng/L) in WWTPs [14, 41].

3.1.7 Summary

Despite that β-lactams are the most consumed antibiotic by humans, they were not detected frequently due to the unstable nature of β-lactams [21]. Many reasons affecting the significant variation of antibiotics' concentrations in wastewater influents such as:

a. the consumption pattern of antibiotics, which may be different in different countries and even in the same country there is a great variation in consumption patterns of antibiotics [21].

b. different sampling procedures; grab/composite sampling results in a great variation in antibiotic concentrations since pharmaceutical loads reached WWTPs are different during the day [43].

c. hourly and seasonal fluctuation; seasonal variation of antibiotic consumption was high between summer and winter since antibiotic consumption is two times higher than consumption in summer which affects the concentrations of antibiotics in wastewater influent samples [44, 45]. Also, during winter, antibiotics in influent samples are diluted by rain [11, 46].

d. wastewater treatment plant scale, which means that higher concentrations of antibiotics in influent samples occurred when WWTP serves a low population and vice versa. Additionally, the concentrations of antibiotics varied in effluent depending on the different sampling methods and the applied wastewater treatment technology [5, 8].

3.2 Presence of antibiotics in sludge samples of WWTP

The analysis process for antibiotics and their transformations in sludge samples is a challenge since antibiotics in sludge have a low detectability rate as well as a low extraction pattern [47]. Therefore, the studies on the detection and presence of antibiotics in sludge are much less than those on wastewater [48]. The results showed the presence of five different classes of antibiotics in both activated and digested sludge samples as shown in **Table 8**.

The most abundant antibiotics were ciprofloxacin (4.8 mg/kg) in digested sludge [40] and norfloxacin (2.7 mg/kg) in activated sludge [14, 49]. Quinolones were detected mostly at the level mg/kg, while most tetracyclines, macrolides, sulfonamides, and others were detected in lower levels (μg/kg) [11]. Despite the wide usage of β-lactams antibiotics in veterinary and human medicine, all sludge samples all over the world showed no presence of β-lactams antibiotics which is mainly attributed to their poor adsorption onto sludge and their unstable characteristics [37].

4. Antibiotics transformation and fate in WWTPs

4.1 Antibiotics removal pathways in WWTPs

Antibiotics in WWTPs can be removed through major pathways including biodegradation, adsorption, membrane separation, and disinfection. Furthermore, there are other pathways for the removal of antibiotics including photolysis, volatilization, and hydrolysis which were eliminated since they have an inconsiderable role in the reduction of antibiotics from WWTPs. For example, β- lactams antibiotics are not stable because of the presence of β- lactam ring which easily can be hydrolyzed. Thus, β-lactams will be hydrolyzed before reaching the WWTPs.

Additionally, some researchers reported that β- lactams have a relatively long half-life due to hydrolysis at neutral pH values (same as in WWTPs) such as more than

5 days for amoxicillin [50] and 52 h for meropenem [51]. Moreover, although β-lactams can be hydrolyzed in WWTP, the contribution of the hydrolysis process for the removal of antibiotics is useless because the wastewater treatment process has a relatively short hydraulic retention time (8–20 h).

In addition to that, there are some antibiotics such as amoxicillin which were degraded by sunlight-photolysis or UV-photolysis [50], macrolides [52], quinolones [53], and tetracyclines [54, 55]. However, this degradation process has a minor significance because wastewater has high concentrations of suspended solids which inhibit the deep penetration of sunlight or UV [5, 8]. Besides, the effect of the photolysis process can be neglected since the hydraulic retention time of WWTPs is much lower than the half-life of most antibiotics in wastewater.

4.2 Antibiotics transformation and fate in conventional WWTP

4.2.1 Primary treatment units

The primary treatment stage in WWTPs consists mainly of screens and primary settling tanks. In some WWTPs, some coagulant chemicals are added in primary treatment units such as ferric ion salts, aluminum salts, or polymers called CEPT (Chemically Enhanced Primary Treatment) [56]. Many previous studies reported that the primary treatment stage in WWTPs has no significant removal for different types of antibiotics including clarithromycin, sulfamethoxazole, ofloxacin, cefalexin, erythromycin, amoxicillin, clindamycin, and cefaclor [13].

Nevertheless, other studies reported that the chemically enhanced primary treatment (CEPT) process can significantly improve the removal efficiencies of some antibiotics such as norfloxacin (67.7%), ofloxacin (55.2%), erythromycin (44.8%), sulfamethoxazole (64.0%), and roxithromycin (76.3%). This effect is due to the destroying effect of coagulants on the chemical chains of some antibiotics [11].

4.2.2 Biological treatment units

Clearly, in biological treatment units, biodegradation and adsorption processes are the main pathways for the transformation of antibiotics in WWTPs. According to the classification of antibiotics, their transformation and fate in biological treatment units can be summarized as the following:

4.2.2.1 β-Lactams

Despite that β-lactams are the most consumed antibiotics for humans and animals, they have not been detected frequently in WWTPs [14, 57], thus there was not much-published data about the fate and transformation of β-lactams in WWTPs. Junker et al. [57] studied the fate of some ^{14}C-labeled antibiotics (benzylpenicillin and ceftriaxone) in the activated sludge process. The results showed that ceftriaxone was not totally mineralized, whereas only about 25% of benzylpenicillin was mineralized. The same results were obtained using biodegradability tests (closed bottle tests method; CBT) at much higher concentrations of β-lactams antibiotics since ceftriaxone was kept unchanged whereas benzylpenicillin was biodegraded up to 27% [58].

The differences in biodegradability between β-lactams antibiotics may be due to the differences in their chemical structures because of diverse side chains [24, 59]. Andreozzi et al. [50] performed a standard batch experiment to study the fate of amoxicillin in the activated sludge process. The results proved that the adsorption and biodegradation processes were responsible for the removal and transformation of amoxicillin.

4.2.2.2 Sulfonamides

Most researchers who studied sulfonamides in WWTPs focused on sulfamethoxazole and N4-acetylsulfamethoxazole (a metabolite of sulfamethoxazole). Batt et al. [41] and Pérez et al. [60] found that sulfonamides were biodegraded to a certain degree (low removal efficiency) in the wastewater biological treatment stage. Sulfonamides were removed during the biological treatment process with an average removal efficiency of 25% [13, 38], as well as sulfamethoxazole showed poor removal efficiency of 20% [42, 61].

Also, some studies reported the resistance of sulfonamides to different treatment processes during wastewater treatment [11, 12, 62]. Nevertheless, some other researchers reported the relatively high removal efficiency of sulfamethoxazole like 55% [34], 56% [63], 66% [64], 67% [65], and 74% [63].

The significant variation in removal efficiencies of sulfonamides during the biological treatment process can be attributed to the following reasons: first, the transformation of some metabolites such as N4-acetylsulfamethoxazole to the parent molecule (sulfamethoxazole) in the influent. Second, the mentioned removal efficiencies depended on grab or composite samples (24 h), which cannot reflect the whole treatment process [12, 66].

Thus, to avoid the previous limitations, some researchers used well-controlled laboratory reactors for studying the fate of sulfonamides and their removal pathway during the activated sludge treatment process [8, 67]. The biodegradation process of three sulfonamides at low concentrations (20 μg/L) using activated sludge reactors was studied by Pérez et al. [60] and they reported that the biodegradation process was so efficient and was able to remove the three sulfonamides for 3 days. Less than 26% of the initial antibiotics' concentrations were present by the third day, whereas by the tenth day, the removal efficiency increased up to 93%. In addition, it was reported that some microorganisms can utilize sulfamethoxazole as a carbon and/or nitrogen source after 3 days lag phase [68]. Despite these studies proving that sulfonamides can be biodegraded, the biodegradation process takes a long time than the usual hydraulic retention time of the biological treatment process at WWTPs.

4.2.2.3 Quinolones

Batt et al. [41] and Xu et al. [12] reported that adsorption followed by biodegradation are the main pathways for the removal of quinolones during biological treatment stages at WWTPs. The removal efficiencies for norfloxacin, ofloxacin, and ciprofloxacin were 87–100%, 75–77%, and 85%, respectively [69, 70]. The adsorption mechanism of quinolones by sludge depends on electrostatic interactions between particles rather than hydrophobic forces [21, 42].

4.2.2.4 Tetracyclines

In the biological treatment process at WWTPs, adsorption is considered the main mechanism for the removal of tetracyclines [41, 71, 72]. Tetracycline (10 μg/L) was removed (up to >95%) rapidly through an adsorption mechanism during 6 h inside activated sludge units. Also, two lab-scale Sequencing Batch Reactors (SBR) were utilized to stimulate the activated sludge process (biological treatment), in these SBRs, the effect of SRT (Sludge Retention Time) and HRT (Hydraulic Retention Time) on transformation and fate of tetracycline were studied (66). The results showed that the removal efficiency of tetracycline in phase 1 (SRT = 10 days; HRT = 24 h) was 86.4 ± 8.7% and phase 2 (SRT = 10 days; HRT = 7.4 h) was 85.1 ± 5.4%, while in phase 3 (SRT = 3 days; HRT = 7.4 h) was 78.4 ± 7.1%. In phase 3, it was clear that the removal efficiency of tetracycline decreased by a reduction in SRT, which indicated that more tetracycline can be adsorbed by old sludge. In addition to that, it was reported that ferrous chloride could enhance the removal of tetracycline through precipitation due to the strong complexation between tetracyclines and ferrous ions [41].

4.2.2.5 Macrolides

All previous studies have indicated that all macrolides were not significantly eliminated, even at low concentrations, during the biological treatment process at WWTPs [10, 67].

4.2.2.6 Trimethoprim

Many studies indicated that trimethoprim was not adsorbed during the Activated Sludge (AS) process [41, 60]. The studies also proved that trimethoprim cannot easily biodegrade during AS with a short sludge retention time [11, 57]. However, Pérez et al. [60] reported that trimethoprim was completely degraded by nitrifying activated sludge with long SRT within 3 days. The nitrifying bacteria present in the nitrifying activated sludge are responsible for trimethoprim degradation since it was noticed that when the activity of nitrifying bacteria is inhibited, the elimination efficiency of trimethoprim decreased from 70 to 25% [42, 73].

4.2.3 Digestion tank

A little number of studies were carried out to study the transformation and fate of antibiotics in digestion tanks because the digestion process is not applied at most of the WWTPs all over the world. Zhang and Bing [5] conducted a two-stage anaerobic sludge digester (SRT = 30 days) to study the fate of fluoroquinolones (norfloxacin and ciprofloxacin) and they reported the stability of these fluoroquinolones. The same results were reported by Lindberg et al. [66] that ciprofloxacin and norfloxacin showed no significant removal under mesophilic sludge digesters (38°C). In contrast, Du et al. [14] utilized anaerobic mesophilic sludge digesters to study the stability of trimethoprim and sulfamethoxazole and they reported the instability of trimethoprim and sulfamethoxazole.

This may be due to that both trimethoprim and sulfamethoxazole have no significant amounts in digested sludge while their concentrations are high in activated sludge. Gartiser et al. [74] studied the biodegradation of nine antibiotics under an anaerobic digestion process (35 ± 1°C) and they reported that the biodegradation process was inefficient for all nine antibiotics except in the case of benzylpenicillin which was biodegraded after 40 days lag phase.

4.3 Antibiotics transformation and fate in advanced treatment processes

4.3.1 Filtration

To improve the quality of treated effluent, some wastewater treatment plants apply advanced treatment units such as Membrane Filtration (MF) or Sand Filtration (SF). During sand filtration, the removal efficiencies of trimethoprim and clarithromycin were 60 and 15%, respectively [8, 34]. Nakada et al. [75] also reported the same results for trimethoprim since the removal efficiency was 55.2% in the sand filtration process. However, sulfamethoxazole and sulphapyridine showed lower removal efficiencies of 26.9 and 14.6%, respectively after the sand filtration process.

Moreover, clarithromycin, azithromycin, roxithromycin, and erythromycin-H_2O showed no removal at all during the sand filtration process which may be due to the presence of highly diverse and effective biofilm on the SF particles [75]. Watkinson et al. [38] reported that about 43 and 94% of total antibiotics were removed by the microfiltration process and reverse osmosis, respectively via eliminating the particles that adsorbed antibiotics on them. Also, the Nanofiltration (NF) process increased the removal efficiencies of antibiotics up to more than 95% for antibiotics such as tetracyclines [42, 76].

4.3.2 Disinfection

Table 9 summarizes the transformation and fate of eight classes of antibiotics after disinfection units at WWTPs. The disinfectants reacted fast with the antibiotics and

Antibiotic	Disinfectant	pH	Performance	
			Contact time (min)	Removal efficiency (%)
β-lactams				
Penicillin G	Ozone	7.7	1	–70
Cefalexin	Ozone	7.7	1	>99
Sulfonamides				
Sulfamethoxazole	Free available chlorine	7.3	1	−95
Sulphapyridine	Ozone	7.95	27	93.9
Fluoroquinolones				
Ciprofloxacin	Free available chlorine	7.7	1	>99
Enrofloxacin	Ozone	7.7	1	>99
Macrolides				
Azithromycin	Ozone	7.95	27	92.6
Erythromycin-H_2O	Ozone	7.95	27	88.7
Clarithromycin	Ozone	7.2	18	>76
Tylosin	Ozone	7.7	1	>99
Roxithromycin	Ozone	7–7.5	8-40-4	>90–99
Tetracyclines				
Tetracycline	Ozone	7.7	1	>99

Antibiotic	Disinfectant	pH	Performance	
			Contact time (min)	Removal efficiency (%)
Amikacin	Ozone	7.7	1	−25
Vancomycin	Ozone	7.7	1	>99
Others				
Trimethoprim	Free available chlorine.	8.1	10	−100
	Ozone	7.7	1	>99
Lincomycin	Ozone	5.5	2	100

Table 9.
Transformation and fate results of antibiotics in the disinfection process.

the removal occurred after 1–27 min. In most studies, ozonation was applied for disinfection purposes. The ozonation oxidized antibiotics either directly by ozone (O_3) or by hydroxyl radicals (•OH). Hydroxyl radicals are generated due to the decay of ozone. Each ozone and hydroxyl radicals have different oxidation mechanisms since O_3 is selective and usually attacks the special functional groups (such as the aromatic structure or a C═C double bond), whereas hydroxyl radicals are non-selective and react with many types of moieties [75].

In addition, Cha et al. [77] studied the ozonation of 14 antibiotics and found that only 4 antibiotics including cephalexin, penicillin G, N(4)-acetylsulfamethoxazole, and amikacin were oxidized by hydroxyl radicals and the other 10 antibiotics reacted mainly with O_3. Moreover, by using other disinfectants such as hypochlorous acid (HOCl), some antibiotics such as trimethoprim showed no degradation, while sulfamethoxazole was degraded after a reaction with HOCl [78]. Liu et al. [10] reported that there was a significant variation in reaction rates of combined chlorine and free available chlorine with antibiotics in wastewater.

5. Conclusion

This review provides insight into the occurrence and fate of antibiotics in domestic wastewater treatment plants. Data was collected about the occurrence of the most widely six groups of antibiotics used for human cure including β-lactams, sulfonamides, quinolones, tetracyclines, macrolides, and others in wastewater and sludge samples of wastewater treatment plants. All previously mentioned groups of antibiotics were detected in wastewater and sludge samples with varied concentrations during the different treatment stages. It was clear that most of the wastewater treatment plants do not have the ability to fully remove these antibiotics. The main removal mechanisms of these antibiotics were biodegradation, adsorption, membrane filtration, and disinfection.

Author details

Raed S. Al-Wasify[1*], Majid M. Alruwaili[2], Fahad S. Aljohani[2], Shimaa R. Hamed[3] and Samar Ragab[4]

1 Water Pollution Research Department, Environmental Research and Climate Changes Institute, National Research Centre, Dokki, Egypt

2 Projects and Technical Services, National Water Company Headquarter, Riyadh, Saudi Arabia

3 Microbial Biotechnology Department, Biotechnology Research Institute, National Research Centre, Dokki, Egypt

4 Biomedical Sciences Department, Center of Microbiology and Phage Therapy, Zewail City of Sciences and Technology, Giza, Egypt

*Address all correspondence to: alwasifyr2013@gmail.com

References

[1] Gorito AM, Ribeiro AR, Almeida CMR, Silva AMT. A review on the application of constructed wetlands for the removal of priority substances and contaminants of emerging concern listed in recently launched EU legislation. Environmental Pollution. 2017;**227**:428-443

[2] Hernández F, Sancho JV, Ibáñez M, Guerrero C. Antibiotic residue determination in environmental waters by LC-MS. Trac-Trends in Analytical Chemistry. 2007;**26**(6):466-485

[3] Amarasiri M, Sano D, Suzuki S. Understanding human health risks caused by antibiotic resistant bacteria (ARB) and antibiotic resistance genes (ARG) in water environments: Current knowledge and questions to be answered. Critical Reviews in Environmental Science and Technology. 2020;**50**(19):2016-2059

[4] Hernando MD, Mezcua M, Fernández-Alba AR, Barceló D. Environmental risk assessment of pharmaceutical residues in wastewater effluents, surface waters and sediments. Talanta. 2006;**69**(2):334-342

[5] Zhang T, Bing L. Occurrence, transformation, and fate of antibiotics in municipal wastewater treatment plants. Critical Reviews in Environmental Science and Technology. 2011;**41**(11): 951-998

[6] Van Boeckel TP, Brower C, Gilbert M, Grenfell BT, Levin SA, Robinson TP, et al. Global trends in antimicrobial use in food animals. Proceedings of the National Academy of Sciences. 2015;**112**:5649-5654

[7] Boyles T, Naicker V, Rawoot N, Raubenheimer P, Eick B, Mendelson M. Sustained reduction in antibiotic consumption in a south African public sector hospital: Four-year outcomes from the Groote Schuur hospital antibiotic stewardship programme. South African Medical Journal. 2017;**107**: 115-118

[8] Al-Isawi R, Ray S, Scholz M. Comparative study of domestic wastewater treatment by mature vertical-flow constructed wetlands and artificial ponds. Ecological Engineering. 2017;**100**:8-18

[9] Giger W, Alder AC, Golet EM, Kohler HPE, McArdell CS, Molnar E, et al. Occurrence and fate of antibiotics as trace contaminants in wastewaters, sewage sludges, and surface waters. Chimia. 2003;**57**(9):485-491

[10] Liu L, Li J, Fan H, Huang X, Wei H, Liu C. Fate of antibiotics from swine wastewater in constructed wetlands with different flow configurations. International Biodeterioration & Biodegradation. 2019;**140**:119-125

[11] Bilgehan N, Taylan D, Serdar K. Behaviour and removal of ciprofloxacin and sulfamethoxazole antibiotics in three different types of full-scale wastewater treatment plants: A comparative study. Water, Air, & Soil Pollution. 2021;**232**: 127

[12] Xu WH, Zhang G, Zou SC, Li XD, Liu YC. Determination of selected antibiotics in the Victoria harbour and the Pearl River, South China using high-performance liquid chromatography-electrospray ionization tandem mass spectrometry. Environmental Pollution. 2007;**145**(3):672-679

[13] Deng Y, Li B, Zhang T. Bacteria that make a meal of sulfonamide antibiotics:

Blind spots and emerging opportunities. Environmental Science & Technology. 2018;**52**(7):3854-3868

[14] Du J, Hongxia Z, Sisi L, Huaijun X, Yan W, Jingwen C. Antibiotics in the coastal water of the south yellow sea in China: Occurrence, distribution and ecological risks. Science of the Total Environment. 2017;**595**:521-527

[15] Zhang T, Zhang M, Zhang XX, Fang HHP. Tetracycline resistance genes and tetracycline resistant lactose-fermenting enterobacteriaceae in activated sludge of sewage treatment plants. Environmental Science & Technology. 2009;**43**(10):3455-3460

[16] Chang XS, Meyer MT, Liu XY, Zhao Q, Chen H, Chen J, et al. Determination of antibiotics in sewage from hospitals, nursery and slaughterhouse, wastewater treatment plant and source water in Chongqing region of three gorge reservoir in China. Environmental Pollution. 2010;**158**(5): 1444-1450

[17] Plósz BG, Leknes H, Liltved H, Thomas KV. Diurnal variations in the occurrence and the fate of hormones and antibiotics in activated sludge wastewater treatment in Oslo. Norway. Science of The Total Environment. 2010; **408**(8):1915-1924

[18] Baquero F, Martínez JL, Cantón R. Antibiotics and antibiotic resistance in water environments. Current Opinion in Biotechnology. 2008;**19**(3):260-265

[19] Díaz-Cruz MS, Barceló D. LC-MS2 trace analysis of antimicrobials in water, sediment and soil. TrAC Trends in Analytical Chemistry. 2005;**24**(7):645-657

[20] Kemper N. Veterinary antibiotics in the aquatic and terrestrial environment. Ecological Indicators. 2008;**8**(1):1-13

[21] Kümmerer K. Antibiotics in the aquatic environment: A review. Part I. Chemosphere. 2009;**75**(4):417-434

[22] Sukul P, Spiteller M. Fluoroquinolone antibiotics in the environment. Reviews of Environmental Contamination and Toxicology. 2007; **191**:131-162

[23] Onesios KM, Yu JT, Bouwer EJ. Biodegradation and removal of pharmaceuticals and personal care products in treatment systems: A review. Biodegradation. 2009;**20**: 441-466

[24] Dodd MC, Buffle MO, Von Gunten U. Oxidation of antibacterial molecules by aqueous ozone: Moiety-specific reaction kinetics and application to ozone-based wastewater treatment. Environmental Science & Technology. 2006;**40**(6):1969-1977

[25] Elnashar MM, Wahba MI, Amin MA, El-Diwany A. Application of Plackett–Burman screening design to the modeling of grafted alginate–carrageenan beads for the immobilization of penicillin G acylase. Journal of Applied Polymers Science. 2014;**131**: 40295. DOI: 10.1002/app.40295

[26] Elnashar MM, Mohamed EH, Ghada EA. Grafted carrageenan gel disks and beads with polyethylenimine and glutaraldehyde for covalent immobilization of penicillin G acylase. Journal of Colloid Science and Biotechnology. 2013;**2**(1):27-33

[27] Elnashar MM, Yassin AM, Kahil T. Novel thermally and mechanically stable hydrogel for enzyme immobilization of penicillin G acylase via covalent technique. Journal of Applied Polymer Sciences. 2008;**109**: 4105-4111

[28] Elnashar MM. Immobilized molecules using biomaterials and nanobiotechnology. Journal of Biomaterials and Nanobiotechnology. 2010;**1**:61-76. DOI: 10.4236/JBNB.2010.11008

[29] Watkinson AJ, Murby EJ, Kolpin DW, Costanzo SD. The occurrence of antibiotics in an urban watershed: From wastewater to drinking water. Science of the Total Environment. 2009;**407**(8):2711-2723

[30] Lin AYC, Yu TH, Lin CF. Pharmaceutical contamination in residential, industrial, and agricultural waste streams: Risk to aqueous environments in Taiwan. Chemosphere. 2008;**74**(1):131-141

[31] Bailón-Pérez MI, García-Campaña AM, Cruces-Blanco C, Iruela MD. Trace determination of β-lactam antibiotics in environmental aqueous samples using off-line and on-line preconcentration in capillary electrophoresis. Journal of Chromatography A. 2008;**1185**(2):273-280

[32] Hirsch R, Ternes T, Haberer K, Kratz KL. Occurrence of antibiotics in the aquatic environment. Science of the Total Environment. 1999;**225**(1–2):109-118

[33] Littlefield NA, Sheldon WG, Allen R, Gaylor DW. Chronic toxicity/carcinogenicity studies of sulphamethazine in Fischer 344/N rats: Two-generation exposure. Food and Chemical Toxicology. 1990;**28**(3):157-167

[34] Göbel A, Thomsen A, McArdell CS, Joss A, Giger W. Occurrence and sorption behavior of sulfonamides, macrolides, and trimethoprim in activated sludge treatment. Environmental Science and Technology. 2005;**39**(11):3981-3989

[35] Xiao Y, Chang H, Jia A, Hu JY. Trace analysis of quinolone and fluoroquinolone antibiotics from wastewaters by liquid chromatography electrospray tandem mass spectrometry. Journal of Chromatography A. 2008;**1214**(1–2):100-108

[36] Seifrtová M, Pena A, Lino CM, Solich P. Determination of fluoroquinolone antibiotics in hospital and municipal wastewaters in Coimbra by liquid chromatography with a monolithic column and fluorescence detection. Analytical and Bioanalytical Chemistry. 2008;**391**(3):799-805

[37] Minh TB, Leung HW, Loi IH, Chan WH, So MK, Mao JQ, et al. Antibiotics in the Hong Kong metropolitan area: Ubiquitous distribution and fate in Victoria Harbor. Marine Pollution Bulletin. 2009;**58**(7):1052-1062

[38] Watkinson AJ, Murby EJ, Costanzo SD. Removal of antibiotics in conventional and advanced wastewater treatment: Implications for environmental discharge and wastewater recycling. Water Research. 2007;**41**(18):4164-4176

[39] Yang SW, Cha JM, Carlson K. Simultaneous extraction and analysis of 11 tetracycline and sulfonamide antibiotics in influent and effluent domestic wastewater by solid-phase extraction and liquid chromatography-electro spray ionization tandem mass spectrometry. Journal of Chromatography A. 2005;**1097**(1–2):40-53

[40] Lindberg RH, Wennberg P, Johansson MI, Tysklind M, Andersson BAV. Screening of human

antibiotic substances and determination of weekly mass flows in five sewage treatment plants in Sweden. Environmental Science and Technology. 2005;**39**(10):3421-3429

[41] Batt AL, Kim S, Aga DS. Comparison of the occurrence of antibiotics in four full-scale wastewater treatment plants with varying designs and operations. Chemosphere. 2007;**68**(3):428-435

[42] Bergeron S, Boopathy R, Nathaniel R, Corbin A, LaFleur G. Presence of antibiotic resistant bacteria and antibiotic resistance genes in raw source water and treated drinking water. International Biodeterioration & Biodegradation. 2015;**102**:370-374

[43] Joss A, Keller E, Alder AC, Göbel A, McArdell CS, Ternes T, et al. Removal of pharmaceuticals and fragrances in biological wastewater treatment. Water Research. 2005;**39**(14):3139-3152

[44] Elseviers MM, Ferech M, Vander Stichele RH, Goossens H. The ESAC project group. Antibiotic use in ambulatory care in Europe (ESAC data 1997–2002): Trends, regional differences and seasonal fluctuations. Pharmacoepidemiology and Drug Safety. 2007;**16**(1):115-123

[45] McArdell CS, Molnar E, Suter MJF, Giger W. Occurrence and fate of macrolide antibiotics in wastewater treatment plants and in the Glatt Valley watershed, Switzerland. Environmental Science and Technology. 2003;**37**(24): 5479-5486

[46] Li B, Zhang T, Xu ZY, Fang HHP. Rapid analysis of 21 antibiotics of multiple classes in municipal wastewater using ultra performance liquid chromatography-tandem mass spectrometry. Analytica Chimica Acta. 2009;**645**(1–2):64-72

[47] O'Connor S, Aga DS. Analysis of tetracycline antibiotics in soil: Advances in extraction, clean-up, and quantification. TrAC Trends in Analytical Chemistry. 2007;**26**(6): 456-465

[48] Kim SC, Carlson K. Quantification of human and veterinary antibiotics in water and sediment using SPE/LC/MS/MS. Analytical and Bioanalytical Chemistry. 2007;**387**(4):1301-1315

[49] Golet EM, Xifra I, Siegrist H, Alder AC, Giger W. Environmental exposure assessment of fluoroquinolone antibacterial agents from sewage to soil. Environmental Science and Technology. 2003;**37**(15):3243-3249

[50] Andreozzi R, Caprio V, Ciniglia C, De Champdoré M, Lo Giudice R, Marotta R, et al. Antibiotics in the environment: Occurrence in Italian STPs, fate, and preliminary assessment on algal toxicity of amoxicillin. Environmental Science and Technology. 2004;**38**(24):6832-6838

[51] Al-Ahmad A, Daschner FD, Kümmerer K. Biodegradability of cefotiam, ciprofloxacin, meropenem, penicillin G, and sulfamethoxazole and inhibition of wastewater bacteria. Archives of Environmental Contamination and Toxicology. 1999; **37**(2):158-163

[52] Vione D, Feitosa-Felizzola J, Minero C, Chiron S. Photo transformation of selected human-used macrolides in surface water: Kinetics, model predictions and degradation pathways. Water Research. 2009;**43**(7): 1-9

[53] Burhenne J, Ludwig M, Nikoloudis P, Spiteller M. Photolytic degradation of fluoroquinolone carboxylic acids in aqueous solution. I.

Primary photoproducts and half-lives. Environmental Science and Pollution Research. 1997;**4**(1):10-15

[54] Jiao SJ, Meng SR, Yin DQ, Wang LH, Chen LY. Aqueous oxytetracycline degradation and the toxicity change of degradation compounds in photoirradiation process. Journal of Environmental Sciences-China. 2008a; **20**(7):806-813

[55] Jiao SJ, Zheng SR, Yin DQ, Wang LH, Chen LY. Aqueous photolysis of tetracycline and toxicity of photolytic products to luminescent bacteria. Chemosphere. 2008b;**73**(3):377-382

[56] Sui Q, Huang J, Deng SB, Yu G, Fan Q. Occurrence and removal of pharmaceuticals, caffeine and DEET in wastewater treatment plants of Beijing, China. Water Research. 2010;**44**(2): 417-426

[57] Junker T, Alexy R, Knacker T, Kümmerer K. Biodegradability of 14C-labeled antibiotics in a modified laboratory scale sewage treatment plant at environmentally relevant concentrations. Environmental Science and Technology. 2006;**40**(1):318-324

[58] Alexy R, Kümpel T, Kümmerer K. Assessment of degradation of 18 antibiotics in the closed bottle test. Chemosphere. 2004;**57**(6):505-512

[59] Kümmerer K, Al-Ahmad A. Biodegradability of the antitumor agents 5-fluorouracil, cytarabine, and gemcitabine: Impact of the chemical structure and synergistic toxicity with hospital effluent. Acta Hydrochimica et Hydrobiolgica. 1997;**25**(4):166-172

[60] Pérez S, Eichhorn P, Aga DS. Evaluating the biodegradability of sulfamethazine, sulfamethoxazole, sulfathiazole, and trimethoprim at different stages of sewage treatment. Environmental Toxicology and Chemistry. 2005;**24**(6):1361-1367

[61] Brown KD, Kulis J, Thomson B, Chapman TH, Mawhinney DB. Occurrence of antibiotics in hospital, residential, and dairy, effluent, municipal wastewater, and the Rio Grande in New Mexico. Science of the Total Environment. 2006;**366**(2–3): 772-783

[62] Miao XS, Bishay F, Chen M, Metcalfe CD. Occurrence of antimicrobials in the final effluents of wastewater treatment plants in Canada. Environmental Science and Technology. 2004;**38**(13):3533-3541

[63] Radjenovic J, Petrovic M, Barceló D. Analysis of pharmaceuticals in wastewater and removal using a membrane bioreactor. Analytical and Bioanalytical Chemistry. 2007;**387**(4): 1365-1377

[64] Clara M, Strenn B, Gans O, Martinez E, Kreuzinger N, Kroiss H. Removal of selected pharmaceuticals, fragrances and endocrine disrupting compounds in a membrane bioreactor and conventional wastewater treatment plants. Water Research. 2005;**39**(19): 4797-4807

[65] Carballa M, Omil F, Lema JM, Llompart M, García-Jares C, Rodríguez I, et al. Behavior of pharmaceuticals, cosmetics and hormones in a sewage treatment plant. Water Research. 2004; **38**(12):2918-2926

[66] Lindberg RH, Olofsson U, Rendahl P, Johansson MI, Tysklind M, Andersson BAV. Behavior of fluoroquinolones and trimethoprim during mechanical, chemical, and active sludge treatment of sewage water and digestion of sludge. Environmental

Science and Technology. 2006;**40**(3): 1042-1048

[67] Joss A, Zabczynski S, Göbel A, Hoffmann B, Löffler D, McArdell CS, et al. Biological degradation of pharmaceuticals in municipal wastewater treatment: Proposing a classification scheme. Water Research. 2006;**40**(8):1686-1696

[68] Drillia P, Dokianakis SN, Fountoulakis MS, Kornaros M, Stamatelatou K, Lyberatos G. On the occasional biodegradation of pharmaceuticals in the activated sludge process: The example of the antibiotic sulfamethoxazole. Journal of Hazardous Materials. 2005;**122**(3):259-265

[69] Radjenovic J, Petrovic M, Barceló D. Fate and distribution of pharmaceuticals in wastewater and sewage sludge of the conventional activated sludge (CAS) and advanced membrane bioreactor (MBR) treatment. Water Research. 2009;**43**(3): 831-841

[70] Vieno NM, Tuhkanen T, Kronberg L. Analysis of neutral and basic pharmaceuticals in sewage treatment plants and in recipient rivers using solid phase extraction and liquid chromatography-tandem mass spectrometry detection. Journal of Chromatography A. 2006;**1134**(1–2): 101-111

[71] Kim S, Eichhorn P, Jensen JN, Weber AS, Aga DS. Removal of antibiotics in wastewater: Effect of hydraulic and solid retention times on the fate of tetracycline in the activated sludge process. Environmental Science and Technology. 2005;**39**(15):5816-5823

[72] Li B, Zhang T. Biodegradation and adsorption of antibiotics the activated sludge process. Environmental Science and Technology. 2010;**44**(9):3468-3473

[73] Batt AL, Kim S, Aga DS. Enhanced biodegradation of iopromide and trimethoprim in nitrifying activated sludge. Environmental Science and Technology. 2006;**40**(23):7367-7373

[74] Gartiser S, Urich E, Alexy R, Kümmerer K. Anaerobic inhibition and biodegradation of antibiotics in ISO test schemes. Chemosphere. 2007;**66**(10): 1839-1848

[75] Nakada N, Shinohara H, Murata A, Kiri K, Managaki S, Sato N, et al. Removal of selected pharmaceuticals and personal care products (PPCPs) and endocrine-disrupting chemicals (EDCs) during sand filtration and ozonation at a municipal sewage treatment plant. Water Research. 2007;**41**(19):4373-4382

[76] Koyuncu I, Arikan OA, Wiesner MRC. Rice, removal of hormones and antibiotics by nanofiltration membranes. Journal of Membrane Science. 2008;**309**(1–2): 94-101

[77] Cha JM, Yang S, Carlson KH. Trace determination of β-lactam antibiotics in surface water and urban wastewater using liquid chromatography combined with electrospray tandem mass spectrometry. Journal of Chromatography A. 2006;**1115**(1–2): 46-57

[78] Dodd MC, Huang CH. Transformation of the antibacterial agent sulfamethoxazole in reactions with chlorine: Kinetics mechanisms, and pathways. Environmental Science and Technology. 2004;**38**(21):5607-5615

Section 4

Biological Decontamination and Nanomembranes

Chapter 9

Biological Seawater Desalination

Enrique O. Martínez

Abstract

More than two billion people worldwide lack access to safe, clean drinking water, and this number is likely to increase due to population growth and rapidly diminishing freshwater supplies. Current seawater desalination methods can effectively provide freshwater and meet the growing demand for this resource. However, they are becoming increasingly controversial owing to their adverse environmental impacts, including high energy consumption and generation of desalination brine. For millions of years, various species of organisms such as plants, microalgae, and bacteria have adapted to environments with variable salinity and have developed mechanisms to eliminate excess intracellular NaCl. This has encouraged scientists to study the possibility of using biological processes for seawater desalination. Biodesalination is an emerging technology for the selective removal of Na + and Cl – from salt water by various halophile organisms, such as macrophytes, microalgae, and cyanobacteria, with very low energy consumption. Microbial desalination cells that allow simultaneous desalination of water in conjunction with wastewater treatment are also included in this category. The direct use of living organisms such as halophile plants, microalgae, and bacteria to desalinate water appears to be a promising field. However, the development and practical applicability of these technologies depend on the living organisms selected for desalinating seawater.

Keywords: seawater, biodesalination, mangroves, microalgae, cyanobacteria, microbial desalination cells

1. Introduction

Globally, increasing freshwater scarcity problems have become more evident owing to continuous population growth, unequal distribution of water resources, mismanagement, and climate change [1, 2]. One approach to address this situation is to use abundant sources, such as seawater and brackish groundwater. However, desalination of these waters occurs at the expense of several environmental impacts due to high energy consumption, mostly fossil fuels, and brine discharge [3–5].

In response to this challenge, biodesalination has emerged as an emerging technology to remove salt from seawater, brine, and wastewater with very low energy consumption [6–8]. This process is an emerging technology for the selective removal of Na + and Cl – from salt water by various halophilic organisms, such as macrophytes, microalgae, and cyanobacteria [9, 10]. Also included in this category are bioelectrochemical systems (BES) that allow simultaneous desalination of water in conjunction with wastewater treatment [11]. Another line of research in the field of biodesalination involves biomimicry and bioinspiration. Within this area, biomimetic membranes based on aquaporins (ABM) have commercial applications [12–18]. Bioinspired and biomass-based solar seawater evaporators (BSSE) are also included in this field of research [19–23].

2. Mangroves and desalination

Halophyte plants, which can survive and thrive under saline conditions, have long been used to restore salt-affected soils. This remediation technology, known as phytodesalination, takes advantage of the capacity of halophytes to absorb large amounts of sodium ions (Na^+) from affected ecosystems and their elimination through accumulation and translocation to different parts of plants [24–26]. A variant of the application of this technology has been the use of halophyte plants in constructed wetlands (CW) for the treatment of wastewater and thus be able to reuse the treated water for irrigation without affecting the soil [27, 28].

Although phytodesalination appears to be a green and promising technology, its application in desalinating water presents technical difficulties [12]. However, a group of halophytes, mangroves have particular adaptations to live in aquatic environments, so they can be used in desalination processes [29].

To survive in saline or brackish environments, mangroves strictly control the absorption of water and ions in their roots, analogous to the reverse osmosis (RO) process, allowing the xylem sap to be almost free of salt [30, 31]. Detailed knowledge of the salt exclusion mechanisms of mangroves has served for the design of filter membranes based on mangrove roots [32]and for artificial trees capable of filtering salt from seawater [33, 34].

Regarding the practical experiences of seawater desalination with halophytes, reed beds with *Avicennia marina* and *Rhizophora mucronata* have been reported [35–38]. This particular phytotechnology is inspired by constructed wetlands (CW) used in wastewater treatment, but in this case, it consists of establishing a flow of seawater through the roots of mangroves to remove NaCl, as shown in **Figure 1**. Another approach uses mangrove plants inside a greenhouse as part of a biothermal desalination process to produce freshwater through transpiration and condensation [35, 37].

3. The use of microalgae and cyanobacteria to desalinate water

Microalgae constitute a heterogeneous group of photosynthetic organisms that can grow in either saline or freshwater and in different climates [36]. These characteristics have led to their direct use as novel and sustainable biological techniques for seawater desalination [12, 39–42]. In fact, it has been found that some common genera of freshwater microalgae, many of which have been widely used in wastewater treatment, such as *Chlorella*, *Scenedesmus* and *Desmodesmus* among others, can survive and grow in water with high NaCl levels [6, 15, 43].

Cyanobacteria, also known as blue-green algae, are prokaryotic (without intracellular structures) photosynthetic organisms with a long evolutionary history [44]. As with microalgae, it has been found that cyanobacteria can develop in environments with high concentrations of NaCl and have the potential to be used in biological desalination processes [45]. Within this group, studies have been conducted on species of the genera *Spirulina*, *Synechococcus*, and *Phormidium*, which are found in hypersaline environments [9, 46–48].

Figure 1.
Conceptual design of constructed wetlands using mangroves as a desalination system. Raw seawater enters the constructed wetland and loses salt as it circulates through the roots of mangroves. The resulting brackish water can be treated using conventional processes such as reverse osmosis.

3.1 Biodesalination mechanisms in microalgae and cyanobacteria

Microalgae and cyanobacteria remove salts from water via bioadsorption and bioabsorption mechanisms [49, 50]. In this sense, it has been shown that microalgae exposed to salt stress respond by increasing biomass and lipid synthesis, which can be converted into biofuels and other high-value products [6, 51].

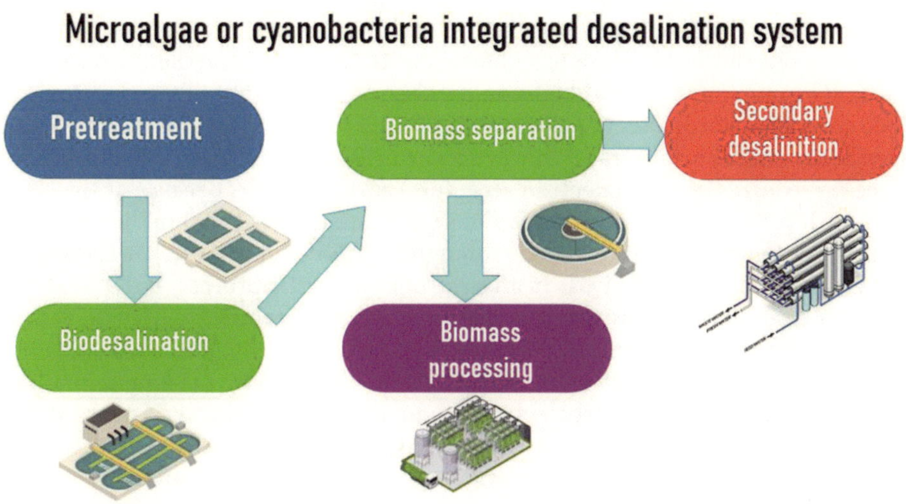

Figure 2.
Schematic representation of microalgae desalination technology integrated with seawater reverse osmosis (SWRO). In this system, pretreated seawater is fed into an algal bioreactor to remove salts from the water. The algae were then removed from the desalinated water, which was then subjected to conventional processes, such as reverse osmosis. The biomass generated in the bioreactor was processed to eliminate salt and recover lipids and other products generated by microalgae.

3.2 Desalination experiences with microalgae and cyanobacteria

A promising application of microalgae is the treatment of produced water (PW), saline wastewater, or the pretreatment of saline water with high nutrient and metal contents [6, 41, 52].

Integration of desalination with algae or cyanobacteria using reverse osmosis (RO) membranes has also been proposed [6]. The idea is to use microalgae or cyanobacteria as a prior coarse desalination step, and then leave the final purification step to reverse osmosis to obtain fresh water with less energy expenditure (**Figure 2**).

However, the use of photosynthetic organisms to desalinate seawater has not advanced beyond the laboratory scale because of unresolved technical difficulties, most notably the separation of biomass from water [53]. Flocculation and sedimentation are the most economical separation technologies that take advantage of the production of extracellular polymeric substances (EPS) [53, 54]. Another approach for the efficient use of microalgae and cyanobacteria in seawater desalination is to immobilize them in natural or synthetic polymer matrices in the form of beads before introducing them into a photobioreactor, which increases the efficiency of desalination and the survival of these microorganisms [55–57].

4. Microbial desalination cells

Microbial desalination cells (MDC) are a variant of bioelectrochemical systems (BES) or microbial electrochemical systems (MES) [7, 58] designed to dilute saline water by electrodialysis simultaneously with the removal of organic matter from wastewater streams [11, 55, 59]. The basic design of an MDC usually consists of three chambers: an anodic chamber inoculated with microorganisms, desalination chamber, and cathodic chamber. An anion exchange membrane (AEM) separates the anode chamber from the intermediate desalination chamber and a cathodic exchange membrane (CEM) separates the intermediate chamber from the cathode chamber (**Figure 3**).

In principle, the desalination mechanism of MDCs depends on the electricity produced by microorganisms that decompose the organic matter of wastewater in the anode chamber and transfer the electrons produced to the anode. The electrons transferred at the anode surface travel through an external wire and an external resistor to reduce the oxidized species (electron acceptors) at the cathode. Because of the electric potential difference in the MDC, the anions (Cl^-) and cations (Na^+) in the desalination chamber move towards the anode through the AEM and towards the cathode through the CEM, respectively, as in an electrodialysis (ED) process [8, 11, 60–62]. These processes treat wastewater, produce electricity, and remove total dissolved solids (TDS) from salt water.

The microorganisms involved in the generation of electricity in bioelectrochemical systems (BES), known as exoelectrogens, are mostly Proteobacteria of the genera *Pseudomonas*, *Shewanella*, and *Geobacter*, although eukaryotes such as yeasts have also been tested [61, 63–67]. However, if natural wastewater is used, biofilms composed of consortia of various species of bacteria develop on the anode [63, 67].

4.1 MDC configurations

The first microbial desalination cell design was based on the prototype proposed by Cao et al. [11], which used potassium ferricyanide as an oxidant. However, owing

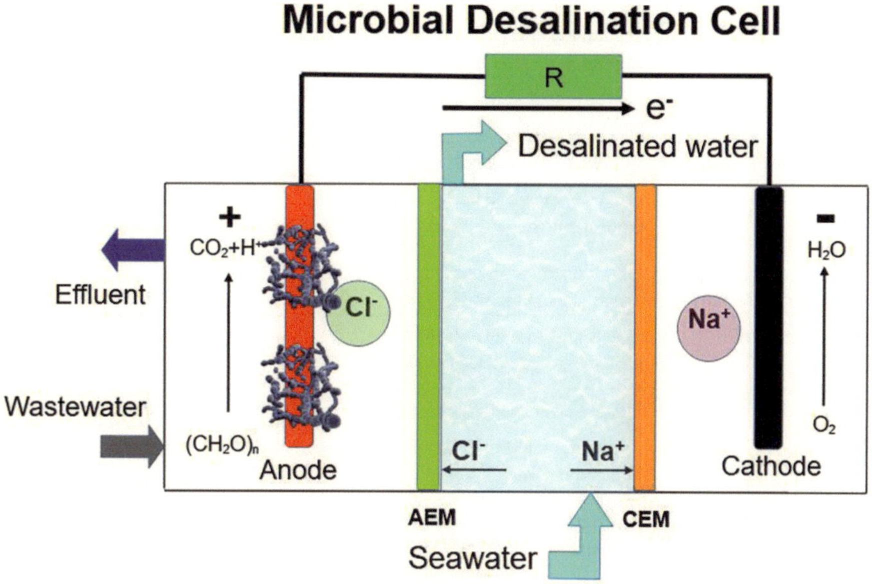

Figure 3.
Schematic representation of a microbial desalination cell. The degradation of dissolved matter in the wastewater occurs in the anode chamber. The electrons produced by exoelectrogenic bacteria travel to the cathode chamber, where the circuit is closed. Simultaneously, seawater circulates between the two electrode chambers separated by ion-exchange membranes. As water circulates, ions are lost, which are directed to the electrode chambers owing to the potential difference between them.

to their toxicity and cost, it is preferable to use air cathodes, but these also have limitations and require expensive catalysts [62, 68–70]. Therefore, a more economical and sustainable alternative was sought, such as cathodes built by living organisms capable of producing oxygen through photosynthesis, known as biocathodes [71–74]. Some of these MDCs with biocathodes have given rise to more complex designs known as photosynthetic microbial desalination cells (PMDCs) [75–79]. From these relatively simple models, more complex experimental configurations have been developed, the descriptions of which can be found in excellent reviews on MDC that study the desalination process in depth [62, 65, 80, 81].

Different microbial desalination cell configurations exhibit different performance and applications [7, 60, 61, 65, 80, 82]. Laboratory-scale studies have shown that stacked MDC (SMDC), osmotic MDC (OsMDC), osmotic fuel cell coupled MDC (OsMFC-MDC), microbial electrolysis desalination and chemical production cells (MEDCC), upflow MDC (UMDC), upflow stacked MDC (USMDC) and microbial reverse electrodialysis cells (MREC) are most appropriate for integration as pretreatment in larger scale reverse osmosis processes [59, 83]. The general characteristics of the most commonly used microbial desalination cells are presented in **Table 1**. More recent developments such as supercapacitive MDC (SC-MDC), quadruple MDC (QMDC), two-chamber tubular MDC (TTMDC), microbial salinity cell (MSC), MDC with integrated osmosis membranes (MDC-FO) and flow electrode MDC (FE-MDC) look promising [84–89].

Another design with perspectives is the photosynthetic MDC (PMDC), which can operate under different conditions according to the purpose pursued, maximizing desalination, wastewater treatment, energy production, or product synthesis [54, 72, 83–85]. There is another variant, the microbial salinity cell (MSC), which is

MDC Configuration	Features	Advantages
Air Cathode MDC	1. Exposure of cathode to air 2. Can be operated in both continuous and batch mode	1. Oxygen used as electron acceptor 2. High reduction in cathode chamber
Bio-cathode MDC	1. Use of bacteria or algae biocathodes 2. Can be operated in both batch and continuous mode	1. The reduction is enhanced by microbes 2. Increases desalination efficiency 3. Reduced start-up time
Stacked MDC (SMDC)	Alternating in AEMs and CEMs	Increases desalination rate
Recirculation MDC (RMDC)	Catholytc and anolyte solutions sequentially re-circulated through cell	1. Increases desalination efficiency 2. Reduces the pH imbalance
Capacitive MDC (CMDC)	1. Double-layer capacitor electrodes 2. Can be operated in both continuous and batch mode	1. Increases desalination efficiency 2. Reduces the pH imbalance 3. Reduced salt contamination
Upflow MDC (UMDC)	1. Tubular reactor containing concentric compartments separated by IEMs 2. Cathodes are exposed to air 3. Operates in continuous mode only	1. Efficient fluid mixing within the chambers 2. Increases desalination efficiency 3. Easier to scale up 4. Can be merged with osmotic fuel cell
Osmotic MDC (OsMDC)	AEM replaced with FO membranes	1. Increases desalination efficiency 2. Cost of FO membrane lower than AEM
Bipolar membrane MDC (BPMDC)	1. AEM and CEM are laminated to form bipolar membrane (BPM) 2. BPM placed next to the anode chamber, making a four-chamber MDC	1. High perm-selectivity 2. Low resistance and voltage drop 3. Increases desalination efficiency 4. Reduces the pH imbalance
Ion-exchange resin coupled MDC	Desalination chamber packed with mixed anions and cations exchange resins	1. Stabilises ohmic resistance 2. Increases desalination rate
Microbial electrodialysis and chemical production cell (MEDCC)	1. An additional chamber of acid production and a BPM is added to the cell 2. Operates in continuous mode only	1. Increases desalination efficiency 2. Reduces the pH imbalance 3. Simultaneous production of acid and alkali
Photosynthetic Microbial Desalination Cells (PMDC)	1. Involves the use of algae as biocathodes 2. Can be operated in batch, continuous and photo-bioreactor modes	1. Relies on the process of photosynthesis 2. Increases desalination efficiency 3. Reduces the pH imbalance

Sources: [7, 59, 60, 62, 65, 68].

Table 1.
Summary of the most common microbial desalination cell configurations.

not a proper MDC [86]. In an MSC, desalination does not occur in the intermediate chamber but in the anode chamber simultaneously with the reduction in chemical oxygen demand (COD). In MSC systems, recirculation is applied to minimize pH imbalance, which is a major bottleneck in MDC systems.

4.2 Operational limitations of MDC technology

Similar to most bioelectrochemical systems, microbial desalination cell technology faces several problems. These include concentration losses, increase in internal resistance, ohmic losses, challenges with MDC architecture, low microbial activity, low desalination rate, organic matter degradation, electron acceptors, electrical conductivity changes, ion transport, fouling on membranes, membrane integrity, pH change issues, and safety of purified water [58, 60, 61, 63, 65, 70, 90–99].

4.3 Scale up of MDCs and integration with other desalination systems

Scaling MDCs for use in real-life scenarios has always been challenging. However, no large-scale studies have been conducted to date [59, 62, 96]. The problems encountered in the laboratory have been expanded and multiplied by increasing the scales [100]. Therefore, the practical use of MDCs requires coupling or integration with other desalination processes [8, 90]. A diagram of the integrated system of an MDC with a reverse osmosis process is shown in **Figure 4**.

MDCs have provided good results at the experimental level when integrated with other desalination technologies [80, 83]. If MDCs are used as part of the pretreatment processes, a significant proportion of the energy used in desalination by reverse osmosis, or by other conventional desalination treatments, is reduced [7, 59, 62, 83, 101]. However, MDCs generate electricity that can be used in desalination.

The largest scale experience to date has been the Microbial Desalination for Low Energy Drinking Water (MIDES) project, accomplished by an international consortium of ten companies and research organizations from seven countries: Austria,

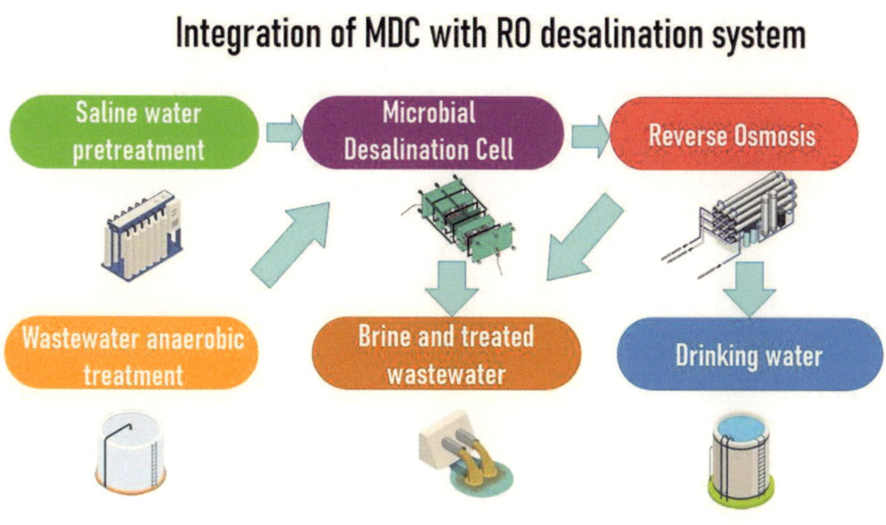

Figure 4.
Schematic representation of microbial desalination cell (MDC) technology integrated with seawater reverse osmosis (SWRO). In this system, pretreated seawater is processed in a wastewater-fed microbial desalination cell bioreactor. The resulting desalinated water is then subjected to conventional processes such as reverse osmosis, while brine and treated effluent are taken for final disposal.

Germany, Hungary, the Netherlands, Portugal, Spain, and Tunisia [102]. Two pilot plants were built, composed of a stack of 15 MDC pilot units, each with an electrode area of 0.4 m2, housed in a 40-foot container with the rest of the peripheral elements, with nominal desalination rates of 4–11 $L{\cdot}m^{-2}{\cdot}h^{-1}$.

5. Conclusions

The growing demand for fresh water, together with concerns about environmental conservation and the water-energy nexus, has led to a growth in research on sustainable desalination technologies. The direct use of living organisms, such as halophile plants, microalgae, and bacteria, to desalinate water appears to be a promising field. However, the development and practical applicability of these technologies depend on the group of living organisms selected to desalinate seawater.

The use of halophyte plants to remove NaCl from soils can be considered the first practical experience of biodesalination. Nonetheless, the use of phytotechnology to desalinate seawater is a nascent field that has not yet been used in practical applications. In any case, it should be noted that the study of mangroves has served as a basis for the development of innovative bioinspired desalination devices.

Regarding the use of microalgae and cyanobacteria, so far it has been shown to be an economical and effective technology to reduce salinity. However, this technology does not allow complete desalination of seawater, although it can serve as an effective pretreatment for reverse osmosis to improve its energy demand, recovery rate, and, therefore, make more sustainable the treatment. Additionally, microalgae can offer additional benefits, such as bioelectricity, biofuels, elimination of nutrients in water, and other products with high added value, making the desalination process more sustainable.

Finally, microbial desalination cells constitute an innovative technology capable of desalinating salt water along with wastewater treatment. Since their introduction in 2009, they have rapidly evolved in design and use, but there is still little evidence that this technology can work in large-scale plants as the sole wastewater treatment and desalination technology. In this sense, the integration of microbial desalination cells in the pretreatment process of reverse osmosis desalination seems to be the most appropriate use of this technology until their efficiency and durability problems are fully resolved.

Author details

Enrique O. Martínez
Ingeoambiente C.A., Caracas, Venezuela

*Address all correspondence to: eomartinez@gmail.com

References

[1] Ahmed FE, Khalil A, Hilal N. Emerging desalination technologies: Current status, challenges and future trends. Desalination. 2021;**517**:115183. DOI: 10.1016/j.desal.2021.115183

[2] He C, Liu Z, Wu J, Pan X, Fang Z, Li J, et al. Future global urban water scarcity and potential solutions. Nature Communications. 2021;**12**:4667. DOI: 10.1038/s41467-021-25026-3

[3] Panagopoulos A, Haralambous K-J. Environmental impacts of desalination and brine treatment–challenges and mitigation measures. Marine Pollution Bulletin. 2020;**161**(Pt. B):111773. DOI: 10.1016/j.marpolbul.2020.111773

[4] Eyl-Mazzega M-A, Cassignolv É. The Geopolitics of Seawater Desalination. In: Études de l'Ifri. Paris: Institut Français des Relations Internationales; 2022. 28 p. ISBN: 979-10-373-0661-6

[5] Shokri A, Sanavi Fard MA. A comprehensive overview of environmental footprints of water desalination and alleviation strategies. International Journal of Environmental Science and Technology. 2022;**20**:2347-2374. DOI: 10.1007/s13762-022-04532-x

[6] Sahle-Demessie E, Aly Hassan A, El Badawy A. Bio-desalination of brackish and seawater using halophytic algae. Desalination. 2019;**465**:104-113. DOI: 10.1016/j.desal. 2019.05.002

[7] Savla N, Pandit S, Khanna N, Mathuriya AS, Jung SP. Microbially powered electrochemical systems coupled with membrane-based technology for sustainable desalination and efficient wastewater treatment. Journal of Korean Society of Environmental Engineers. 2020;**42**(2):360-380. DOI: 10.4491/ksee.2020.42.7.360

[8] García-Rodríguez L, Delgado-Torres AM. Renewable energy-driven desalination: New trends and future prospects of small capacity systems. PRO. 2022;**10**(4):745. DOI: 10.3390/pr10040745

[9] Ghobashy MOI, Bahattab O, Alatawi A, Aljohani MM, Helal MMI. A novel approach for the biological desalination of major anions in seawater using three microalgal species: A kinetic study. Sustainability. 2022;**14**(2):7018. DOI: 10.3390/su14127018

[10] Ahmed ME, Zafar AM, Hamouda MA, Aly Hassan A, Arimbrathodi S. Biodesalination research trends: A bibliometric analysis and recent developments. Sustainability. 2023;**15**(1):16. DOI: 10.3390/su15010016

[11] Cao X, Huang X, Liang P, Xiao K, Zhou Y, Zhang X, et al. A new method for water desalination using microbial desalination cells. Environmental Science & Technology. 2009;**43**(18):7148-7152. DOI: 10.1021/es901950j

[12] Taheri R, Razmjou A, Szekely G, Hou J, Ghezelbash GR. Biodesalination - On harnessing the potential of nature's desalination processes. Bioinspiration & Biomimetics. 2016;**11**(4):041001. DOI: 10.1088/1748-3190/11/4/041001

[13] Abdelrasoul A, Doan H, Lohi A. Biomimetic and Bioinspired Membranes for New Frontiers in Sustainable Water Treatment Technology. London: IntechOpen; 2017. 234 p. DOI: 10.5772/65691

[14] Werber JR, Elimelech M. Permselectivity limits of biomimetic

desalination membranes. Science Advances. 2018;**4**(6):eaar8266. DOI: 10.1126/sciadv.aar8266

[15] Fuwad A, Ryu H, Malmstadt N, Kim SM, Jeon T-J. Biomimetic membranes as potential tools for water purification: Preceding and future avenues. Desalination. 2019;**458**:97-115. DOI: 10.1016/j.desal.2019.02.003

[16] Porter CJ, Werber JR, Zhong M, Wilson CJ, Elimelech M. Pathways and challenges for biomimetic desalination membranes with sub-nanometer channels. ACS Nano. 2020;**14**(9):10894-10916. DOI: 10.1021/acsnano.0c05753

[17] Beratto-Ramos A, Dagnino-Leone J, Martínez-Oyanedel J, Aranda M, Bórquez R. Fabrication and filtration performance of aquaporin biomimetic membranes for water treatment. Separation & Purification Reviews. 2022;**51**(3):340-357. DOI: 10.1080/15422119.2021.1948865

[18] Zhang L, Liu F, Wang J, Lin H, Han Q. Bioinspired nanobubble water channel membranes for ultrafast osmosis desalination. Journal of Membrane Science. 2023;**675**:121554. DOI: 10.1016/j.memsci.2023. 121554

[19] Liu X, Tian Y, Wu Y, Caratenuto A, Chen F, Cui S, et al. Seawater desalination derived entirely from ocean biomass. Journal of Materials Chemistry A. 2021;**39**:22313-22324. DOI: 10.1039/d1ta05068k

[20] Li Z, Wei S, Ge Y, Zhang Z, Li Z. Biomass-based materials for solar-powered seawater evaporation. Science of the Total Environment. 2023;**858**(Pt. 3):160003. DOI: 10.1016/j.scitotenv.2022.160003

[21] Dong Y, Tan Y, Wang K, Cai Y, Li J, Sonne C, et al. Reviewing wood-based solar-driven interfacial evaporators for desalination. Water Resesearch. 2022;**223**:119011. DOI: 10.1016/j.watres.2022.119011

[22] Zhang H, Li L, He N, Wang H, Wang B, Dong T, et al. Bioinspired hierarchical evaporator via cell wall engineering for highly efficient and sustainable solar desalination. EcoMat. 2022;**4**(5):e12216. DOI: 10.1002/com2.12216

[23] Zhang Y, Wang F, Yu Y, Wu J, Cai Y, Shi J, et al. Multi-bioinspired hierarchical integrated hydrogel for passive fog harvesting and solar-driven seawater desalination. Chemical Engineering Journal. 2023;**466**:143330. DOI: 10.1016/j.cej.2023.143330

[24] Rahman MM, Mostofa MG, Keya SS, Siddiqui MN, Ansary MMU, Das AK, et al. Adaptive mechanisms of halophytes and their potential in improving salinity tolerance in plants. International Journal of Molecular Sciences. 2021;**22**(19):10733. DOI: 10.3390/ijms221910733

[25] Ghazaryan KA, Harutyunyan A, Khachatryan HE, Singh A, Minkina TM, Rajput V, et al. Study of tolerance and phytodesalination potential of wheat, oat, emmer, and barley for sustainable saline agriculture. Applied Ecology and Environmental Research. 2023;**21**(5):4853-4882. DOI: 10.15666/aeer/2105_48534882

[26] Khan S, Masoodi TH, Pala NA, Murtaza S, Mugloo JA, Sofi PA, et al. Phytoremediation prospects for restoration of contamination in the natural ecosystems. Water. 2023;**15**(8):1498. DOI: 10.3390/w15081498

[27] Freedman A, Gross A, Shelef O, Rachmilevitch S, Arnon S. Salt uptake and evapotranspiration under arid

conditions in horizontal subsurface flow constructed wetland planted with halophytes. Ecological Engineering. 2014;**70**:282-286. DOI: 10.1016/j.ecoleng.2014.06.012

[28] Vaudry N, Sun Y, Afolabi O. Exploiting constructed wetlands for industrial effluent phytodesalination in Jing-Jin-Ji urban agglomeration, China. International Journal of Phytoremediation. 2023;**2**(7):851-867. DOI: 10.1080/15226514.2022.2115005

[29] Kim K, Seo E, Chang S-K, Park TJ, Lee SJ. Novel water filtration of saline water in the outermost layer of mangrove roots. Scientific Reports. 2016;**6**:20426. DOI: 10.1038/srep20426

[30] Reef R, Lovelock CE. Regulation of water balance in mangroves. Annals of Botany. 2015;**115**(3):385-395. DOI: 10.1093/aob/mcu174

[31] Sudhir S, Arunprasath A, Vel VS. A critical review on adaptations, and biological activities of the mangroves. Journal of Natural Pesticide Research. 2022;**1**:100006. DOI: 10.1016/j.napere.2022.100006

[32] Kim K, Kim H, Lim JH, Lee SJ. Development of a desalination membrane bioinspired by mangrove roots for spontaneous filtration of sodium ions. ACS Nano. 2016;**10**(12):11428-11433. DOI: 10.1021/acsnano.6b07001

[33] Wang Y, Lee J, Werber JR, Elimelech M. Capillary-driven desalination in a synthetic mangrove. Science Advances. 2020;**6**(8):eaax5253. DOI: 10.1126/sciadv.aax5253

[34] Lee J. Perspectives and design considerations of capillary-driven artificial trees for fast dewatering processes. Scientific Reports. 2021;**11**:8631. DOI: 10.1038/s41598-021-88006-z

[35] Chimayati RL, Titah HS. Removal of salinity using interaction mangrove plants and bacteria in batch reed bed system reactor. Journal of Ecological Engineering. 2019;**20**(4):84-93. DOI: 10.12911/22998993/102792

[36] Puspaningrum TC, Titah HS. The removal of salinity in a reed bed system using mangroves and bacteria in a continuous flow series reactor. Journal of Ecological Engineering. 2020;**21**(6):212-223. DOI: 10.12911/22998993/124075

[37] Finck C. A bio-thermic seawater desalination system using halophytes. Water Supply. 2014;**14**(4):657-663. DOI: 10.2166/ws.2014.020

[38] Sahoo D, Seckbach J, editors. The algae world. In: Cellular Origin, Life in Extreme Habitats and Astrobiology. 1st ed. Vol. 26. New York: Springer; 2015. 610 p. DOI: 10.1007/978-94-017-7321-8

[39] El Nadi MHA, Waheb A, Saad ISA. Desalination using algae ponds under nature Egyptian conditions. Journal of Water Resources and Ocean Science. 2014;**3**(6):69-73. DOI: 10.11648/j.wros.20140306.11

[40] Figler A, B-Béres V, Dobronoki D, Márton K, Nagy SA, Bácsi I. Salt tolerance and desalination abilities of nine common green microalgae isolates. Water. 2020;**11**(12):2527. DOI: 10.3390/w11122527

[41] Moayedi A, Yargholi B, Pazira E, Babazadeh H. Investigation of bio-desalination potential algae and their effect on water quality. Desalination and Water Treatment. 2021;**212**:78-86. DOI: 10.5004/dwt.2021.26638

[42] Gautam S, Kapoor D. Application of halophilic algae for water desalination.

In: El-Sheekh M, Abomohra AE-F, editors. Handbook of Algal Biofuels. Aspects of Cultivation, Conversion, and Biorefinery. Amsterdan: Elsevier; 2022. pp. 161-179. DOI: 10.1016/b978-0-12-823764-9.00002-9.ch8

[43] Al Bazedi G, Ismail MM, Mugwanya M, Sewilam H. Desalination concentrate microalgae cultivation: Biomass production and applications. Sustainable Water Resources Management. 2023;**9**:108. DOI: 10.1007/s40899-023-00887-2

[44] Demoulin CF, Lara YJ, Cornet L, François C, Baurain D, Wilmotte A, et al. Cyanobacteria evolution: Insight from the fossil record. Free Radical Biology and Medicine. 2019;**140**:206-223. DOI: 10.1016/j.freeradbiomed.2019.05.007

[45] Minas K, Karunakaran E, Bond T, Gandy C, Honsbein A, Madsen M, et al. Biodesalination: An emerging technology for targeted removal of Na^+ and Cl^- from seawater by cyanobacteria. Desalination and Water Treatment. 2015;**55**(10):2647-2668. DOI: 10.1080/19443994.2014.940647

[46] Kirkwood AE, Buchheim JA, Buchheim MA, Henley WJ. Cyanobacterial diversity and halotolerance in a variable hypersaline environment. Microbial Ecology. 2008;**55**:453-465. DOI: 10.1007/s00248-009291-5

[47] Amezaga JM, Amtmann A, Biggs CA, Bond T, Gandy CJ, Honsbein A, et al. Biodesalination: A case study for applications of photosynthetic bacteria in water treatment. Plant Physiology. 2014;**164**(4):1661-1676. DOI: 10.1104/pp.113.233973

[48] Ghalhari MA, Mafigholami TA, Khoshmaneshzadeh B. Optimization of the biological salt removal process from artificial industrial wastewater with high TDS by *Spirulina* microalga using the response surface method. Water Science & Technology. 2022;**86**(5):1168-1180. DOI: 10.2166/wst.2022.270

[49] Patel AK, Tseng Y-S, Singhania RR, Chen C-W, Chang J-S, Dong CD. Novel application of microalgae platform for biodesalination process: A review 2021. Bioresource Technology. 2021;**337**:125343. DOI: 10.1016/j.biortech.2021.125343

[50] Zafar AM, Javed MA, Aly Hassan A, Sahle-Demessie E, Harmon S. Biodesalination using halophytic cyanobacterium *Phormidium keutzingianum* from brackish to the hypersaline water. Chemosphere. 2022;**307**(Pt. 4):136082. DOI: 10.1016/j.chemosphere.2022.136082

[51] Shirazia SA, Rastegary J, Aghajani M, Ghassemi A. Simultaneous biomass production and water desalination concentrate treatment by using microalgae. Desalination and Water Treatment. 2018;**135**:101-107. DOI: 10.5004/dwt.2018.23163

[52] Nadersha S, Aly HA. Immobilized algae for produced water treatment and desalination. International Journal of Environmental Science and Development. 2022;**13**(6):264-269. DOI: 10.18178/ijesd.2022.13.6.1403

[53] Zafar AM, Javed MA, Aly Hassan A, Mehmood K, Sahle-Demessie E. Recent updates on ions and nutrients uptake by halotolerant freshwater and marine microalgae in conditions of high salinity. Journal of Water Process Engineering. 2021;**44**:102382. DOI: 10.1016/j.jwpe.2021.102382

[54] Madkour AGME, Ibrahim HAH, El-Sayed WMM, El-Moselhy KM. Bioflocculation technique for microalgal harvesting and wastewater nutrient

recovery. Iranian Journal of Fisheries Sciences. 2020;**19**(4):1780-1794. DOI: 10.22092/ijfs.2018.117674

[55] Shaikhoun S, Zafar AM, Wu Y-Hb, Aly HA. Biodesalination performance of Phormidium keutzingianum concentrated using two methods (immobilization and centrifugation). Arabian Journal of Chemistry. 2022;**15**(11):104282. DOI: 10.1016/j.arabjc.2022.104282

[56] Zafar AM, Al Mosteka H, Aly HA. Performance of immobilized microalgal strains for biodesalination of real seawater. Desalination. 2023;**561**:116704. DOI: 10.1016/j.desal.2023.116704

[57] Zafar AM, Aly HA. Seawater biodesalination treatment using Phormidium keutzingianum in attached growth-packed bed continuous flow stirred tank reactor. Environmental Research. 2023;**236**(Pt2):116784. DOI: 10.1016/j.envres.2023.116784

[58] Wang H, Ren ZJ. A comprehensive review of microbial electrochemical systems as a platform technology. Biotechnology Advances. 2013;**31**(8):1796-1807. DOI: 10.1016/j.biotechadv.2013.10.001

[59] Salehmin MNI, Lim SS, Satar I, Daud WRW. Pushing microbial desalination cells towards field application: Prevailing challenges, potential mitigation strategies, and future prospects. Science of the Total Environment. 2021;**759**:143485. DOI: 10.1016/j.scitotenv.2020.143485

[60] Pandit S, Sarode S, Das D. Fundamentals of microbial desalination cell. In: Das D, editor. Microbial Fuel Cell - A Bioelectrochemical System that Converts Waste to Watts. New York: Springer; 2018. pp. 353-371. DOI: 10.1007/978-3-319-66793-5_18.ch18

[61] Imoro AZ, Mensah M, Buamah R. Developments in the microbial desalination cell technology: A review. Water-Energy Nexus. 2021;**4**:76-87. DOI: 10.1016/j.wen.2021.04.002

[62] Saeed HM, Husseini GA, Yousef S, Saif J, Al-Asheh S, Fara AA, et al. Microbial desalination cell technology: A review and a case study. Desalination. 2015;**359**:1-13. DOI: 10.1016/j.desal.2014.12.024

[63] Luo H, Xu P, Ren Z. Long-term performance and characterization of microbial desalination cells in treating domestic wastewater. Bioresource Technology. 2012;**120**:187-193. DOI: 10.1016/j.biortech.2012.06.054

[64] Guang L, Ato-Koomson D, Jingyu H, Ewusi-Mensah D, Miwornunyuie N. Performance of exoelectrogenic bacteria used in microbial desalination cell technology. International Journal of Environmental Research and Public Health. 2020;**17**(3):1121. DOI: 10.3390/ijerph17031121

[65] Jatoi AS, Hashmi Z, Mazari SA, Mubarak NM, Karri RR, Ramesh S, et al. A comprehensive review of microbial desalination cells for present and future challenges. Desalination. 2022;**535**:115808. DOI: 10.1016/j.desal.2022.115808

[66] Mardiana U, Innocent C, Cretin M, Buchari B, Gandasasmita S. Yeast fuel cell: Application for desalination. IOP Conference Series: Materials Science and Engineering. 2016;**107**:012049. DOI: 10.1088/1757-899X/107/1/012049

[67] Yuan H, Sun S, Abu-Reesh IM, Badgley B, He Z. Unravelling and reconstructing the nexus of salinity, electricity and microbial ecology for bioelectrochemical desalination. Environmental Science & Technology.

2017;**51**(21):12672-12682. DOI: 10.1021/acs.est.7b03763

[68] Jingyu H, Ewusi-Mensah D, Norgbey E. Microbial desalination cells technology: A review of the factors affecting the process, performance and efficiency. Desalination and Water Treatment. 2017;**87**:140-159. DOI: 10.1007/978-3-319-66793-5_18

[69] Ramírez-Moreno M, Rodenas P, Aliaguilla M, Bosch-Jimenez P, Borràs E, Zamora P, et al. Comparative performance of microbial desalination cells using air diffusion and liquid cathode reactions: Study of the salt removal and desalination efficiency. Frontiers in Energy Research. 2019;7:135. DOI: 10.3389/fenrg.2019.00135

[70] Al Balushi NJ, Nayak JK, Rahman S, Sana A, Al-Mamun A. A comprehensive study on air-cathode limitations and its mitigation strategies in microbial desalination cell - A review. Energies. 2022;**15**(20):7459. DOI: 10.3390/en15207459

[71] Gude VG, Kokabian B, Gadhamshetty V. Beneficial bioelectrochemical systems for energy, water, and biomass production. Journal of Microbial & Biochemical Technology. 2013;**S6**:005. DOI: 10.4172/1948-5948.S6-005

[72] Kokabian B, Gude VG. Photosynthetic microbial desalination cells (PMDCs) for clean energy, water and biomass production. Environmental Science: Processes & Impacts. 2013;**15**:2178-2185. DOI: 10.1039/c3em00415e

[73] Arana TJ, Gude VG. A microbial desalination process with microalgae biocathode using sodium bicarbonate as an inorganic carbon source. International Biodeterioration & Biodegradation. 2018;**130**:91-97. DOI: 10.1016/j.ibiod.2018.04.003

[74] Koltysheva D, Shchurska K, Kuzminskyi Y. Microalgae and cyanobacteria as biological agents of biocathodes in biofuel cells. Biotechnologia. 2021;**102**(4):437-444. DOI: 10.5114/bta.2021.111108

[75] Kokabian B, Ghimire U, Gude VG. Water deionization with renewable energy production in microalgae-microbial desalination process. Renewable Energy. 2018;**122**:354-361. DOI: 10.1016/j.renene.2018.01.061

[76] Kokabian B, Smith R, Brooks JP, Gude VG. Bioelectricity production in photosynthetic microbial desalination cells under different flow configurations. Journal of Industrial and Engineering Chemistry. 2018;**58**:131-139. DOI: 10.1016/j.jiec.2017.09.017

[77] Hui WJ, Ewusi-Mensah D, Huang J. Using C. *vulgaris* assisted microbial desalination cell as a green technology in landfill leachate pre-treatment: A factor-performance relation study. Journal of Water Reuse and Desalination. 2020;**10**(1):1-16. DOI: 10.2166/wrd.2019.073

[78] Barahoei M, Hatamipour MS, Khosravi M, Afsharzadeh S, Feghhipour SE. Salinity reduction of brackish water using a chemical photosynthesis desalination cell. Science of the Total Environment. 2021;**779**:146473. DOI: 10.1016/j.scitotenv.2021.146473

[79] Ewusi-Mensah D, Huang J, Chaparro LK, Rodenas P, Ramírez-Moreno M, Ortiz JM, et al. Algae-assisted microbial desalination cell: Analysis of cathode performance and desalination efficiency assessment. Processes. 2021;**9**(11):2011. DOI: 10.3390/pr9112011

[80] Zahid M, Savla N, Pandit S, Thakur VK, Jung SP, Gupta PK, et al. Microbial desalination cell: Desalination through conserving energy. Desalination. 2022;**521**:115381. DOI: 10.1016/j.desal.2021.115381

[81] Odunlami OA, Vershima DA, Tagbo CV, Ogunlade S, Nkongho S. Microbial desalination cell technique - A review. South African Journal of Chemical Engineering. 2023;**46**:312-329. DOI: 10.1016/j.sajce.2023.07.011

[82] Al-Mamun A. Microbial desalination cell based wastewater treatment and resource recovery: Various challenges. In: Mungray AK, Mungray AA, Sonawane SS, Sonawane SH, editors. Novel Approaches Towards Wastewater Treatment and Resource Recovery Technologies (P). Amsterdam: Elsevier; 2022. pp. 73-107. DOI: 10.1016/B978-0-323-90627-2.00015-0.ch5

[83] Shivakumar T, Razaviarani V. An integrated approach to enhance the desalination process: Coupling reverse osmosis process and microbial desalination cells in the UAE. Water Supply. 2021;**21**(3):1127-1143. DOI: 10.2166/ws.2020.375

[84] Santoro C, Abad FB, Serov A, Kodali M, Howe KJ, Soavic F, et al. Supercapacitive microbial desalination cells: New class of power generating devices for reduction of salinity content. Applied Energy. 2017;**208**:25-36. DOI: 10.1016/j.apenergy.2017.10.056

[85] Jafary T, Al-Mamun A, Alhimali H, Baawain MS, Rahman MS, Rahman S, et al. Enhanced power generation and desalination rate in a novel quadruple microbial desalination cell with a single desalination chamber. Renewable and Sustainable Energy Reviews. 2020;**127**:109855. DOI: 10.1016/j.rser.2020.109855

[86] Jafary T, Al-Mamun A, Alhimali H, Baawain MS, Rahman S, Tarpeh WA, et al. Novel two-chamber tubular microbial desalination cell for bioelectricity production, wastewater treatment and desalination with a focus on self-generated pH control. Desalination. 2020;**481**:114358. DOI: 10.1016/j.desal.2020.114358

[87] Al Hinai A, Jafary T, Alhimali H, Rahman S, Al-Mamun A. Desalination and acid-base recovery in a novel design of microbial desalination and chemical recovery cell. Desalination. 2022;**525**:115488. DOI: 10.1016/j.desal.2021.115488

[88] Elnahas M, Elawwad A, Ghallab A, Ettouney R, El-Rifai M. An integrated MDC-FO membrane configuration for simultaneous desalination, wastewater treatment and energy recovery. RSC Advances. 2023;**13**(25):17038-17050. DOI: 10.1039/d3ra00149k

[89] Yang Z, Li Y, Zhan Z, Song Y, Zhang L, Jin Y, et al. Enhanced power generation, organics removal and water desalination in a microbial desalination cell (MDC) with flow electrodes. Science of the Total Environment. 2023;**858**(Pt. 1):159914. DOI: 10.1016/j.scitotenv.2022.159914

[90] Kokabian B, Gude VG. Role of membranes in bioelectrochemical systems. Membrane Water Treatment. 2015;**6**(1):53-75. DOI: 10.12989/mwt.2015.6.1.053

[91] Alhimali H, Jafary T, Al-Mamun A, Baawain MS, Vakili-Nezhaad GR. New insights into the application of microbial desalination cells for desalination and bioelectricity generation. Biofuel Research Journal.

2019;**6**(4):1090-1099. DOI: 10.18331/brj2019.6.4.5

[92] Ragab M, Elawwad A, Abdel-Halim H. Evaluating the performance of microbial desalination cells subjected to different operating temperatures. Desalination. 2019;**462**:56-66. DOI: 10.1016/j.desal.2019.04.008

[93] Liu F, Moustafa H, Hassouna MS, He Z. Resource recovery from wastewater can be an application niche of microbial desalination cells. Environment International. 2020;**142**:105855. DOI: 10.1016/j.envint.2020.105855

[94] Yuliasni R, Kadier A, Zen N, Setianingsih NI, Kurniawan SB, Ma P-C. Water desalination and bioelectricity generation using three chambers microbial salinity cell reactor with electrolyte recirculation. Journal of Ecological Engineering. 2020;**21**:129-136. DOI: 10.12911/22998993/126983

[95] Khaled H, Zrelli A, Hamed M, Chaouachi B. Modelisation and optimization of a microbial desalination cell system. Journal of Sustainable Bioenergy Systems. 2021;**11**(3):118-130. DOI: 10.4236/jsbs.2021.113009

[96] Rahman S, Al-Mamun A, Jafary T, Alhimali H, Baawain MS. Effect of internal and external resistances on desalination in microbial desalination cell. Water Science & Technology. 2021;**83**(10):2389-2403. DOI: 10.2166/wst.2021.145

[97] Abd-almohi HH, Alismaeel ZT, M-Ridha MJ. Study of microbial desalination cell performance; power generation and desalination efficiency using pure oxygen in a cathode chamber. Al-Khwarizmi Engineering Journal. 2022;**18**(3):37-47. DOI: 10.22153/kej.2022.07.002

[98] Nawaz A, Ul-Haq I, Qaisar K, Gunes B, Raj SI, Mohyuddin K, et al. Microbial fuel cells: Insight into simultaneous wastewater treatment and bioelectricity generation. Process Safety and Environmental Protection. 2022;**161**:357-373. DOI: 10.1016/j.psep.2022.03.039

[99] Ramírez-Moreno M, Esteve-Núñez A, Ortiz JM. Study of microbial desalination cell performance with different saline streams: Analysis of current efficiency and freshwater production. Journal of Environmental Chemical Engineering. 2023;**11**(1):109240. DOI: 10.1016/j.jece.2022.109240

[100] Rahman S, Jafary T, Al-Mamun A, Baawain MS, Choudhury MR, Alhimali H, et al. Towards upscaling microbial desalination cell technology: A comprehensive review on current challenges and future prospects. Journal of Cleaner Production. 2021;**288**:125597. DOI: 10.1016/j.jclepro.2020.125597

[101] ElMekawy A, Hegab HM, Pant D. The near-future integration of microbial desalination cells with reverse osmosis technology. Energy & Environmental Science. 2014;7:3921-3933. DOI: 10.1039/c4ee02208d

[102] Salinas-Rodríguez SG, Arévalo J, Ortiz JM, Borràs-Camps E, Monsalvo-Garcia V, Kennedy MD, et al. Microbial Desalination Cells for Low Energy Drinking Water. London: IWA Publishing; 2021. 231 p. DOI: 10.2166/781789062120

Chapter 10

Nanostructures in Water Purifications

Selcan Karakuş and Magdy M.M. Elnashar

Abstract

Effective water purification strategies are essential in addressing the serious global concern of scarce water. Nanomaterials have recently garnered considerable interest due to their excellent chemical, mechanical, physical, and biological properties, making them promising candidates for use in water filtration systems. Nanomaterials, with their high hydrophilicity, surface area, and variable surface characteristics, have shown potential in removing water-based contaminants. This review provides an overview of current developments in the development of nano-membrane materials for filtration systems. We discuss various kinds of nanoplatforms, such as polymeric nanocomposites, MXene nanosheets, metal/metal oxide nanoparticles (NPs), carbon nanotubes, metal–organic frameworks, nanofibers, and nanotubes, and their mechanisms of action in removing impurities. Furthermore, we summarize the possibilities and challenges associated with the use of nano-membrane systems, including potential environmental impacts and the need for sustainable and affordable production technologies. Overall, the application of nanomaterials in purifying water shows great potential for providing safe and clean drinking water to people around the world.

Keywords: water purification, water pollutants, desalination, nano-membranes, water quality index, MXene, nano-structure

1. Introduction

Global warming and water pollution cause difficulties in finding clean water, and this crisis gets worse as the world economy and population grow [1]. Several contaminants caused by industrial or natural activity are detected as impurities in water, rendering it non-potable or unsafe for drinking.

Understanding that water covers 70% of the earth, but only 3% of the water on Earth is drinkable and most of this water (2%) is not accessible as it is locked in ice caps, glaciers or the soil. Over 1.1 billion individuals do not have access to freshwater, and 2.7 billion people globally experience water scarcity for at least 1 month of the year. Poor sanitation affects 2.4 billion people because it puts them at risk for typhoid, water-borne illnesses, and diarrhea. Every year, diarrheal diseases alone claim the lives of 2 million people, predominantly youngsters. Droughts might affect two-thirds of the worldwide people by 2025. Furthermore, environments across the globe will experience increased damage [2]. Thus, development in water filtration systems is critical and urgently needed to maintain the supply of healthy water.

The types of water pollutants can be classified as:

a. Physical pollutants

Mostly refer to water contamination brought on by rocks and sediments. These contaminants cause liquids' color, opacity, and texture to alter visibly.

b. Chemical pollutants

Chemical pollutants can be manmade or natural. Bleach, insecticides, and pipeline corrosion are examples of man-made chemical pollution. Whereas natural chemical pollutants include toxins, nitrogen, and arsenic produced by bacteria.

c. Biological pollutants

Viruses, bacteria, protozoa, parasites and microorganisms are examples of biological contaminants.

d. Radiological pollutants

Radiological pollutants can be manmade or natural. They are the least frequent pollutants. However, there is a strong correlation between radioactive substances (e.g. cesuim and arsenic) and major health issues, including cancer. Nearly every ionized molecule can be radioactive, although cesium is the most prevalent element. A water supply can become contaminated by the improper disposal of radioactive waste, damaged reactors, or nuclear fallout. Radiological contaminants are growing in size as a result of heavier reliance on nuclear power.

In another classification as shown in **Figure 1**, the main sources of water pollution are sewage, industrial waste, global warming, atmospheric deposition, marine dumping, underground storage leakage, radioactive waste and oil pollution. More specifically, wastewater, pigments, heavy metals, radioactive elements, drugs, solvents, and textile dyes are typically the primary pollutants that seriously pollute the world's largest freshwater resources, having damaging effects on both the health of humans and aquatic

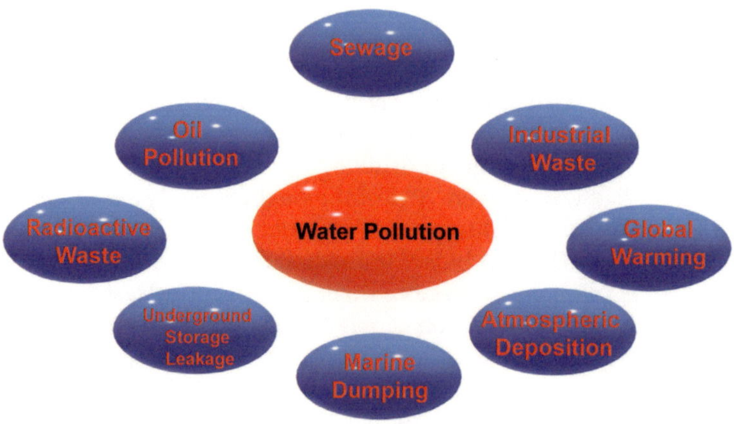

Figure 1.
Sources of water pollution.

species [3]. For examples, the presence of specific metal ions, primarily chlorides, sulphates, and bicarbonates of magnesium and calcium result in the accumulation of lime scale can clog plumbing and encourage galvanic corrosion. Iron is another example, even in minute amounts, as low as 0.3 ppm, iron alters the flavors of foods and beverages, can discolor flooring, sinks, and other fixtures, and can cause pipe blockages [4].

2. Water treatment processes

Water treatment processes (WTPs) can be divided into two categories: conventional WTPs and non-conventional WTPs as shown in **Table 1**. The most fundamental and often used traditional methods in WTP include sedimentation, coagulation, flocculation, and filteration.

In Australia, coagulation; flocculation; sedimentation; filtration and disinfection as a conential WTPS and they use biological activated carbon treatment as a non-convential WTPS.

2.1 Water quality index

The WQI method employs six criteria to determine the quality of river water: pH, ammonia-nitrogen (AN), suspended solids (SS), and biochemical oxygen demand (BOD) [5]. The Department of Environment Malaysia (DOE) developed the idea of evaluating water quality based on pollutant loading and river water categories under the National Water Quality Standards (NWQS) [5]. Nine variables are combined by the WQI: oxygen that has been broken down, nitrate, pH, temperature, phosphate, fully broken down solids, coliform bacteria, and natural oxygen interest (Body). The Water Quality Index (WQI) utilizes a single worth to represent the overall water

Conventional WTPs	Non-conventional WTPs
• Coagulation • Flocculation • Electrocoagulation • Filtration • Chlorination • Fluorination • Membrane bioreactors • Sand Filtration • Disinfection • Pulsatube Clarifier Actiflo' Technology • Dissolved Air Floatation (DAF) • Sedimentation (Membrane filtration technology) • Reverse osmosis • Ion exchange • Adsorption	• Ozone in water treatment • Applications of Ultraviolet (UV) Light • Biological Activated Carbon (BAC) concept • Ultrafiltration (UF) and Microfiltration in Membrane Technology

Table 1.
Comparison of various water treatment processes.

quality of a specific source at a given time, based on selected water quality criteria. The ranking is based on the values of Q and W, where Q represents the degree of water quality relative to any given boundary and W refers to the general significance of the individual boundary to the overall quality of water. The five classifications for the positional measures of water quality are as follows [5, 6]:

The WQI is termed,

- extremely bad when it is less than 25.
- awful between 26 and 50.
- moderate between 51 and 70.
- acceptable between 71 and 90.
- great between 91 and 100.

For example, in Australia, the average WQI values for Port Kembla (76.78), Port Jackson (79.07), Port Yamba (77.67), Port Eden (67.42), and Port Newcastle (75.15) were calculated. The water quality ratings for these ports during high and low tides were assessed as good, except for Port Botany and Eden [6].

3. Drawbacks of traditional water purification techniques

Water treatment for consumption and other applications has traditionally been done using conventional methods including chlorination, sedimentation, and membrane processes. Nevertheless, these strategies have several limitations that reduce their potency in tackling the escalating worldwide shortage of drinking water. For example, the application of chlorine as a cleaner may generate hazardous disinfection by - products that may be harmful to people's health. Chemical precipitation and filtration strategies might also fall short in their capabilities to get removal of some pollutants, like organic toxic elements, which can be dangerous to both the environment and the health of humans. These solutions can also be costly, energy-intensive, and complicated to set up, which makes them less practicable for adoption in remote. Therefore, it is necessary to develop novel and ecologically friendly water purification solutions that can get beyond these limitations and offer everyone a source of pure and secure water. Several techniques are examined for this purpose individually or in combinations as below:

Reverse osmosis: By using this procedure, harmful minerals are removed from the water, making the treated water acidic. Chemicals, chlorine, chloramines, volatile organic compounds, and medicines cannot be removed with this technique. Demands a lot of energy.

Distillation: The majority of toxins are left behind and need a lot of water and energy. Pollutants having boiling points above 100°C are challenging to eliminate.

Ultraviolet/treatment: Expensive approach that is deactivated by water turbidity and cloudiness. Removing heavy metals and other nonliving pollutants is inefficient.

Chemical transformation: Reagents are needed in excess. The finished product can be a poor mixture that cannot be released into the environment. Inactive amid difficult circumstances. This approach is not very discriminating.

Fluctuation and coagulation: To attain the ideal pH, this procedure is difficult and less effective and calls for alkaline additions.

Biological process: Microorganisms are challenging to regulate and sensitive to environmental conditions. The microbial cells are damaged by intermediates. Cost-wise, this is inefficient. It takes time.

Ultrafiltration: Dissolved inorganics cannot be removed using this technique. Demands a lot of energy. Prone to particulate clogging and challenging to clean.

Microfiltration: nitrates, fluoride, metals, salt, volatile organics, color, and other contaminants cannot be removed. It requires routine cleaning. Membrane fouling is inevitable. Less vulnerable to microorganisms, particularly viruses.

Carbon filtration: Nutrients like salt, fluoride, metals, and nitrates cannot be removed. Solids that cannot be dissolved cause clogging. Susceptible to mold. Filters must be changed frequently.

Thus, there was a need to find more advanced techniques to improve the efficiency and decrease the price of filtration systems.

4. Membrane filtration technology

Membrane filtration is a separation method that covers a wide variety of pollutants, from NPs to microparticles, and a wide range of water filtration systems are produced to use a simple and effective way in the filtration of industrial and domestic wastewater. Micro/nanoparticles can be produced using a variety of technologies, such as the Innotech encapsulator, interphase, ionic-gelation methods, vibrational jet-flow, and dripping techniques [7, 8]. As an example of microparticles, Ali et al. (2017), prepared three biopolymeric formulations in the micro size, 2 mm ± 10%, based on chitosan, alginate and carrageenan to chelate heavy metals [9]. The results showed that most of biopolymeric beads can chelate up to 85–100% of cations (500 mg/L).

5. An overview of the nano-membrane systems

Nanomaterials on the other hand can be categorized as zero-dimensional (0D), one-dimensional (1D), two-dimensional (2D), or three-dimensional (3D) based on their dimensions. Quantum dots and nanodispersions are examples of 0D nanomaterials that have all of their dimensions at the nanoscale. Whereas 1D nanomaterials, such as nanotubes and nanowires, have two dimensions at the nanoscale. 2D nanomaterials, such as films with nanosized coatings and nanosheets, are characterized as nanostructures. Whereas mesoporous bulk powders and nanoclusters are a feature of 3D nanomaterials. The design of sustainable and biocompatible membranes and the use of Euglena, nanomaterial-based, carbon- and nanostructure-based membranes for water purification systems are just a few samples of the achievements that have already been produced [10, 11]. This popularization of membrane materials is due mainly to filtration systems' substantially smaller carbon footprint, cost-effectiveness, and high performance as compared to other methods of purifying water [12].

Today, a wide range of combined mixtures, including volatiles, small particles, and organic solvents, have been effectively removed using membranes with controllable pore sizes based on ultrafiltration, microfiltration, nanofiltration, solvent extraction, gas separation technology, and reverse osmosis. The functional use of membrane separation methods rapidly requires the production of high-performance

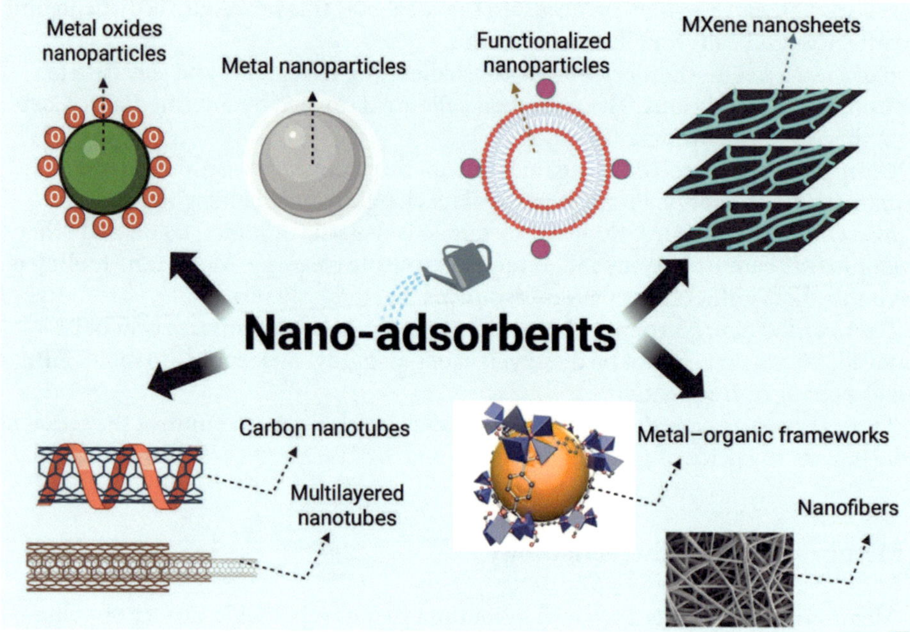

Figure 2.
Utilization of nanomaterials with varied dimensions in the development of nano-adsorbents.

nanomembranes with excellent stability, high selectivity, permeability, and protective coating properties.

To solve the above problems, nanomaterials of various sizes have been used to develop nano-based membranes. These nanomaterials include polymeric nanocomposites [13], metal/metal oxide NPs [14], polymer/metal/metal oxide NPs [15], zeolites [16], carbon nanotubes [17], graphene/graphene oxide (GO) [18], transition metal dichalcogenide-based nanosheets [19], two-dimensional (2D) MXene nanosheets [20], and metal–organic frameworks [21]. In **Figure 2**, the utilization of nanomaterials with varied dimensions in the development of nano-adsorbents is presented. With the quick advancement of materials science, several nanostructures, including zwitterionic colloids NPs [22], hyperbranched polymer dendrimer nanostructures [23], biomass-based metallic NPs [24], TiO_2 NPs [25], $CoFe_2O_4$/FeO/CoFe nanocomposites [26], CuO/CN [27], ZnO nanoneedles-Ti_3C_2 MXene/PVDF [28] and poly (vinyl alcohol) (PVA)-based nanofibers [29], have been employed to develop membrane materials for water purification applications.

5.1 Polymeric nanocomposites

Polymeric nanocomposites are nanosized formed from NPs and polymers. They are employed in water purification to remove pollutants such as microorganisms, organic molecules, and toxic metals. The matrix phase provides durability and mechanical strength, while the nanomaterials offer enhanced reactivity and a large surface area. The adsorption process of polymeric nanocomposites is based on chemical and physical interactions between pollutants and adsorbent. The physical adsorption involves weak intermolecular interactions and Van der Waals forces between pollutants and adsorbent. Chemical adsorption involves the generation of covalent or ionic interactions between pollutants and adsorbent [30].

Two different types of polymers nanobased membranes have been developed:

i. direct coating or self-assembly of polymer nanomaterials into homogeneous nanobased membranes.

ii. incorporation of polymer-based nanostructures into a polymer matrix to fabricate mixed matrix polymer nanomaterial-based membranes [31].

5.2 Metal/Metal oxide nanoparticles

Metal/metal oxide NPs are commonly used in water treatment to remove pollutants, including microorganisms, pathogens, toxic metals, and organic compounds. These nanostructures are highly reactive and efficient at removing pollutants due to their unique morphologies. The adsorption processes of metal and metal oxide NPs are based on the chemical/physical interactions between adsorbent and impurities. The surface characteristics of the NPs, such as their structure, morphology, and charge density, play a major role in determining their selectivity and adsorption efficiency. Adsorption mechanisms of these NPs involve both chemical and physical interactions between the adsorbent and adsorbed species. The physical adsorption is based on hydrogen bonding, dipole–dipole attraction, and Van der Waals forces, between pollutants and adsorbents. The high surface area and reactivity of meal-based nanoparticles enhance physical adsorption by improving contact area between adsorbents and adsorbates. Chemical adsorption, on the other hand, involves the formation of covalent or ionic bonds between adsorbents and impurities. This can occur through various mechanisms, including ion exchange and complex formation. Metal and metal oxide NPs can form chemical bonds with a variety of contaminants, including toxic metals and organic compounds. Using advanced oxidation processes, such as photocatalytic activity and Fenton processes, metal and metal oxide NPs can also catalyze the removal of contaminants. Reactive oxygen species, which can oxidize and degrade a wide range of chemicals, are produced during these interactions. In addition, both physical and chemical interactions with impurities play a role in the adsorption mechanism of metal and metal oxide NPs. The adsorption efficiency and specificity of NPs are strongly affected by their surface morphology [32].

5.3 Carbon nanotubes

They form hollow and cylindric structures called carbon nanotubes. They are used in the elimination of pollutants including pathogens, germs, and organic compounds from water. Carbon nanotubes have a broad surface and are reactive, which makes them excellent at removing toxins. The unique electronic properties of CNTs enable them to facilitate the adsorption of polar and ionic pollutants via hydrophobic interactions, hydrogen bonding, π-π interfaces, electrostatic interactions. In contrast, chemical adsorption of CNTs involves the generation of covalent or ionic bonds through various mechanisms, such electrostatic interaction, and hydrogen bonding. Overall, CNTs are well-suited for removing pollutants in water treatment due to their excellent surface area, distinctive electronic properties, and strong adsorption capacity [33].

5.4 MXene nanosheets

MXene nanosheets are effective adsorbents for removing impurities because of their two-dimensional layered structure. The unique surface chemistry and shape of

MXene nanosheets form the basis for their adsorption mechanism. Surface functional groups such as -O, -OH, and -F, on surface of MXene nanosheets enable the adsorption of various impurities through hydrogen bonds, electrostatic interactions, and covalent bonds. The high surface area and aspect ratio of the nanosheets create a large number of adsorption sites, enhancing their capacity and selectivity for adsorption [34].

5.5 Metal–organic frameworks

Metal–organic frameworks (MOFs) have an excellent porous structure that makes them promising adsorbents for a variety of pollutants. The adsorption mechanism of MOFs is based on a combination of physical and chemical interactions. The porous structure provides a large surface area and many open metal sites that can interact with the adsorbate. The chemical interactions involve the coordination of metal centers with the adsorbate through various chemical processes such as complexation, ion exchange, electrostatic interaction, pore filling, hydrogen bonding, redox reaction, covalent attachment, encapsulation adsorption, and chelation. The organic ligands in MOFs can also interact with the adsorbate through hydrogen bonding, van der Waals forces, and π-π interactions. Moreover, the structure, size, shape, and functional groups of MOFs can be modified to increase their affinity for specific pollutants, making them an appealing choice for removing a wide range of pollutants [35].

5.6 Nanofibers

Nanofibers are flexible adsorbent materials that find wide applications in water treatment. Although the adsorption mechanism of nanofibers can vary depending on the specific material and impurity, there are general mechanisms that apply to multiple nanofiber-based adsorbents. One of these mechanisms is physical adsorption, where impurities are trapped within the pores and channels of the nanofiber structure through van der Waals forces, reduction–oxidation adsorption, weak intermolecular interactions. Due to their high surface area and porosity, nanofibers provide numerous adsorption sites, making them highly effective. In addition to physical adsorption, some nanofiber-based adsorbents also use chemical adsorption mechanisms. These mechanisms involve the formation of covalent or ionic bonds between impurities and the functional groups on surface of nanofibers or the use of electrostatic interactions to attract and capture charged impurities. Overall, nanofibers' combination of high surface area, porosity, and chemical reactivity make them highly effective for a wide range of impurities in water [36].

5.7 Nanomembranes

In light of the creation of an effective purification mechanism, membrane systems manufactured with different types of nanostructures inside the polymer matrix typically show a significant enhancement in permeability. Despite the improved purification efficiency of these inorganic NPs-based membrane systems, the majority of inorganic NPs tend to agglomerate and have poor biocompatibility with the biopolymer matrix. A wide range of 2D nanostructured materials, including GO, tungsten disulfide (WS2), molybdenum disulfide (MoS2), 2D oxides, and hexagonal boron nitride, have shown potential as components for innovative separation and purification in the areas of desalination and water treatments, since they possess both the structural performance of coatings and the adaptability of polymeric materials [37]. In brief, using 2D nanosized materials to create these advanced membrane systems represents a potential way to advance water filtration methods. As an example of these membranes is Mxene-based membranes.

5.7.1 MXene- based separation membranes

MXene, or 2D transition metal carbides, has just entered the market as a new 2D material for separation and purification systems. The general formula for MXene, a 2D transition metal carbide and nitride, is Mn + 1AXn (n = 1, 2, and 3). These ceramics were named MAX phases by researchers. d-block transition metal M (V, Cr, Ni, Ti, Sc, etc.). The primary group element is A. Either N or C atoms constitute X. 2D transition metal carbide/nitride (MXene) 2D nanosheets were produced by methodically etching the layers of main group elements from the bulk precursor of the ternary MAX phase [38]. 2D MXene and MXene-composites have been shown to be suitable membrane solutions for oil/water separation [39], sustainable water desalination [40], hydrogen purification [41], and dye/salt separation [42] due to their hydrophilic nature, simple scale-up manufacturing, good biocompatibility, and nontoxicity. Moreover, MXene has considerable benefits for adjustable tuning of the d-spacing of MXene films, simple surface functionalization, and improved adhesion with enabling surfaces due to the chemical functional groups on the surface.

The impacts of several influencing parameters and various synthesis techniques of 2D MXene-nanosheets have been investigated and reported in the literature [43–45]. With a strong emphasis on filtration mechanisms, the applications of developed membranes for oil/water separation have been widely researched [46]. 2D MXene-nanosheet-based membranes have been reported to exhibit excellent filtration performances in a variety of applications, including the membrane separation of seawater, waste management, and the purification and separation of petroleum and oil products. The membrane stability and regeneration characteristics of the previously employed membranes have also been evaluated [47]. Many difficulties with the development, use, and application of MXene-based membranes have been highlighted, and recommendations for further study have been presented. Uniform 2D MXene-nanosheets-based membranes have been developed because of their varied shapes, adjustable functionality, excellent water solubility, and compatibility with biopolymeric matrices. Furthermore, from 0D to 2D nanostructures can interact with contaminants in a wide range of ways, depending on the kind of pollutants being investigated, including Van der Waals interactions, Lewis's acid–base, chemical bonding, π–π stacking, hydrophobicity, surface complexation, electrostatic interaction, and hydrogen bonding [48]. Recent studies focused on controlling the hydrophilicity and hydrophobicity, charge property, pore size, surface integrity, and morphology of MXene-based membranes to increase the membrane's performance.

Table 2 offers a comprehensive list of current examples of adsorption-based nanostructures and their distinctive attributes for efficient pollution removal. These experimental findings, showcasing the diverse array of nanostructures and their applications in combatting industrial and environmental contamination challenges, serve as a general resource for researchers and professionals in the field.

Adsorbent	Adsorbate	Characterization	Adsorption performance	Ref.
Silver NPs combined with yttrium oxide (Ag-Y_2O_3) nanocomposite	Cu(II) and Cr(VI)	Particle size of 65 ± 3 nm	Adsorption capacity: 815 mg $g^{-1}$and 867.85 mg g^{-1} for Cu(II) and Cr(VI)	[49]
Bimetallic Fe/Pd NPs	As (III)	Particle sizes ranging from 30 to 60 nm. 1.176	Adsorption capacity: 1.176 mg g^{-1}	[50]
Fe_3O_4 NPs	Indigo carmine	Nanoparticles with diameters from 20 to 30 nm	Adsorption capacity: 11.51 mg g^{-1} to 17.45 mg g^{-1}	[51]
Fe_3O_4/MgO NPs	Amaranth dye	Nanorods with diameters of 14 nm and length of 22	Adsorption capacity: 17.94 mg g^{-1} to 37.98 mg g^{-1}	[52]
CuO NPs	Methylene blue dye	Average crystalline size of 19.28 nm	Adsorption capacity: 464.24 mg g^{-1}	[53]
Functionalization of chitosan biopolymer with SiO_2 NPs	Acid red 88 dye	Final consistent particle size of 250 μm	Adsorption capacity: 252.4 mg g^{-1}	[54]
Alginate/ poly(acrylamide)/ carbon nanotubes	Cationic dyes (methylene blue, MB, and crystal violet, CV)	Total porosity was estimated to be 76.6 ± 0.6%	MB and CV adsorption on the composite increased by 1.5 and 1.6 times.	[55]
Tri-n-Butyl Phosphate functionalized Multi Walled Carbon Nanotubes (MWCNTs-COO-TBP)	Ruthenium	Thickness of entire wall measured to be ~7.5 nm	Adsorption capacity: 151.51 mg g^{-1}	[56]
Iron nitrate modified carbon nanotube composites	Arsenic	The flocculation of FCNTs has increased, with more and more uniform pores	Uptake: 99.1%	[57]
UiO-66-based MOFs	Congo red (CR), and cationic dyes of methylene blue (MB) and rhodamine B (Rh B)	Angular morphology	Adsorption capacity of CR: 93.05 mg g^{-1} and 94.20 mg g^{-1}	[58]
MXene-based 2D $Ti_3C_2T_x$ nanosheets	Cadmium (Cd^{2+})	Lamella-like structure	Adsorption capacity: 475.6 mg g^{-1}	[59]
MXene@Ag/cryogel composites	Sulfamethoxazole and mercury	Composite with Ag particle size of 100–200 nm.	Uptake: (98% mercury) and degradation (97% sulfomethoxazole)	[60]

Table 2.
Examples of adsorption-based nanostructures from recent times and their characteristics in the removal of pollutants.

DOI: http://dx.doi.org/10.5772/intechopen.113893

6. Conclusion

Several water treatment methods were discussed in this paper. Sedimentation, filtration, flocculation, and coagulation are some of the most essential and popular classical processes employed in WTP. By weighing the advantages and disadvantages of each technique, it is simple to decide which type of filtration technology is best for which kind of region. The 2D MXene-nanosheet-based membrane has diverse forms, adaptable functionality, good water solubility, and compatibility with biopolymeric matrices, and it has made significant strides, however there are still many obstacles to be overcome before more sophisticated membranes can be created.

Author details

Selcan Karakuş[1] and Magdy M.M. Elnashar[2,3]*

1 Faculty of Engineering, Department of Chemistry, Istanbul University-Cerrahpaşa, Istanbul, Turkey

2 School of Medicine, Curtin University, Perth, Australia

3 Centre of Scientific Excellence, Polymers Department, Advanced Materials and Nanotechnology Laboratory, National Research Centre, Cairo, Egypt

*Address all correspondence to: magmel@gmail.com; magdyelnashar@curtin.edu.au

References

[1] Khatua A, Jana S, Kar TK. A fuzzy rule-based model to assess the effects of global warming, pollution and harvesting on the production of Hilsa fishes. Ecological Informatics. 2020;**57**:101070. DOI: 10.1016/J.ECOINF.2020.101070

[2] Duda AM, El-Ashry MT. Addressing the global water and environment crises through integrated approaches to the management of land. Water and Ecological Resources. 2009;**25**:115-126. DOI: 10.1080/02508060008686803

[3] Chowdhary P, Bharagava RN, Mishra S, Khan N. Role of industries in water scarcity and its adverse effects on environment and human health. Environmental Concerns and Sustainable Development. 2020:235-256. DOI: 10.1007/978-981-13-5889-0_12

[4] Das B, Hazarika P, Saikia G, et al. Removal of iron from groundwater by ash: A systematic study of a traditional method. Journal of Hazardous Materials. 2007;**141**:834-841. DOI: 10.1016/J.JHAZMAT.2006.07.052

[5] El-Serehy HA, Abdallah HS, Al-Misned FA, et al. Assessing water quality and classifying trophic status for scientifically based managing the water resources of the Lake Timsah, the lake with salinity stratification along the Suez Canal. Saudi Journal of Biological Sciences. 2018;**25**:1247-1256. DOI: 10.1016/J.SJBS.2018.05.022

[6] Jahan S, Strezov V. Water quality assessment of Australian ports using water quality evaluation indices. PLoS One. 2017;**12**:e0189284. DOI: 10.1371/JOURNAL.PONE.0189284

[7] Danial EN, Elnashar MMM, Awad GEA. Immobilized Inulinase on grafted alginate beads prepared by the one-step and the two-steps methods. Industrial and Engineering Chemistry Research. 2010;**49**:3120-3125. DOI: 10.1021/IE100011Z

[8] Elnashar MMM, Danial EN, Awad GEA. Novel carrier of grafted alginate for covalent immobilization of Inulinase. Industrial and Engineering Chemistry Research. 2009;**48**:9781-9785. DOI: 10.1021/IE9011276

[9] Ali KA, Hassan ME, Elnashar MMM. Development of functionalized carrageenan, chitosan and alginate as polymeric chelating ligands for water softening. International Journal of Environmental Science and Technology. 2017;**14**:2009-2014. DOI: 10.1007/S13762-017-1298-Y/METRICS

[10] Ma W, Ding Y, Zhang M, et al. Nature-inspired chemistry toward hierarchical superhydrophobic, antibacterial and biocompatible nanofibrous membranes for effective UV-shielding, self-cleaning and oil-water separation. Journal of Hazardous Materials. 2020;**384**:121476. DOI: 10.1016/J.JHAZMAT.2019.121476

[11] Teng L, Yue C, Zhang G. Epoxied SiO_2 nanoparticles and polyethyleneimine (PEI) coated polyvinylidene fluoride (PVDF) membrane for improved oil water separation, anti-fouling, dye and heavy metal ions removal capabilities. Journal of Colloid and Interface Science. 2023;**630**:416-429. DOI: 10.1016/J.JCIS.2022.09.148

[12] Priya AK, Gnanasekaran L, Kumar PS, et al. Recent trends and advancements in nanoporous membranes for water purification. Chemosphere. 2022;**303**:135205. DOI: 10.1016/J.CHEMOSPHERE.2022.135205

[13] Cheng Y, Xia C, Garalleh HAL, et al. A review on optimistic development of polymeric nanocomposite membrane on environmental remediation. Chemosphere. 2023;**315**:137706. DOI: 10.1016/J.CHEMOSPHERE.2022.137706

[14] Cheng X, Li P, Liu W, et al. Activation of peroxymonosulfate by metal oxide nanoparticles for mitigating organic membrane fouling in surface water treatment. Separation and Purification Technology. 2020;**246**:116935. DOI: 10.1016/J.SEPPUR.2020.116935

[15] Hao L, Li L, Yu S, Liu J. Humic acid-coated hydrated ferric oxides-polymer nanocomposites for heavy metal removal in water. Science of the Total Environment. 2022;**834**:155427. DOI: 10.1016/J.SCITOTENV.2022.155427

[16] Li Z, Jiang WT, Jean JS, et al. Combination of hydrous iron oxide precipitation with zeolite filtration to remove arsenic from contaminated water. Desalination. 2011;**280**:203-207. DOI: 10.1016/J.DESAL.2011.07.009

[17] Thamaraiselvan C, Lau WJ, Dosoretz CG. Coupled electrochemical transformation and filtration of water pollutants by cathodic-carbon nanotube membranes. Journal of Environmental Chemical Engineering. 2022;**10**:107670. DOI: 10.1016/J.JECE.2022.107670

[18] Fernández-Márquez M, Pla R, Oliveira AS, et al. Improvement of water filtration performance of graphene oxide membranes on nylon support by UV-assisted reduction treatment: Control of molecular weight cut-off. Chemical Engineering Journal. 2022;**449**:137807. DOI: 10.1016/J.CEJ.2022.137807

[19] Peng H, Wang R, Mei L, et al. Transition metal dichalcogenide-based functional membrane: Synthesis, modification, and water purification applications. Matter. 2023;**6**:59-96. DOI: 10.1016/J.MATT.2022.09.019

[20] Xiang J, Wang X, Ding M, et al. The role of lateral size of MXene nanosheets in membrane filtration of dyeing wastewater: Membrane characteristic and performance. Chemosphere. 2022;**294**:133728. DOI: 10.1016/J.CHEMOSPHERE.2022.133728

[21] Fang M, Cambedouzou J, Cot D, et al. Facile membrane preparation from colloidally stable metal-organic framework-polymer nanoparticles. J Memb Sci. 2022;**657**:120669. DOI: 10.1016/J.MEMSCI.2022.120669

[22] Ma R, Ji YL, Guo YS, et al. Fabrication of antifouling reverse osmosis membranes by incorporating zwitterionic colloids nanoparticles for brackish water desalination. Desalination. 2017;**416**:35-44. DOI: 10.1016/J.DESAL.2017.04.016

[23] Hooshyar SP, Panahi HA, Moniri E, Farsadrooh M. Tailoring a new hyperbranched PEGylated dendrimer nano-polymer as a super-adsorbent for magnetic solid-phase extraction and determination of letrozole in biological and pharmaceutical samples. Journal of Molecular Liquids. 2021;**338**:116772. DOI: 10.1016/J.MOLLIQ.2021.116772

[24] Islam S, Bairagi S, Kamali MR. Review on green biomass-synthesized metallic nanoparticles and composites and their photocatalytic water purification applications: Progress and perspectives. Chemical Engineering Journal Advances. 2023;**14**:100460. DOI: 10.1016/J.CEJA.2023.100460

[25] Davarikia Y, Aroujalian A, Salimi P. Immobilization of TiO_2 nanoparticles on PES substrate via dopamine and

poly (vinyl alcohol) for long-term oil/water purification. Process Safety and Environmental Protection. 2022;**166**:656-668. DOI: 10.1016/J.PSEP.2022.08.067

[26] Tan SS, Wang N, Yang CQ, et al. Exchange bias and exchange spring in $CoFe_2O_4$/FeO/CoFe nanocomposites. Journal of Magnetism and Magnetic Materials. 2022;**556**:169428. DOI: 10.1016/J.JMMM.2022.169428

[27] Zuo S, Xia D, Guan Z, et al. Dual-functional CuO/CN for highly efficient solar evaporation and water purification. Separation and Purification Technology. 2021;**254**:117611. DOI: 10.1016/J.SEPPUR.2020.117611

[28] Sahu S, Dhar PD. 1D/2D ZnO Nanoneedles/Ti_3C_2 MXene enrobed PVDF electrospun membrane for effective water purification. Applied Surface Science. 2023;**622**:156905. DOI: 10.1016/J.APSUSC.2023.156905

[29] Chokwe RC, Kebede TG, Dube S, Nindi MM. Fabrication of electrospun Mondia whitei/PVA nanofibres: Application in the removal of acidic drugs. Heliyon. 2022;**8**:e11075. DOI: 10.1016/J.HELIYON.2022.E11075

[30] Hnamte M, Pulikkal AK. Clay-polymer nanocomposites for water and wastewater treatment: A comprehensive review. Chemosphere. 2022;**307**:135869. DOI: 10.1016/J.CHEMOSPHERE.2022.135869

[31] Elnashar MM. Review article: "Immobilized molecules using biomaterials and Nanobiotechnology". Journal of Biomaterials and Nanobiotechnology. 2010;**1**:61-76. DOI: 10.4236/JBNB.2010.11008 (Q4, 113)

[32] Qumar U, Hassan JZ, Bhatti RA, et al. Photocatalysis vs adsorption by metal oxide nanoparticles. Journal of Materials Science and Technology. 2022;**131**:122-166. DOI: 10.1016/J.JMST.2022.05.020

[33] Mishra S, Sundaram B. Efficacy and challenges of carbon nanotube in wastewater and water treatment. Environmental Nanotechnology Monitering & Management. 2023;**19**:100764. DOI: 10.1016/J.ENMM.2022.100764

[34] Xu N, Wang W, Zhu Z, et al. Recent developments in photocatalytic water treatment technology with MXene material: A review. Chemical Engineering Journal Advances. 2022;**12**:100418. DOI: 10.1016/J.CEJA.2022.100418

[35] Xu HJ, Hu PY. Progress on fundamentals of adsorption transport of metal-organic frameworks materials and sustainable applications for water harvesting and carbon capture. Journal of Cleaner Production. 2023;**393**:136253. DOI: 10.1016/J.JCLEPRO.2023.136253

[36] PGS A, Irani M, Rad LR. Mechanisms of the removal of the metal ions, dyes, and drugs from wastewaters by the electrospun nanofiber membranes. Journal of the Taiwan Institute of Chemical Engineers. 2023;**143**:104625. DOI: 10.1016/J.JTICE.2022.104625

[37] Alam I, Chowdhury I. Applications of two-dimensional nanostructures for water filtration. Separation Science and Technology (New York). 2022;**15**:281-286. DOI: 10.1016/B978-0-323-90763-7.00014-7

[38] Srivastava A, Verma A, Prajapati YK. Effect of 2D, TMD, perovskite, and 2D transition metal carbide/nitride materials on performance parameters of SPR biosensor. Handbook of Nanomaterials for Sensing Applications. 2021:57-90. DOI: 10.1016/B978-0-12-820783-3.00005-1

[39] Hou K, Zhou H, Zhu K, et al. Superwetting $Ti_3C_2T_X$ MXene membranes intercalated with sodium alginate for oil/water separation. Carbohydrate Polymer Technologies and Applications. 2023;**5**:100278. DOI: 10.1016/J.CARPTA.2022.100278

[40] Solangi NH, Mubarak NM, Karri RR, et al. Applications of advanced MXene-based composite membranes for sustainable water desalination. Chemosphere. 2023;**314**:137643. DOI: 10.1016/J.CHEMOSPHERE.2022.137643

[41] Li R, Fu X, Liu G, et al. Room-temperature in situ synthesis of MOF@MXene membrane for efficient hydrogen purification. Journal of Membrane Science. 2022;**664**:121097. DOI: 10.1016/J.MEMSCI.2022.121097

[42] Wang H, He Z, Yang Q, et al. Fabrication of 2D/2D composite membrane via combining functionalized MXene and MoS_2 nanosheets for dye/salt separation. Journal of Environmental Chemical Engineering. 2022;**10**:108365. DOI: 10.1016/J.JECE.2022.108365

[43] Xie H, Ma S, He Z. Facile preparation of PANI/MoO_x nanowires decorated MXene film electrodes for electrochemical supercapacitors. Electrochimica Acta. 2023;**448**:142173. DOI: 10.1016/J.ELECTACTA.2023.142173

[44] Ahouei MA, Syed TH, Bishop V, et al. $Ti_3C_2T_X$ MXene framework materials: Preparation, properties and applications in energy and environment. Catalysis Today. 2023;**409**:162-172. DOI: 10.1016/J.CATTOD.2022.11.001

[45] Xu J, You J, Wang L, et al. MXenes serving aqueous supercapacitors: Preparation, energy storage mechanism and electrochemical performance enhancement. Sustainable Materials and Technologies. 2022;**33**:e00490. DOI: 10.1016/J.SUSMAT.2022.E00490

[46] Tang X, Guo B, Zhang S, et al. Layer-by-layer repaired lamellar membrane for low stacking defect of MXene nanosheets and efficient separation performance in water purification. Journal of Environmental Chemical Engineering. 2023;**11**:109450. DOI: 10.1016/J.JECE.2023.109450

[47] Ihsanullah I, Bilal M. Recent advances in the development of MXene-based membranes for oil/water separation: A critical review. Applied Materials Today. 2022;**29**:101674. DOI: 10.1016/J.APMT.2022.101674

[48] Sajid M, Asif M, Baig N, et al. Carbon nanotubes-based adsorbents: Properties, functionalization, interaction mechanisms, and applications in water purification. Journal of Water Process Engineering. 2022;**47**:102815. DOI: 10.1016/J.JWPE.2022.102815

[49] Kumar Pradhan S, Pareek V, Panwar J, Gupta S. Synthesis and characterization of ecofriendly silver nanoparticles combined with yttrium oxide ($Ag-Y_2O_3$) nanocomposite with assorted adsorption capacity for Cu(II) and Cr(VI) removal: A mechanism perspective. Journal of Water Process Engineering. 2019;**32**:10049. DOI: 10.1016/j.jwpe.2019.100917

[50] Lin Y, Jin X, Islam Khan N, et al. Efficient removal of As (III) by calcined green synthesized bimetallic Fe/Pd nanoparticles based on adsorption and oxidation. Journal of Cleaner Production. 2020;**286**:124987. DOI: 10.1016/j.jclepro.2020.124987

[51] Souza Damasceno B, Felipe Viana Da Silva A, Cláudia Vaz De Araújo A. Dye adsorption onto magnetic and superparamagnetic Fe_3O_4 nanoparticles:

A detailed comparative study. Journal of Environmental Chemical Engineering. 2020;**8**:103994.DOI: 10.1016/j.jece.2020.103994

[52] Salem ANM, Ahmed MA, El-Shahat MF. Selective adsorption of amaranth dye on Fe_3O_4/MgO nanoparticles. Journal of Molecular Liquids. 2016;**219**:780-788. DOI: 10.1016/J.MOLLIQ.2016.03.084

[53] Prajapati AK, Kumar MM. Comprehensive kinetic and mass transfer modeling for methylene blue dye adsorption onto CuO nanoparticles loaded on nanoporous activated carbon prepared from waste coconut shell. Journal of Molecular Liquids. 2020;**307**:112949. DOI: 10.1016/j.molliq.2020.112949

[54] Wu R, Saud Abdulhameed A, Kong Yong S, et al. Functionalization of chitosan biopolymer with SiO_2 nanoparticles and benzaldehyde via hydrothermal process for acid red 88 dye adsorption: Box-Behnken design optimization. International Journal of Biological Macromolecules. 2023;**247**:125806. DOI: 10.1016/j.ijbiomac.2023.125806

[55] Da Costa JS, De Souza JF, Dos Santos DRS, Fajardo AR. Composite aerogels of alginate/poly(acrylamide)/carbon nanotubes with enhanced performance for cationic dyes adsorption. Journal of Sol-gel Science and Technology. 2023;**84**:124586. DOI: 10.1016/j.mseb.2023.116820

[56] Mohanty B, Kumar Parida P, Abinash Bhuyan C, et al. Adsorption behavior of ruthenium in tri-n-butyl phosphate functionalized multi walled carbon nanotubes. Diamond and Related Materials. 2023;**139**:110336. DOI: 10.1016/j.diamond.2023.110336

[57] Li Y, Li Z, Qu G, et al. Mechanism, behaviour and application of iron nitrate modified carbon nanotube composites for the adsorption of arsenic in aqueous solutions. Chinese Journal of Chemical Engineering. 2023;**60**:26-36. DOI: 10.1016/j.cjche.2023.01.010

[58] Lei Y, Zhao J, Song H, et al. Enhanced adsorption of dyes by functionalized UiO-66 nanoparticles: Adsorption properties and mechanisms. Journal of Molecular Structure. 2023;**1292**:136111. DOI: 10.1016/j.molstruc.2023.136111

[59] Ilyas M, Younas M, Ul M, et al. MXene-based 2D $Ti_3C_2T_x$ nanosheets for highly efficient cadmium (Cd^{2+}) adsorption. Journal of Water Process Engineering. 2023;**55**:104131. DOI: 10.1016/j.jwpe.2023.104131

[60] Megbenu HK, Daulbayev C, Nursharip A, et al. Photocatalytic and adsorption performance of MXene@Ag/cryogel composites for sulfamethoxazole and mercury removal from water matrices. Environmental Technology and Innovation. 2023;**32**:103350. DOI: 10.1016/j.eti.2023.103350

Section 5

Industrial Process and Energy Generation

Chapter 11

VoxSton – A New Water Clarification System

Juhani Pylkkanen

Abstract

A new water clarification system for industrial and communal wastewater is presented here. The system consists of two centrifugal separators, a screw separator and two vortex flow flotation units. It can separate the solids like fibers out of the sludge and press them to dry matter content of up to 30 percent and clean the sludge water for recycling and discharge to nature. The clarification system can, in most cases, treat process and communal sludge continuously or daily basis directly from outlets of the wastewater to be treated to finish without equalization and precipitation. The objective is to recycle the process and wastewater, and to collect different ingredients for reuse as much as possible, and to prevent microbe growth. The new system saves in energy and space as well as in operational costs by roughly 50 percent compared to present advanced systems. The capital costs are estimated to be lower too. The new treatment or part of it has been applied for purification of various water matrices from city and industrial wastewaters to natural lake waters.

Keywords: water treatment, water purification, flotation, particle removal, recycling, separation, clarification, integrated clarification, pharmaceutic residues removal, radon removal

1. Introduction

A new industrial process and communal wastewater clarification system VoxSton developed by SansOx Ltd. and Rannan Teollisuuskone Oy in Finland is presented in this chapter. The system consists of two centrifugal separators, a screw separator and two vortex flow flotation units. It can separate the solids like fibers out of the sludge and press them to dry matter content of 30 percent, and clean the sludge water for recycling and discharge to nature.

The first test runs have been conducted process phase by phase with real pulp and paper process sludge successfully. The treatment or part of it has been applied for purification of various water matrices from city and industrial wastewaters to refreshing of natural lake waters. The special feature of the system is that the industrial and communal sludge can be treated in most cases continuously or daily basis directly from the liquid outlet to finish without equalization and precipitation in terms of hours instead of days and weeks. Separation and recycling potential is presented in **Figure 1**. The water treatment technology discussed here is able to fulfill the most separation demands presented in the figure. Removal of the separated ingredients

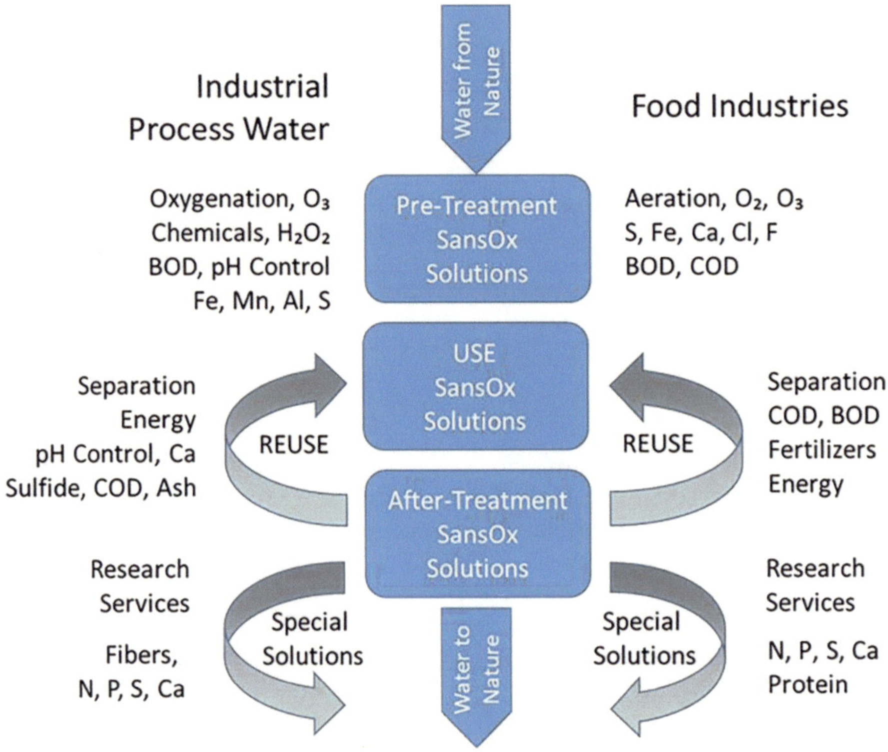

Figure 1.
Industrial recycling economy.

for recycling might need dedicated filtering. The most challenging separation and removal cases have been dealt with radon gas, pharmaceutic residues and disinfection of a hospital wastewater.

The new system saves in energy and chemicals, and in space as well as in operational costs roughly 50 percent compared to present advanced systems. The treatment process is integrated and fast. Use of chemicals has been reduced significantly due to molecular activation and efficient mixing in hermetic tube condition. For example CO_2 consumption was reduced to 1/5 by OxTube in a Ca removal case. The capital costs are estimated to be lower too. The main benefits of the new treatment system are presented in **Table 1**.

2. Flotation theory in brief

Solids removal of the new process water treatment system VoxSton is based on centrifugal and screw separation. Separation and removal of soluble ingredients are based on OxTube separation, molecular activation, and flotation in tube and cell condition. Flotation bubbles are generated by OxTube in VoxFlot water intake and fed them to bottom of the cell. The design principles of the VoxFlot are based on the flotation theory that is preferred here in brief [1–3]. The OxTube treatment and VoxFlot flotation are described in Chapter 3.

P&P Process Water of 5000 m^3/d			**Cost Estimation 2016**	
Item	**Conventional Process**		**Novel VoxSton Process**	
Investment	2.9 to 4.3 M€	580 to 860 €/m^3	2.5 to 3.4 M€	500 to 680 €/m^3
Operation Costs	1050 €/d	0.21 €/m^J	680 €/d	0.14 €/m^3
Maintenance Costs	340 €/d	0.07 €/m^3	170 €/d	0.04 €/m^3
Less Energy Cons.	7000 kWh/d	1.4 kWh/m^{3}d	3600 kWh/d	0.72 kWh/m^{3}d
Space	3850 m^2		1400 m^2	
Less CO_2 + removal	High energy consumption		Low energy consumption	
Odor Control	Difficult		In control by Efficient Oxygenation	
COD and BOD	Slow, high energy consumption		Fast and low energy consumption	
Disinfection	Chlorine and UV used		Ozone with air into OxTube	
Microbe Growth	Difficult to eliminate		Easy to eliminate with aeration	
Chemicals Removal	Difficult		Possible, verified with drugs residues	

Table 1.
Conclusions on the comparison of the novel VoxSton and present industrial process water treatment systems of the same capacity.

The mechanisms for the bubble-particle attachment are very complex and consists of three steps; collision, attachment (adsorption) and detachment (desorption). The collision is achieved by particles being within the collision tube of a bubble and this is affected by the velocity and radius of the bubble. The collision tube corresponds to the region in which a particle will collide with the bubble.

The attachment of the particle to the bubble is controlled by the induction time of the particle and bubble. The particle and bubble need to bind and this occurs if the time in which the particle and bubble are in contact with each other is larger than the required induction time. This induction time is affected by

- the fluid viscosity,
- particle and bubble size, and
- the forces between the particle and bubbles.

The detachment of a particle and bubble occurs when the force exerted by the surface tension is exceeded by shear and gravitational forces. These forces are complex and vary within the cell. High shear will be experienced close to the impeller of a mechanical flotation cell and mostly gravitational force in the collection and cleaning zone of a flotation column.

The attachment of the bubbles to the surface is determined by the interfacial energies between the solid, liquid, and gas phases. This is determined by the Young-Duprè equation [1]

$$'\Upsilon_{lv}\cos\theta = ('\Upsilon_{sv} - '\Upsilon_{sl}) \quad (1)$$

where:

- Υ_{lv} is the surface energy of the liquid/gas vapor interface,
- Υ_{sv} is the surface energy of the solid/gas vapor interface,
- Υ_{sl} is the surface energy of the solid/liquid interface,
- θ is the contact angle, the angle formed at the junction between gas, solid, and liquid phases.

A common quantity used to describe the collection efficiency of a flotation process is *flotation recovery R*. This quantity incorporates the probabilities of collision and attachment of particles to gas flotation bubbles [2, 3].

$$R = \frac{N_c}{\left(\frac{\pi}{4}\right)\left(d_p + d_b\right)^2 H_C} \tag{2}$$

where:

- $N_c = PN_c^i$ is the probability of the particle being collected, where *P* is probability coefficient and N_c^i the number of possible particle collisions
- d_p is particle diameter
- d_b is bubble diameter
- *H* is a specified height within the flotation at which the recovery was calculated
- *c* is the particle concentration

3. Description of VoxSton

The concept of the new industrial and communal sludge and water treatment called VoxSton consists of the following phases and devices:

1. SaoxFuge centrifugal separator with a screen shelter flange for enhancement of solids content from 0.5 to 5 percent in concentrate fed to the screw separator.
2. Milston screw separator for solids separation of dry matter content up to 30 percent.
3. VoxFlot vortex flotation process for floggulation, flocs separation and cleaning of the water coming from SaoxFuge and Milston separation. The flotation foam can be fed to the screw separator or to a drying area.

4. Final phase might be filtering, e.g., sand filtering. OxTube Treatment is identified by measures to enhance filtering, e.g., iron and manganese removal, efficiency. Reaction gases can be removed by VoxFlot Flotation or by a separate gas remover.

The entire concept is presented in **Figure 2**. The devices can be installed according to an integrated treatment process principles. Number of pools, tanks and pumps can be reduced, and energy and chemicals saved. Microbe growth, smell and uncertain phenomena can be eliminated by speeding up the process lead time. Of course, the process quality and recycling will improve significantly.

Present processes can be improved by the pretreatment and the secondary treatment that could be similar to the concept in **Figure 2**. These kinds of retrofits have been completed in present wastewater treatment and water intake facilities even without additional pumps and energy consumption. The secondary treatment from a sediment pool or from any other still water vessel requires normally an additional pump.

3.1 SaoxFuge centrifugal separator

Separation efficiency of SaoxFuge is based on centrifugal force and sediment effect that depends on the liquid viscosity and the specific gravity difference. The liquid flows a spiral path among the SaoxFuge pipe and a strong centrifugal force is created. The centrifugal acceleration depends on the flow speed and curve radius; it could be up to 1000 G. The liquid flow in the SaoxFuge can be pressurized up to 10 bar that makes the flow smoother, and reasonable high flow speed without turbulences is possible. The liquid flow turns to a spinning mode which is compensated by a counter vortex flow in SaoxFuge intake. The SaoxFuge Separator is presented in **Figure 3**.

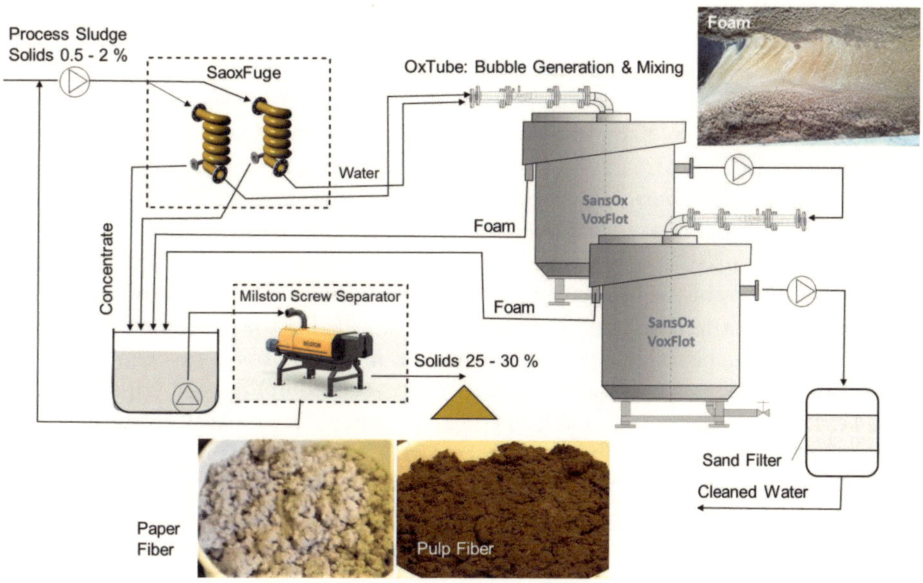

Figure 2.
Conceptual illustration of VoxSton process water treatment. In the most cases the process water can be treated from process outlet to finish continuously or daily basis without equalization and precipitation in terms of hours.

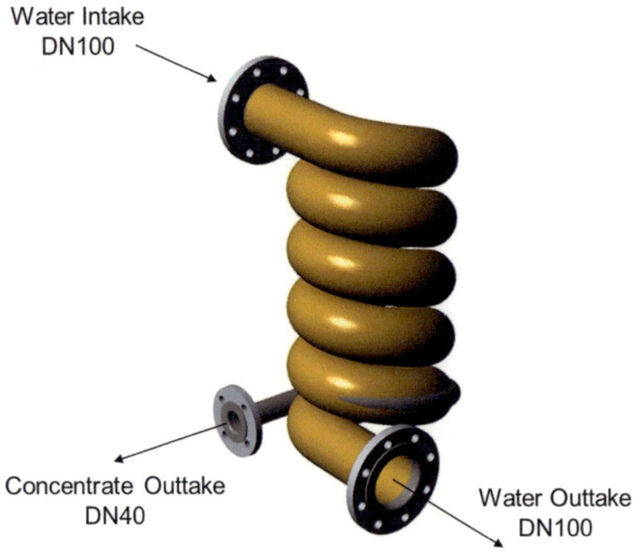

Figure 3.
SaoxFuge DN100 centrifugal separator is simple and reliable, no moving parts. Separation is a continuous and even process.

3.2 OxTube water treatment and bubble generation

A new hermetic water treatment called OxTube Integrated Clarification and some of its applications are presented briefly here. There are many ways and combinations to apply this new water treatment technology. In order to understand the OxTube phenomena from industrial water clarification point of view the basic principles and some related examples are presented in this context.

OxTube treats the water in hermetic flowing condition. It separates soluble ingredients from the water, splits molecular and ionic structures, activates molecules, and clarifies the water. These treatment functions happen within a second or less in its seamless process of the hermetic tube condition. OxTube sucks air in its nozzle zone, generates bubbles and mixes them evenly in the water to be treated. The other gases and chemicals can be mixed evenly with the water flow in the nozzle zone. The meeting or collision probability of the molecules, particles and bubbles is high. Desirable chemical reactions follow immediately and they are well controlled in the hermetic condition. Compressed air can be fed into the nozzle or after, if more bubbles are needed.

The principle of OxTube water treatment is presented in **Figure 4** with a functional prototype visually. The entire clarification process is very well controlled due to its hermetic procedure. Its foam generation efficiency in flotation is illustrated in **Figure 5** in raw water intake from a lake to industrial processes. **Figure 8** shows visible bubble generation, biomass split and flotation in a lake refreshing which is similar to purification of industrial waters. Measured aeration and oxygenation efficiency is presented in **Figure 6**.

OxTube efficiency on separation of soluble ingredients, molecular activation and clarification is shown visibly with Swiss Alpen tap water in **Figure 7**. Separation and activation can be illustrated by OxTube treatment without air suction, and the integrated clarification within a second with air suction left and right, respectively.

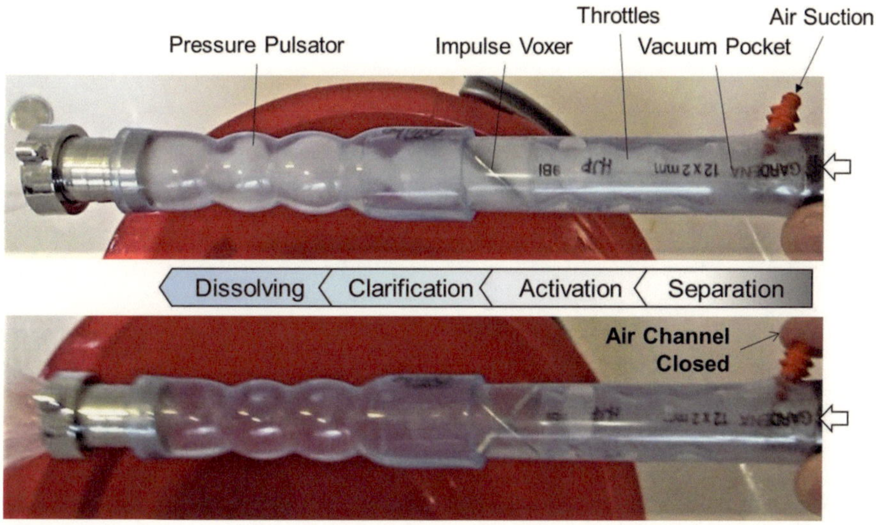

Figure 4.
OxTube water treatment process illustrated visually with air suction. The seamless process of separation with bubbles, activation, and clarification is performed within a second. The water is all the way an even mixture of small air bubbles, water and separated substances with air suction (upper picture). Collision probability of molecules, particles and bubbles is extreme high.

Figure 5.
Result of the efficient reaction of OxTube aerated lake water and related foam generation on the surface without additional flotation bubbles.

The water temperature and pressure are kept exactly the same in this trial. Turbidity of 0.0 NTU has been achieved in clarifications.

OxTube water treatment technology has been applied successfully in removal of pharmaceutics, radon gas, calcium, manganese and iron from various water matrices, and disinfection with ozone of a hospital wastewater. More than 40 different pharmaceutical residues in total over 24 µg/l were removed 90 percent by OxTube Integrated

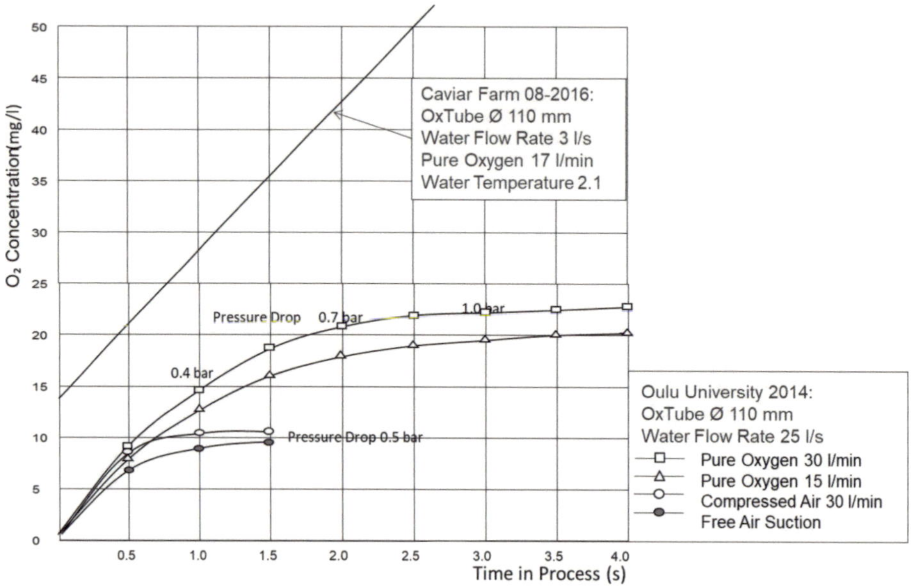

Figure 6.
Gas dissolving performance of the OxTube DN100. O_2 concentration of 100% was reached continuously within a second/Oulu university [4]. The highest O_2 concentration achieved continuously in an OxTube DN100 installation 2016 is 55 mgO_2/l at 2.1°C.

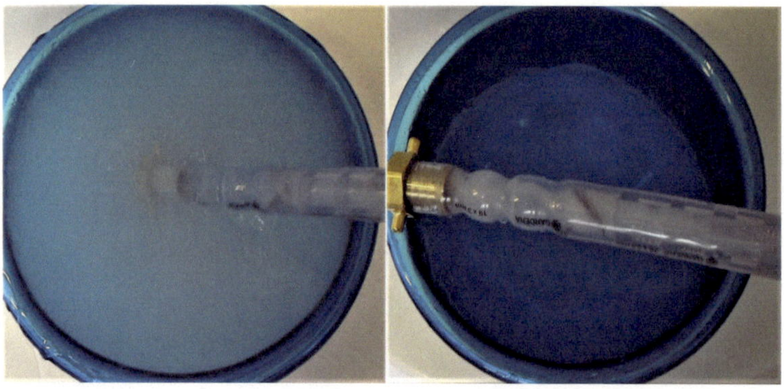

Figure 7.
Efficiency of OxTube water treatment with air; left Swiss Alpen tap water after the soluble ingredients separation and molecular activation, right the integrated clarification at the same condition. The separation, activation, clarification and replacement dissolving are performed seamlessly within a second in OxTube.

Clarification within a second in one through run [5]. Radon removal from 57 to 8 Bq/L is achieved continuously in five installations of ground water intakes at Water Plants in The Philippines. Disinfection of a hospital wastewater with ozone is shown in **Table 2** [5, 6]. Viruses were reduced over 60 percent without any gas and chemicals feed. Microbe growth can be reduced or even eliminated by OxTube oxygenation with clean air. Further, microbes and particles can be removed by VoxFlot flotation.

OxTube can be installed in various positions in present treatment processes and facilities. Calcium removal was improved up to 100 percent and CO_2 consumption reduced to 1/5 just by installing one OxTube in the present industrial water treatment facility of a can manufacturer.

Disinfection of Hospital Waste Water/THL Summary April 3, 2018

Sample	*E.coli* PMY/ ml	Log-red	Reduction %	MS2 PFU/ ml	Log-red	Reduction %	T °C	EC µS/ cm	pH	Turbity NTU	O_3 Res mg/l	O_2 mg/l
Test Water	2			1			12,8	283	7,73	0,68	0,03	10,26
Inoculum	2,500,000			870,000								
0 mgO_3/l	1,700,000		32,00	340,000		60,92	12,4	285	7,88	0,46	0,02	11,65
1,43 mgO_3/l	<0,001 (MR.)	>9,23	99,99,999,994	<2 (MR.)	>5,27	>99,9995	13,2	285	7,81	0,31	0,10	>20
1,27 mgO_3/l	<0,001 (MR.)	>9,23	99,99,999,994	<2 (MR.)	>5,27	>99,9995	12,6	294	7,83	0,27	0,14	>20
1,11 mgO_3/l	<0,001 (MR.)	>9,23	99,99,999,994	<2 (MR.)	>5,27	>99,9995	12,9	288	7,73	0,44	0,10	>20
0,95 mgO_3/l	<0,001 (MR.)	>9,23	99,99,999,994	<2 (MR.)	>5,27	>99,9995	14,3	279	7,84	0,37	0,10	>20
0,79 mgO_3/l	<0,001 (MR.)	>9,23	99,99,999,994	<2 (MR.)	>5,27	>99,9995	14,0	289	7,85	0,44	0,08	>20
0,61 mgO_3/l	<0,001 (MR.)	>9,23	99,99,999,994	<2 (MR.)	>5,27	>99,9995	14,0	287	7,74	0,40	0,07	>20

Reduction of 100% of E. coli and MS2 achieved with O_3 feed within 0.7 second [5, 6].

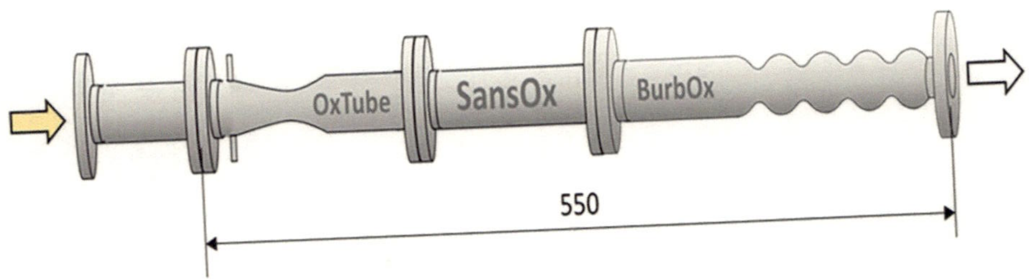

Table 2.
High molecule and microbe collision probability can be shown by the disinfection tests completed at Savonia University of Applied Sciences in Kuopio Finland.

Figure 8.
Natural water refreshing by OxTube integrated water clarification visibly. OxTube separates soluble ingredients, activates molecules, and generates air bubbles for the high collision flotation in tube. Split bio mass attaches on bubbles, and mass floats are formed on the surface easy to scribe out. The turbid lake water turned clear within 5 days. Turbidity of 0.0 NTU has been achieved in some purification cases.

OxTube clarification and flotation capabilities have been applied in refreshing of natural waters and their ecosystem. OxTube creates in its tube condition an efficient flotation with high collision of bubbles and bio particles. Turbidity of 0.0 NTU has been achieved in some wastewater purification cases with retention of some hours after the OxTube treatment. There is a flotation effect in the tube flow after bubble generation. This OxTube Flotation is a new innovation and it can be named as Tube Flotation. The natural water clarification by OxTube is shown visibly in **Figure 8**.

3.3 Milston screw separator

Milston is a compact and efficient screw separator verified with various sludge and slurry separations. It is available as a fixed installation and mobile applications, manual or completely automated. Particularly, Milston products are robust and reliable, easy to maintain and keep clean by internal washing system. Milston Hero 1 and its test results of P&P sludge separation are presented in **Figure 9**. The water, after solids separation, can be fed into the water treatment process as presented in **Figure 2**.

Milston Screw Separator can be used in various positions of industrial as well as other sludge processing. Content of solids should be at least around 5 percent. The concentration enhancement can be performed by SaoxFuge centrifugal separator.

3.4 VoxFlot flotation

Vortex Flotation Separation developed by SansOx is a unique separation method in which the liquid flows with the particles continuously in a vortex path, the gas bubbles are formed in inlet of the liquid flow by OxTube, and in the vortex flotation flow before the flotation cell. The bubbles are evenly distributed in the flow. In the flotation cell bubbles spin and flow cross the vortex liquid flow in optimal angle, the particles in the vortex

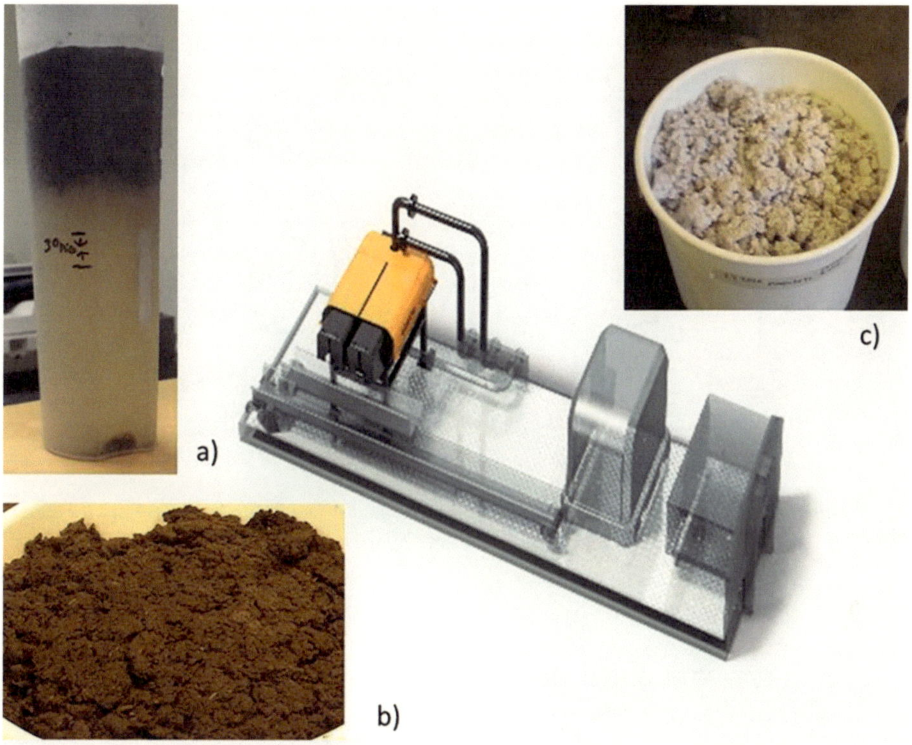

Figure 9.
Milston HERO 1 slurry separation capacity up to 75 m^3/h; a) pulp sludge left above, b) separated pulp sludge left beneath, c) separated paper sludge right above.

flow stick gently on the spinning gas bubbles and are transferred to the cell surface. The Vortex Flotation Separation by SansOx is named as VoxFlot. The concept and principle of the VoxFlot are presented in **Figures 10–12**. The flotation cell is compact and small in size. The bubbles are generated evenly by OxTube into the water flow before entrance to the cell and they function all the way with high collision probability and attachment.

The liquid is treated before the vortex flotation phase in the VoxFlot cell by the OxTube of SansOx, **Figures 10** and **11**. The OxTube treatment consists of ingredients separation, molecular activation, bubble generation, chemicals feed and mixing. Aeration and additional compressed air feed if needed for additional air bubbles can be performed. The treatment is done continuously and fast in terms of seconds by the OxTube process that mixes chemicals, e.g., coagulants, dissolves gases, e.g., oxygen into the liquid, and generates lots of flocs and bubbles of air and reaction gases directly into the flowing liquid. Meeting or collision probability of bubbles and particles is high in the tube condition of OxTube. The treatment supports flotation theory referred briefly in Chapter 2 in all its aspects.

3.4.1 Superior performance of VoxFlot

The design principles of the VoxFlot are based on the flotation theory presented in Chapter 2, and experiences of the new and present systems, and particularly verified separation and dissolving performance of the OxTube. A special focus has been set on the collision probability, bubble generation, ingredients separation, molecular

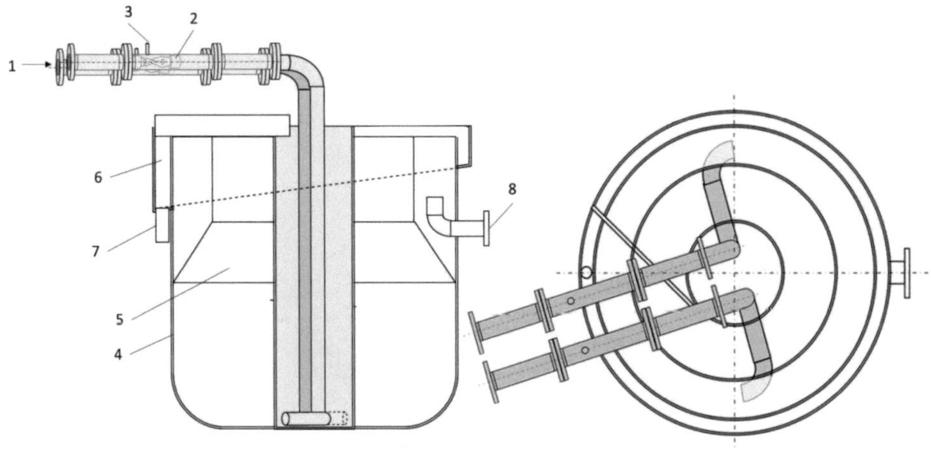

Figure 10.
VoxFlot flotation; 1. Liquid intake, 2. OxTube bubble generator and dissolver, 3. Air intake, 4. Flotation cell, 5. Bubble trap, 6. Foam collector, 7. Foam outtake, 8. Cleaned water outtake.

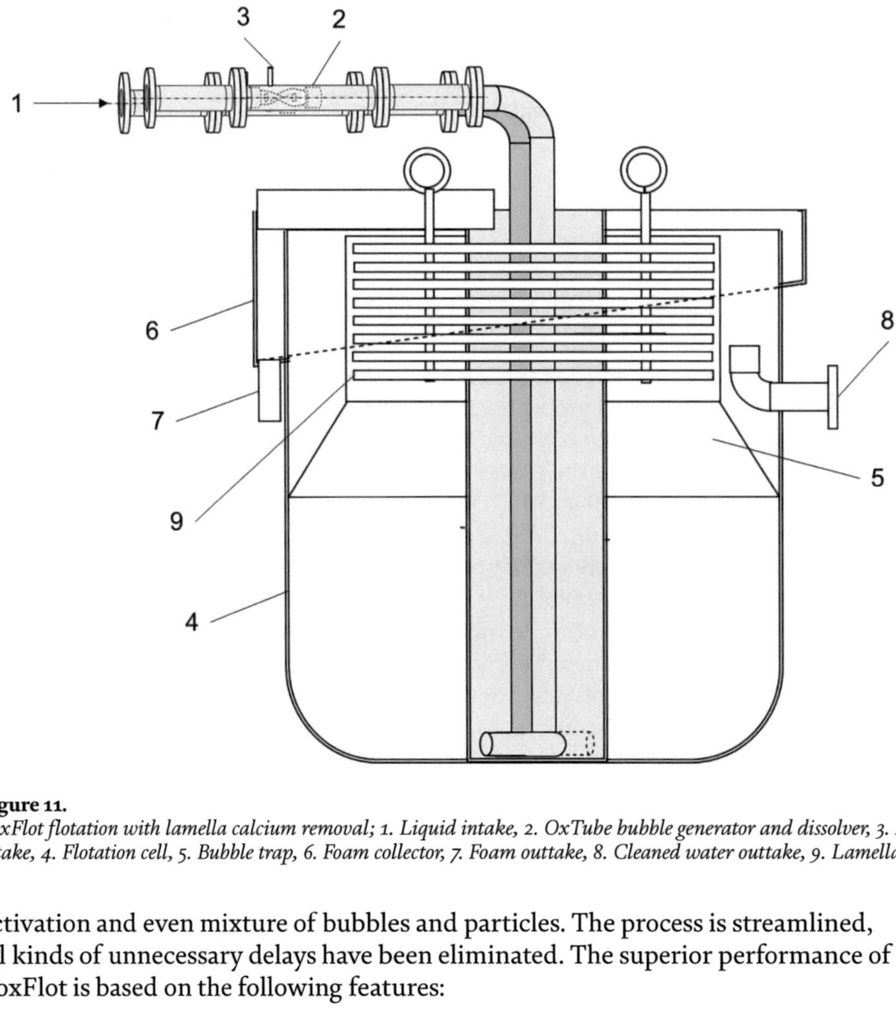

Figure 11.
VoxFlot flotation with lamella calcium removal; 1. Liquid intake, 2. OxTube bubble generator and dissolver, 3. Air intake, 4. Flotation cell, 5. Bubble trap, 6. Foam collector, 7. Foam outtake, 8. Cleaned water outtake, 9. Lamella.

activation and even mixture of bubbles and particles. The process is streamlined, all kinds of unnecessary delays have been eliminated. The superior performance of VoxFlot is based on the following features:

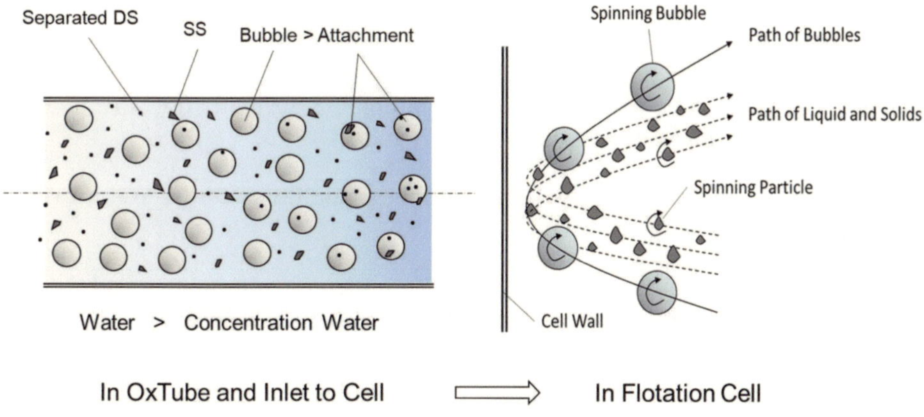

Figure 12.
OxTube generates bubbles and even mixture in the hermetic tube condition. Meeting probability of particles and bubbles is high. The vortex flow into the cell through the bubble trap causes a spinning effect of bubbles and solids that increases attachment.

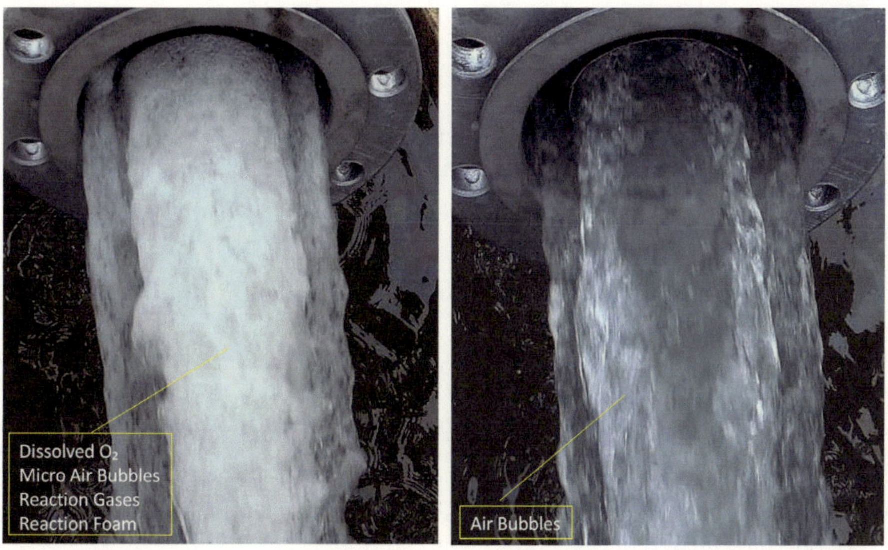

Figure 13.
Left OxTube aerated water flow and right non-OxTube aerated water flow. OxTube aeration creates high concentration of dissolved O_2, immediate reaction and huge amount of bubbles of 0.1 mm diameter in average. The surface tension and viscosity of the liquid is reduced significantly [7].

- Bubbles are generated in OxTube in the liquid flow evenly which enhances the meeting or collision probability of bubbles and particles, **Figures 4, 12**, and **13**.
- OxTube separates dissolved ingredients and activates molecules which increases performance, and decreases energy and chemicals consumption, **Figure** 7.
- Chemicals if needed can be fed in OxTube together with air or separately.

- The bubbles are evenly distributed, number of the bubbles is huge and their size is small and controllable, **Figures** **8**, **12**, and **13**.

- There are mixture of air and reaction gas bubbles in the liquid flow that increases attachment performance; **Figures 4**, **8**, **12**, and **13**.

- Additional compressed air can be fed into the liquid flow in order to generate larger bubbles.

- Surface energy of the solid/gas interface Υ_{sv} is increased by bubbles and the spinning effect caused by the vortex flow; the attachment of solid to gas is increased; Eq. (1) and **Figure 12**.

- Surface energy of the solid/liquid interface Υ_{sl} is reduced by gas dissolving; surface tension and viscosity of the liquid are lowered by 10 to 14 percent, and the attachment of solid to gas is increased; Eq. (1), test runs and **Figures 5–12**.

- The vortex flow turns bubbles and solids toward the main stream that reduces impact of shear and gravitational forces on the detachment; **Figure 12**.

- The collection efficiency of the flotation process (flotation recovery R) is increased by high collision probability N_c, small bubble size d_b, and high density of air and reaction gas bubbles; **Figures 12** and **13**, and Eq. (2).

- The vortex flow in the flotation cell generates the spinning effect of the bubbles and solids that increases the attachment and collision performance, **Figure 12**.

3.4.2 Main economic benefits of VoxFlot

VoxFlot flotation system has significant economic benefits compared to the present systems as follows:

- Bubble generation is performed in the main flow in OxTube just after the water intake; compressed air is reduced or not necessary which reduces energy consumption.

- The flotation cell size is reduced into a half or more due to the vortex flow and bubble trap which reduces capital costs and floor space, **Figures 10–12**.

- The vortex flow keeps the flotation cell clean for longer which reduces service costs.

- There are no rotational nor moving elements, e.g., impellers in mixing/dissolving.

- OxTube treatment and the vortex flow prevent growth of bacteria population (**Table 2**), BOD, fermentation and sediment retention on the bottom and wall.

- COD is reduced fast by the efficient aeration, reasonable high oxygen concentration, and by the continuous treatment process.

- Lamella for Ca removal can be integrated into VoxFlot Unit, as shown in **Figure 11**.

- The main economic benefits are summarized in **Table 1**.

4. Conclusions

Water itself is a clean element and present everywhere in the natural processes and ecosystems. It is heavily loaded in process and wastewaters as well as in the nature on earth and ecosystems. The water can purify itself or leave its load by evaporation powered by sun energy. Water circulation can be said as the largest transportation system globally.

The VoxSton system removes many ways the entire or partial load of various water matrices and treats the water for recycling and discharge to nature. It is an integrated water clarification system that separates soluble ingredients, activates molecules, generates flotation bubbles evenly in the water flow, presses solids for recycling and treats the water for recycling and discharging to nature.

The VoxSton industrial process and wastewater treatment system is efficient and it can in the most cases treat the liquid from the process to finish continuously or daily basis. It can be installed directly into the liquid flow without equalization and precipitation.

The VoxSton or its parts can be applied in purification of various water matrices. It has been applied in water intake, industrial processes, various wastewater purifications, fish farms, refreshing of natural waters and pretreatment at water plants. The most challenging cases have been removal of pharmaceutical residues from city wastewater and radon from intake of ground water at water plants, and disinfection of a hospital wastewater.

The new process water treatment system VoxSton is an integrated and fast. It saves in energy, chemicals and space as well as in operational costs roughly by 50 percent compared to present advanced systems. The capital costs are estimated to be lower too.

The main benefits of the VoxSton compared to present industrial water treatment systems are presented in **Table 1**.

Note

VoxSton concept for process waters of P&P Industries has been published first time in 2016 in China [8, 9]. The basic concept remains the same but its subsystems of OxTube and VoxFlot are developed further applicable for many other water matrices.

Author details

Juhani Pylkkanen
SansOx Ltd., Lahti, Finland

*Address all correspondence to: juhani.pylkkanen@t2o.fi

References

[1] Kawatra SK. Flotation Fundamentals. Michigan Technological University, Department of Chemistry. MTU Chemistry; 2015. p. 6

[2] Nguyen AV. On modelling of bubble–particle attachment probability in flotation. International Journal of Mineral Processing. 1996;**53**:225-249

[3] Shahbazi B, Rezaib B, Koleinic SMJ. Bubble–particle collision and attachment probability on fine particles flotation. Chemical Engineering and Processing. 2010;**49**:622-627

[4] Laakso V. Test Results on Diploma [Thesis]. Oulu University, Faculty of Technology; 2014

[5] Pylkkanen J. OxTube integrated water clarification for removal of pharmaceutical residues and disinfection. In: IWA The International Water Association 'Ecotechnologies for Wastewater Treatment' EcoSTP2021, Milan. 21-25 June 2021

[6] Ikonen J et al. Presence and reduction of anthropogenic substances with UV light and oxidizing disinfectants in wastewater—A case study at Kuopio, Finland. Water. 2021;**13**:360. DOI: 10.3390/w13030360

[7] Laakso V. Sompasaari, Finland: Customer Project of SansOx Ltd; 2015

[8] Pylkkanen J. New Pulp and Paper Process Water Treatment System - VoxSton. Vol. 35. No. 12. China Pulp & Paper; 2016. pp. 66-70 in Chinese with English abstract

[9] Pylkkänen JH. VoxSton – New process water treatment system. China Paper Newsletters. 2016;7, New Products/ Technology:49-51 in Chinese

Chapter 12

Microorganisms and Microbial Communities in Bioelectrochemical Systems for Wastewater Bioremediation and Energy Generation

Lina María Agudelo-Escobar and Santiago Erazo Cabrera

Abstract

Water resource sustainability is a critical global concern, leading to extensive scientific research. Proposed alternatives for wastewater effluent use include the promising Bioelectrochemical Systems (BES) that not only treat wastewater effectively but also generate electricity, produce biofuels, and synthesize valuable compounds through integrated microbial and electrochemical processes. BES research aims to enhance device design and develop superior electrochemical materials for optimal performance. The efficiency of treatment and energy co-generation depends on the metabolic characteristics of microbial communities responsible for oxidation-reduction processes in wastewater. The diversity of these communities, along with electron transport mechanisms and metabolic pathways, significantly impacts BES functionality and effectiveness. This study focuses on microorganisms in various BES setups, presenting their electrochemical performance. It compiles data on microbial ecology, emphasizing controlled communities and model microorganisms from wastewater treatment systems. The study highlights the scarce research on native microbial communities for agroindustrial wastewater. Its main goal is to consolidate information on microorganisms with electrogenic capacity, demonstrating their potential in different bioelectrochemical systems. These applications can transform wastewater bioremediation and enable the production of green energy, biofuels, and high-value compounds.

Keywords: bioelectrosystem, microbial community bioelectrosystem, microbial community MFC, microbial community MEC, native microbial community bioelectrosystem

1. Introduction

The development of humanity has led to an exponential increase in the consumption of natural resources and the generation of waste derived from anthropogenic activities. One of the most consumed resources is water. It is estimated that in the world the consumption for domestic, agricultural, and industrial activities is greater than 4000 km^3 of water per year [1]. As a consequence, a large amount of wastewater with a high environmental impact has been generated. According to UNESCO: "in low-income nations only 8% of domestic and industrial wastewater is treated" [2]. Wastewater treatment systems are classified into: (I) *mechanical or primary processes:* settling, screening, sedimentation; (II) *biological or secondary processes*: aerobic and anaerobic microbial metabolism of organic compounds; and (III) *physicochemical or tertiary processes*: advanced oxidations, filtrations with ionic charges, chlorination, ozonation, among others [3]. One of the most widely applied strategies for the purification of organic matter in wastewater is carried out through Wastewater Treatment Plants (WWTP), which combine primary and secondary methods [4]. The implementation of WWTPs is costly and low-income countries, whose productive activities correspond mainly to the exploitation of raw materials, and do not incorporate WWTPs into their agricultural or industrial production processes. These treatments are carried out on domestic and industrial wastewater, mainly in urban areas, and according to recent data, only an average of 20% of the wastewater produced are treated in the world before being returned to the ecosystems [5]. In recent years, research has been carried out for the non-conventional biological treatment of wastewater using bioelectrochemical systems (BES) [6–9].

The BES can be wastewater treatment systems adapted and designed based on a microbial metabolism that, through reactions of oxidation and reduction of organic matter, carry out the purification of contaminated water, producing at the same time electrical energy, biofuels, or other chemical compounds of interest [6, 8, 10]. There is a general classification of BES based on their application; (a) *microbial fuel cells (MFCs)*: developed mainly for electricity production, but have recently been used for wastewater treatment and as biosensors for the detection of toxic chemicals; (b) *microbial electrolysis cells (MEC)* developed for the electrochemical production of biofuels such as hydrogen or methane, and finally, (c) *microbial electrochemical technologies (MET)*: used to desalinate water or produce chemical compounds such as hydrogen peroxide [11]. The configuration of the BES and the origin and differentiation of microbial communities to be used in these systems depend on the composition of the substrate and the compounds of interest that can be generated from a specific cellular metabolism. The different types of BES are designed and implemented according to the objective of wastewater treatment, and some are coupled to conventional secondary treatments, such is the case of wetlands connected to MFC, in which symbiotic microorganism-plant relationships are used as a strategy of optimization in the efficiency of reduction of organic matter content [12, 13]. BES have also been implemented for the treatment of residual water or a fixed effluent, with cascade feeding or simultaneous feeding under continuous operation, in which different microbial communities have been used in each stage or system to increase the production rate energy and decontamination levels [14, 15]. Moreover, various conventional wastewater treatment technologies have been effectively combined with BES. For instance, MFCs have been coupled with stabilization ponds, while BES have been integrated into aerobic or anaerobic treatment systems specifically designed for nutrient (nitrites/nitrates, sulfates) or heavy metal (Fe, Cu, Cd, etc.) removal purposes [7, 16–23].

For the design and adaptation of BES technologies in the unconventional treatment of wastewater, it is necessary to know individually the metabolic responses and the enzymatic machinery of the microorganisms present in the system. Currently, microorganisms of different genera with the ability to degrade organic matter and with potential for application in bioelectrochemical systems have been described. Mainly bacteria and archaea have been studied, although there are reports of electrogenic activity in eukaryotes, such as yeasts [10, 24, 25]. On the other hand, in a significant amount of research, detailed descriptions of the mechanisms of mass and energy transport in the cell have been made; and in recent years, some studies have been registered that seek to adapt diverse microbial communities to BES [12, 23, 25–30]. However, in the case of microbial communities, due to the complexity of the interactions, it is difficult to establish the metabolic processes and mass and energy transport mechanisms involved in bioelectrochemical processes. This review presents a consolidation of the information on the main microorganisms used in BES, the metabolic processes they carry out for the degradation of organic matter and the use of substrates for the cogeneration of energy and the production of other metabolites of interest, and the possible interactions between the microorganisms of the microbial communities in these bioelectrochemical systems.

2. Microorganisms in bioelectrochemical systems

Microorganisms in bioelectrochemical systems are responsible for the transformations of the substances or compounds present in the organic matter or substrate, since through their metabolic processes they carry out oxidation and reduction reactions, which in turn constitute the mechanisms of energy generation and high value compounds. Its implementation for wastewater treatment and cogeneration of energy or compounds of interest depend on the interaction between the physicochemical characteristics of the substrates and the metabolic capacity associated with microbial diversity [7, 14, 19, 23, 31–36]. In the case of MFCs, the main metabolism described corresponds to the redox (electrogenic) degradation of organic matter. In MEC systems, microorganisms with a metabolism of methanogenesis and hydrogenogenesis related to the production of biofuels have been recorded [28, 37–39].

The biotechnological capacity of cells to produce energy called "electrogenic potential" was first described in 1911 by Potter, who evaluated the electrogenic potential of *Escherichia coli* and *Saccharomyces cerevisiae* [40]. Advances in molecular biology and microscopic characterization techniques have allowed the identification of multiple microorganisms and their metabolism, as well as their ability to generate energy. Some main microbial groups have been identified, with species known as model electrogenic microorganisms, which present adaptive characteristics of great interest in BES, such as the presence of some cellular structures (pili and nanotubes) and the excretion of mediating substances that act as direct and indirect electron transport mechanisms in the different bioelectrochemical systems [41–44]. The predominant microorganisms found in these systems belong to the Bacteria domain, mainly the phyla Proteobacteria, Chloroflexi and Firmicutes [13, 26, 33, 45–47]; and the genera *Geobacter* sp., *Shewanella* sp., and *Pseudomonas* sp.

3. Model electrogenic microorganisms

The Electrogenic model microorganisms have a remarkable electron transfer capacity, exhibiting physical and metabolic characteristics that allow the production of usable energy from the consumption of organic matter [10, 11, 48]. The bacterial genus *Geobacter* sp. it has been reported as the one with the highest electricity production rates in bioelectrochemical systems [32, 43, 49, 50]. Several studies have reported this electrogenic capacity; J Chen and his collaborators established that in a double-chamber MFC fed with synthetic wastewater, in which power density values of 71.6 mW/m^2 and 65.4 mW/m^2 were achieved at the anode and the cathode, respectively; the predominant genus in the microbial community developed from activated sludge from a WWTP was *Geobacter* sp. [51]. Similarly, Shen et al. reported that in an MFC designed for the co-degradation of phenolic compounds, a production of 267.2 mW/m^2 was reached with the predominance of *Geobacter* sp. within the identified bacterial genera [52]. Also, Paitier et al., in a bottle-type open cathode MFC, characterized the microbial community and identified the predominance of *Geobacter* in the formation of the electrogenic biofilm on the anode during one month of operation, reaching an energy performance of approximately 50 mW/m^2. The authors reported a predominant exponential growth of *Geobacter* during the first 5 days of biofilm formation [50]. In the report by McAnulty et al., an MFC that converts methane directly into an electric current, the capacity of *Geobacter sulfurreducens* to produce electrons from acetate was evaluated. A maximum power density of 160 mW/m^2 was obtained [53]. On the other hand, considering different types of ESP, in a MEC dedicated to methane production, Zhao et al. evaluated the performance and interaction between *Geobacter* sp. and *Methanosaeta* sp. in the reduction of CO_2 to CH_4, reaching 3017.6 mL of CH_4 in 51 days of operation; the authors suggest from their results that *Geobacter* develops a key metabolism as a precursor to methanogenesis in bioelectrochemical systems [41]. However, although *Geobacter* sp. is a microorganism with a great electrogenic potential, several studies indicate that it is not the only relevant microorganism in BES [11, 19, 42, 45, 52, 54–57]. For example, for the development of electrochemical technologies with unconventional materials, such as ceramic biocathodes as those made of terracotta, the predominance of fermentative bacteria has been identified, which exhibits metabolic capacities different from *Geobacter*, proving that there is an influence on the interaction between the microorganisms and materials used in BES. It has been suggested that *Geobacter* interacts more efficiently with conductive materials such as carbon felt [58].

For the bacterial genus *Pseudomonas sp*. that it is present in almost all ecosystems, the electrogenic potential associated with its metabolism has also been identified; and it has been established that it plays an important role in electron transport in BES [33, 42, 52, 59]. *P. aeruginosa* stands out as the model microorganism. However, other species within this same genus with high electrogenic capacity have also been reported. In the study by Ilamathi et al. of a single-chamber MFC, the power densities determined for a *P. aeruginosa* strain were 601 μW/m^2 and 323 μW/m^2 for a *P. fluorescens* strain [60]. Arkatkar et al. obtained a current of 0.036 mA in a double-chamber MFC inoculated with *P. aeruginosa* [42]. In turn, non-conventional MFC designs have been tested using this microorganism. For example, Zhang et al. reported a power density of 3322 mW/m^2 for an MFC with a cylindrical reactor design [61]. Similarly, in the comparative study by Bagchi and Behera, this microorganism was used to evaluate the influence of the manufacturing material on an MFC, reaching power density values of 0.96 W/m^3 for an MFC made of plastic and 0.69 W/m^3 for the MFC built with ceramic materials [52]. Also, alternative strategies for the operation of bioelectrochemical systems using *P. aeruginosa* have been evaluated. For example, Qiao et al. evaluated the effect of the implementation of inhibitors such as sodium hyaluronate, on the formation of biofilm on the anode in an MFC inoculated with *P. aeruginosa*, indicating a current density of 4.8 μA/cm^2 [62]; Yong et al. evaluated the mixed aerobic-anaerobic operation of an MFC inoculated with *P. aeruginosa*, obtaining a current density of 99.8 mA/cm^2 [16].

Similarly, *Shewanella sp.*, recognized as a facultative bacterium, has been reported as an exoelectrogenic microorganism [32, 57, 60, 63]. The presence of these microorganisms in nutrient-limited cultures has been described, a characteristic that may favor their implementation in BES [26]. This bacterial genus has been widely identified in studies with bioelectrochemical systems in which various microbial communities have been used as inoculum. In a comparative study carried out by Hassan et al., in a double-chamber MFC, they compared microbial communities in petrochemical and domestic wastewater, reaching current densities of 156 mA/m^2 and 123 mA/m^2, respectively. The authors reported the active presence of *Shewanella* sp. in the cells in which wastewater of domestic origin was used [64]. In particular, the *Shewanella oneidensis* MR-1 strain has been identified as an electrogenic reference, as reported by Wang et al., in their study of a non-conventional laminar flow MFC fed with synthetic wastewater and glycerol, in which a potential for 16.05 mW [65]. In another work by Wang et al., the mutualistic interaction between *Shewanella oneidensis* and *Escherichia coli* in a double-chamber MFC fed with a synthetic substrate, a current density of 2.0 μA/cm^2 was achieved [44]. Likewise, Hirose et al. reported a power density of approximately 250 mW/m^2 in a single-chamber electrolysis cell fed with a synthetic substrate, electrogenic activity that was attributed to the metabolic response of *S. oneidensis* MR-1. As stated by the authors, this microorganism possesses molecular mechanism that enables it to detect electrode potentials and effectively regulate its catabolic pathways, so this capability allows the microorganism to adapt and optimize its metabolic activities in response to the electrical conditions presented by the electrodes [66]. This molecular mechanism provides valuable insights into the intricate processes underlying the bioelectrochemical interactions and further enhances our understanding of the microbial behavior within the context of electrochemical systems.

On the other hand, another variety of electrogenic bacteria has been described in BES, and some genera are distributed depending on the origin of the wastewater or even depending on the differentiation of the microbial communities through the operation time of the system, within the frequently described bacterial genera, *Clostridium* sp. [4, 28, 49, 67, 68]. As an illustration, the *Clostridium beijerinckii* strain was subjected to evaluation in both a H_2 fermentation reactor and an open cathode MFC, with the feedstock being port drainage sediment, in this study yielded impressive results, including the production of 104 mmol/L of H_2, 5 mmol/L of acetate, 33 mmol/L of butyrate, 3 mmol/L of lactate, and 1 mmol/L of ethanol; additionally, the power density achieved was recorded at 1.2 W/m^2 [28]. These findings demonstrate the potential of this strain and highlight the feasibility of harnessing its capabilities for efficient and sustainable bioenergy production. Another bacterial genus studied has been *Desulfovibrio* sp.; that in a double-chamber MFC it reached an electrical production of 185 mW/m^2 and in a MEC, where it was the predominant bacterial genus, a hydrogen production of 0.28 m^3-H_2/m^3-d was reached [26]. Also, reports have been made of the electrogenic activity of *Klebsiella* sp. in a double-chamber MFC, and from finding in a symbiotic relationship of *Klebsiella variicola* and *P. aeruginosa*, a power density of 14.78 W/m^3 was reached [69]. Additionally, the presence of *Acinetobacter* sp., *Bacillus* sp., *Thiobacillus* sp., *Desulfomonas* sp., among other genera, has been reported. Some of the archaea studied according to their electrogenic potential are *P. furiosus* with an electrical production, in terms of power density, of 225 mW/m^2. The current density produced by *Ferroglobus placidus* and

Geoglobus ahangari has been reported to be 680 mA/m^2 and 570 mA/m^2, respectively. The study of some archaea in BES has also been carried out in mixed communities including bacteria, where volume power density values between 2138 mW/m^3 and 6924 mW/m^3 have been recorded [4, 11, 53, 62]. Recently, it has been reported that some genera of archaea, such as *Methanosaeta* and *Methanosarcina*, have symbiotic interactions with other electrogenic microorganisms, such as *Geobacter* sp. [4] which generates a particular interest in its presence and functionality in the microbial interactions of ecosystems in BES.

The microbial community present in bioelectrochemical systems is not exclusively constituted by electrogenic microorganisms, they coexist with a great variety of microorganisms that through their metabolic processes and enzymatic machinery carry out the degradation of complex substrates such as polymers, carbohydrates, proteins, hydrocarbons, fats and oils, among others; contributing to obtaining of precursor substances for electrogenic metabolism, which significantly influence energy productivity in BES [15, 22, 49, 64, 70–75]. The presence of different microorganisms in bioelectrochemical systems indicates the important ecological role they play in nature, as well as suggesting their participation in bioanode formation and electron transport in bioelectrochemical systems [47, 68, 76, 77]. However, a considerable amount of BES studies focusses primarily on the design and configuration of devices developed as MFCs or MECs, and research on microorganisms and their interactions is relatively minor. The study of microbial communities can be considered more complex due to the interactions that occur between microorganisms when they are active in BES. In these systems, ecological successions or interdependent mixed metabolisms could occur between electrogenic microorganisms and those that do not present electrogenic activity. Knowledge of the symbiotic relationships of microbial communities in BES would allow the adaptation of this technology for the treatment of a wide variety of substrates and environments, including wastewater from multiple industries. **Table 1** presents a list of electrogenic microorganisms reported by various authors, which were evaluated in the different BES using multiple substrates.

Microorganisms	**Type of BES**	**Substrate**	**Electrochemical response**	**References**
Geobacter sp.	MFC	Activated sludge	71.6 mW/m^2	[51]
Geobacter sp.	MFC	Phenolic waste water	267.2 mW/m^2	[75]
Geobacter sp.	Bottle-type air cathode single chamber MFC	Waste water + acetate	50 mW/m^2	[50]
Geobacter sulfurreducens	MFC-MEC (CH_4)	Synthetic wastewater (sodium acetate)	160 mW/m^2	[53]
Geobacter sp. and *Methanosaeta* sp.	MEC (CH_4)	Activated sludge + glucose	3017.6 mL CH_4	[41]
Pseudomonas aeruginosa	Single-chamber MFC.	Modified King's B médium + azo dye: Orange 16 (RO-16) and black 5 (RB-5)	601 μW/m^2	[78]
Pseudomonas fluorescens			323 μW/m^2	

Microorganisms	Type of BES	Substrate	Electrochemical response	References
Pseudomonas aeruginosa	Two-chamber MFC.	Synthetic wastewater (glucose)	0.036 mA	[42]
Pseudomonas aeruginosa	Non-conventional MFC (cylinder).	Anaerobic activated sludge + glucose	3322 mW/m^2	[9]
Shewanella oneidensis MR-1	MFC.	Synthetic wastewater (glycerol)	16.05 mW	[79]
Shewanella sp.	Two-chamber MFC.	Industrial (petrochemical) and domestic waste water	156 mA/m^2 123 mA/m^2	[64]
Shewanella oneidensis MR-1	Single-chamber MFC.	LMM + NaCl	250 mW/m^2	[66]
Shewanella sp. and *E. coli*	Two-chamber MFC (U shaped).	Saline solution + glucose	2,0 $\mu A/cm^2$	[44]
Clostridium beijerinckii	MEC	Port drainage sediment	104 mmol/L H_2 1.2 W/m^2	[28]
Desulfovibrio sp. (*Comunidad mixta*)	Two-chamber MFC.	Activated sludge + sodium acetate	185 mW/m^2	[26]
	MEC.		0.28 m^3-H_2/m^3-d	
Klebsiella variicola and *P. aeruginosa*	Two-chamber MFC.	Domestic waste water + POME	14.78 W/m^3	[69]
Geobacter sp. *Petrimonas* sp. *Erysipelotrichia* sp. *Escherichia* sp. *Shigella* sp.	Two-chamber MFC	Food waste sludge	422 mW/m^2	[56]
Clostridium sp. *Chthonomonas* sp.	Double-chamber BES	Active sludge	99 mW/m^2	[54]
Chloroflexi sp. *Thiobacillus* sp. *Desulfovibrio* sp. *Desulfomicrobium* sp. *Desulfobulbus* sp. *Desulfuromonas* sp.	GW-MFC	Active sludge	24.7 mW/m^2	[19]
Native bacteria	Compost soil MFC (CSMFC)	Compost (soil)	3.2 mW/m^2	[21]
Clostridium sp. *Paracoccus* sp. *Pseudomonas* sp. *Arcobacter* sp.	Pilot scale MFC (1-year operation)	Domestic sewage	2 mA	[80]
Desulfovibrio sp.	Two-chamber MFC	Simulated textile wastewater	258 mW/m^2	[30]
Citrobacter sp. *Thiobacillus* sp. *Thioclava* sp. *Corynebacterium* sp. *Arcobacter* sp. *Desulfovibrio* sp. *Geobacter* sp.	Two-chamber MFC	Domestic anaerobic and aerobic sludges and simulated wastewater (with RB19 dye)	85.2 mW/m^2	[18]

Microorganisms	Type of BES	Substrate	Electrochemical response	References
Methylobacterium extorquens	MEC	Synthetic wastewater (16 g/L succinate)	2.53 mM/g-wet cell/h (formate production)	[39]
Rummeliibacillus sp. *Burkholderia* sp. *Enterococcus* sp. *Clostridium* sp.	Cylinder-type air cathode single chamber MFC.	Food waste hydrolysate.	0.97 kWh/kg COD	[81]
Azospirillum sp. *Petrimonas* sp. *Pseudomonas* sp. *Geobacter* sp.	Three-electrode single chamber BES	Synthetic wastewater (1 g/L acetate)	2.81 mW/cm^2	[74]
Native bacteria	Two-chamber MFC	Domestic waste water	1856 mW/m^2	[82]
Native bacteria	Single-chamber air-cathode Ceramic MFC	Activated sewage sludge	189 mW/m^2	[83]
Thauera sp. *Pseudomonas* sp. *Candidatus Methylomirabilis oxyfera*	Two-chamber MFC	Anaerobic granular sludge	98.7 mW/m^2	[47]
A. woodii *M. maripaludis*	Two-chamber MEC	Synthetic medium	Electromethanogenesis 0.1–0.14 μmol/cm^2 h^1 (–400 mV) Acetate 0.21–0.23 μmol/cm^2 h^1 (–400 mV)	[84]
Parabacteroides sp. *Proteiniphilum* sp. *Catonella* sp. Clostridium sp.	Air-cathode MFC	Cow manure	44 mW/m^2	[85]
Cloacibacillus sp. *Proteiniphilum* sp. *Clostridium* sp. *Sphaerochaeta* sp.	Two-chamber MFC	Biogas slurry (corn stover)	296 mW/m^2	[70]
Native bacteria	Fixed-bed air-cathode MFC	Synthetic medium + activated sludge	1.149 mA	[7]
Azoarcus sp. *Nitratireductor* sp. *Ochrobactrum* sp. *Serratia* sp.	Single-chamber air-cathode MFC	Granular anaerobic sludge	22.8–112.7 mA/m^3	[86]
Native bacteria (rhizobacteria)	Integrated drip hydroponics MFC	Domestic waste water	31.88 mW/m^2	[21]
Anammox *Planctomyces* *Bacillus* sp. *Desulfovibrio* sp. *Geobacter* sp. *Nitrospira* sp. *Thauera* sp.	CW-MFC	Domestic waste water (sludge)	3714.08 mW/m^2	[12]

Microorganisms	Type of BES	Substrate	Electrochemical response	References
Geobacter sp. *Pseudomonas* sp.	1000 L modularized MFC system	Municipal waste water	7.58 W/m^2	[87]

Table 1.
Summary of electrogenic microorganisms reported.

4. Electron transport mechanisms in electrogenic microorganisms

The energy production in the BES is directly related to the chemical transformations of the organic matter present in the substrates fed in these systems [19, 75, 88, 89]. In the chemical reactions of oxidation and reduction, the mechanisms of electron transport (ET) are essential to ensure that the substances that are oxidized give up or transfer electrons to the substances that are reduced, that is, substances that gain electrons and process that is used in BES to achieve power generation [10, 35, 41, 46]. **Figure 1** shows a representation of the flow of electrons and protons resulting from cellular metabolism present in a bioelectrochemical system. Direct and indirect ET mechanisms have been identified and used for the general characterization and classification of electrogenic microorganisms [6, 11, 18]. Direct electron transport (DET) mechanisms refer to the physical contact between electrogenic microorganisms and the electrode, and direct contact between the outer membrane of the bacteria and the anode surface [24]. The indirect electron transport (IET) mechanism, also known as mediated electron transport (MET), occurs due to the presence of mediating substances that facilitate the flow of electrons toward the surface of the anode electrode [33, 35, 57]. These substances are produced by some microorganisms in their cellular metabolism, as occurs in the oxidation of compounds and the generation of fermentative products, such as hydrogen [22, 90, 91]; or they can be added to the systems to

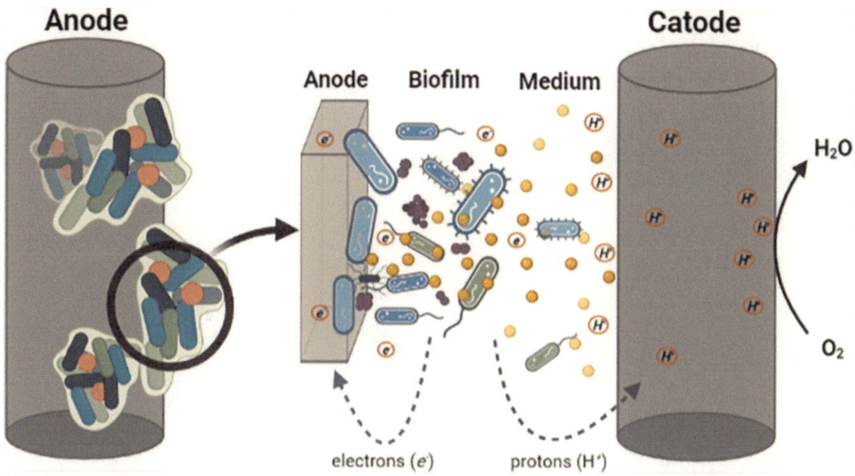

Figure 1.
Representation of the flow of electrons and protons resulting from cellular metabolism present in a BES (MFC double chamber).

facilitate the electron transport process [25]. The development of direct or indirect mechanisms for electron transport in BES depends on the enzymatic machinery of the electrogenic microorganisms used and on the experimental conditions such as the composition and concentration of substances of the substrate used in the operation of the bioelectrochemical system [42]. The indirect electron transport mechanism presents a lower efficiency in the production of electricity in the BES [6].

In electrogenic microorganisms such as *Geobacter* sp. and *Shewanella* sp., mainly DET mechanisms have been identified, associated with the formation of membrane-extending structures known as pili [26, 42]. The pili are short membrane extensions that establish direct contact with the outside of the cell or other microorganisms for the exchange, mainly, of ions, nutrients, and genetic material [10, 41, 52, 72, 75, 92]. **Figure 2** shows a schematic representation of the direct mechanism of electron transfer through a pili of *Geobacter* sp. The formation of these membrane extensions has been described as a mechanism of ecological adaptation of cells to the environment. Different studies have established that the genetic machinery for the appearance of pili is based on the expression of cytochrome C (OmcS) [85, 86]. Similar to *Geobacter* sp., the bacterial genera *Shewanella* sp. and *Rhodoferax* sp. possess protein complexes of type C cytochromes [8, 93–95].

On the other hand, for *Shewanella* sp. the formation of nanotubes that allow the direct transport of electrons between the cell and its environment has been reported [96, 97]; and also, indirect transport mechanisms have been described with the presence of the group of quinones and quinol in the cytoplasmic membrane, which transfer electrons to extracellular acceptors such as minerals containing Fe (III), mediating substances that finally carry out the transport of electrons toward the electrodes of the BES [60]. **Figure 3** shows a schematic representation of both electron transport mechanisms for *Shewanella* sp.

Bacteria such as *Pseudomonas* sp. are included in microorganisms with indirect electron exchange mechanisms, as they excrete biomolecules to facilitate the transfer or transport of electrons [32, 33, 92]. For the bacteria *Pseudomonas* sp., and *Lactococcus*

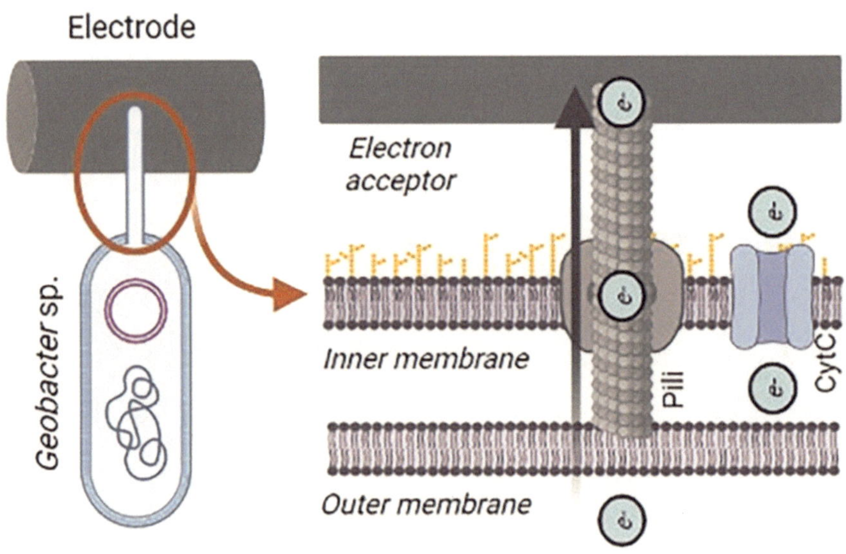

Figure 2.
Schematic representation of DET mechanism of Geobacter *sp. (pili and cytC).*

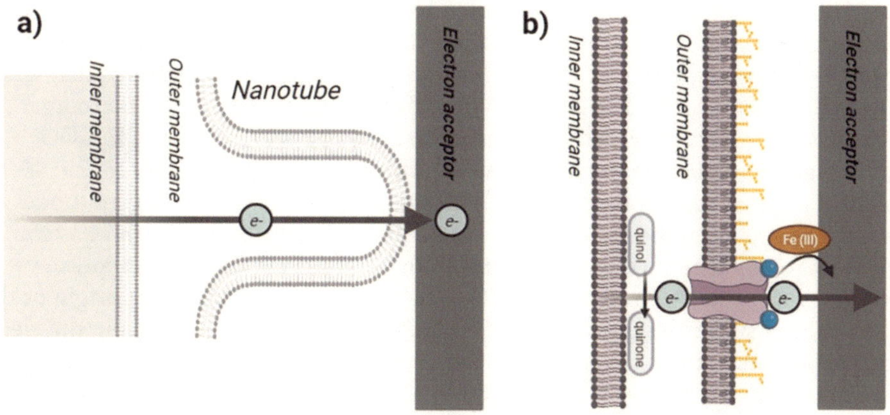

Figure 3.
Schematic representation of (a) DET (nanotubes) and (b) IET from Shewanella *sp.*

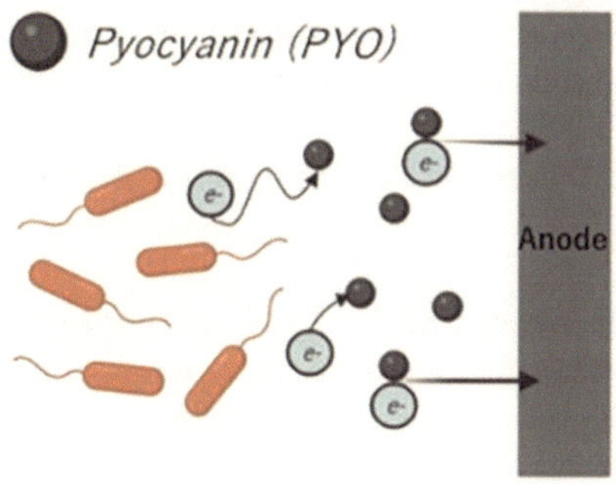

Figure 4.
Schematic representation of IET through phenazines (PYO) in Pseudomonas *sp.*

sp. the presence of phenazine-type pigments (cyclic hydrocarbons with the presence of two N groups), riboflavins and quinones, produced by secondary metabolic pathways, has been identified. Some examples of these pigments are pyocyanin and 2-amino-3-carboxy-1,4-naphthoquinone (ACNQ), mediating substances that carry out the electron transport process in BES systems [35, 52]. The metabolic performance of *P. aeruginosa* depends, to a large extent, on the presence of phenazines, since these substances have been reported to be involved in virulence, signaling, iron metabolism, and electron transport [35]. The phenazine of greatest interest in the study of bioelectrochemical systems is pyocyanin (PYO) due to its participation in the transfer of electrons at high catalytic reaction rates and low overpotential [61, 96, 98]. **Figure 4** shows a schematic representation of the indirect electron transfer mechanism for *Pseudomonas* sp.

5. Microbial metabolism in bioelectrochemical systems

Some main metabolic routes that electrogenic microorganisms carry out to achieve the degradation of compounds present in the substrates in BES have been identified [66, 92, 99]. In general terms, it can be considered that the oxidation of organic matter is carried out by microorganisms for cell growth and maintenance. In this process, ions associated with the energy transfer mechanisms are generated, as described in **Figure 5**, and correspond respectively to the anode and cathode reactions that take place in the BES [48, 89]. The metabolism and electrogenic activity of microorganisms on controlled substrates, especially synthetic substrates with acetate ($CH_3COO—$) or simple sugars as carbon source, has been extensively investigated [6, 9, 18, 20, 49, 85]. Different authors have evaluated the electrogenic capacity of the so-called model microorganisms from these substrates to establish efficiencies in electron transfer processes and to identify the reaction stoichiometry present in the metabolism of microbial consortia [4, 64, 84, 100, 101]. However, the diversity in the origin of the wastewater or substrates used in the BES generate significant variations in the electrogenic potential of the system due to the composition of the substances present in these substrates; for example, it has been established that in domestic wastewater predominates oxidizable organic matter [9, 55, 80, 102]. In industrial wastewater, the composition depends on the productive activity from which it comes; and components such as fats, dyes, petroleum derivatives and a large number of recalcitrant contaminants have been identified [14, 17, 30, 56, 73, 75, 89, 103]. In substrates of agricultural origin, a variety of complex carbon compounds and nitrogen compounds derived from pesticides and fertilizers have been found [47, 89].

$$CH_3COO^- + 2O_2 \rightarrow 2HCO_3^- + H^+ \ (\Delta G = -847{,}6\ kJ/mol)\ (E^\circ = 0{,}805\ V\ vs\ SHE) \tag{1}$$

The usable energy in bioelectrochemical systems depends on organic compounds that can be biochemically transformed through redox reactions, giving rise to different substances and metabolic intermediates. Eq. (1) presents a simplified reaction of the transformation of a simple substrate such as acetate, which is achieved through the electrogenic metabolism of microbial communities in a BES [47, 93]. However, in a complex substrate such as domestic or industrial wastewater, in which the presence of multiple compounds has been identified, there are many more intermediate reactions, as a product of the metabolic activity of the microorganisms involved in the oxidation

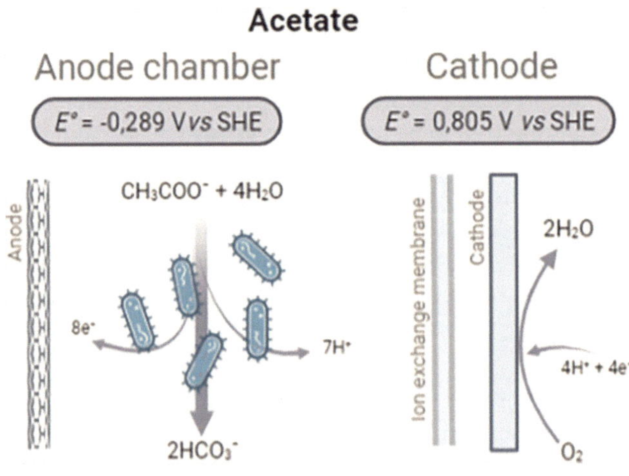

Figure 5.
Schematic representation of IET through phenazines (PYO) in Pseudomonas *sp.*

and reduction processes. For example, for MEC-type systems used for the production of methane or hydrogen, a series of complementary reactions known as methanogenesis and hydrogenogenesis occur [28, 37, 39, 41, 53, 70]. The reactions present some variations, since what is intended in these BES is the production of biofuels, so the oxidation of organic matter is carried out until obtaining methane gas (CH_4) or until obtaining biohydrogen (H_2) [7, 11]. In this type of system, there is generally a symbiotic relationship of anaerobic microorganisms, especially bacteria and archaea, which degrade organic matter and produce biofuels with an interdependent metabolism; that is, the microorganisms in the present community mutually benefit from the metabolic intermediates and by-products produced by the successive substrate transformation reactions [11, 25, 53, 84]. **Figure 6** presents the stoichiometry of redox and methanogenesis of organic compounds in a BES using CO_2 as carbon source [47].

Nutrients derived from nitrogen and sulfur from wastewater represent a particular interest for bioremediation processes due to their high impact on ecosystems. These compounds have been described as the main cause of the effects of eutrophication and pollution of natural water sources [19, 22, 71, 104]. The use of BES for the purification of nitrites, nitrates, and sulfates in wastewater has been shown to be a promising strategy to reduce the degree of pollution and co-generation of energy. An example is the study carried out by Mahmoudi and collaborators in which the performance of a bioelectrochemical reactor with two chambers (anaerobic and aerobic) was evaluated, in which a current generation of 0.841 mA and an ammonium and COD removal of 54.6% and 87.2% were observed. The authors concluded that the COD/N ratio is one of the parameters with the greatest influence on the nitrification process to control the growth of autotrophic and heterotrophic bacteria [7]. In some cases, the wastewater may contain other nutrients derived from sulfur, so a series of stoichiometric adaptations of electrogenic metabolism for sulfate reduction have been proposed [89]. Simplified metabolic reactions for nitrates and sulfates in BES are presented in **Figure 7**.

In nature, there are ecological successions and symbiotic relationships between microorganisms in which different metabolisms are expressed for the assimilation of complex nutrients and pollutants [23, 46, 84]. The expression of specific metabolic pathways could represent a strategy for the optimization of bioelectrochemical systems, since the electrogenic metabolism of a diverse microbial community would

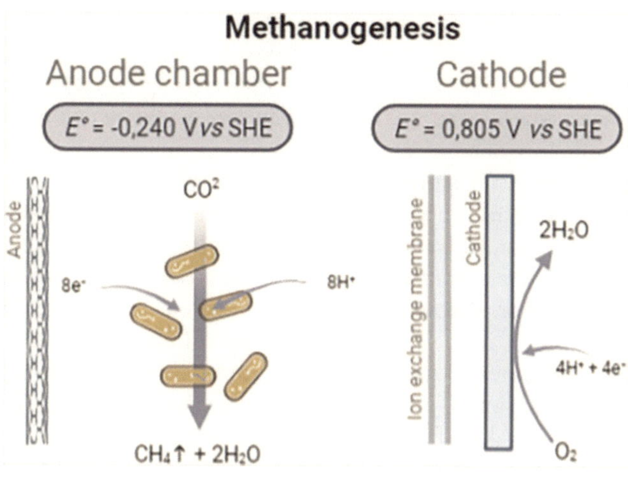

Figure 6.
Metabolic reactions of methanogenesis in BES.

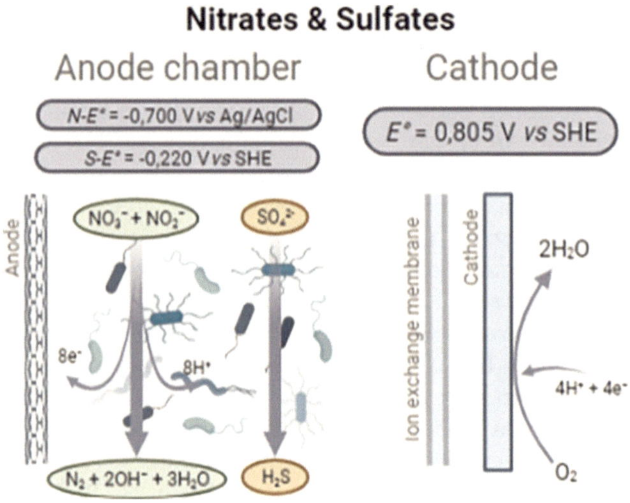

Figure 7.
Metabolic redox reactions of nitrites/nitrates (N) and sulfates (S) in BES.

allow an increase in the efficiency and scope of these systems for the treatment of different contaminated environmental matrices, which are constituted by recalcitrant organic and inorganic compounds [17, 48, 64, 72, 74, 75, 105].

6. Microbial communities in bioelectrochemical systems

Numerous studies report BES in which a great diversity of consortiums and microbial communities are used for the treatment of domestic and industrial wastewater for the degradation or reduction of polluting compounds such as organic waste, dyes, pesticides, phenolic compounds and petrochemicals; as well as for the production of metabolites with high added value such as volatile fatty acids, organic acids, biofuels, among others [8, 41, 50, 64, 80, 89, 106–108]. In some cases, the presence of "model" microorganisms is indicative of electrogenic activity. As reported by Liang et al., there is a presence of *Geobacter* sp. and *Pseudomonas* sp. in a 1000-L reactor, in which a production of 7.68 W/m^2 was obtained with synthetic wastewater and 3.64 W/m^2 with domestic wastewater [87]. However, the diversity of the microbial communities used in bioelectrochemical systems depends fundamentally on the origin or nature of the inoculum or the substrate used.

In the treatment of domestic wastewater in these systems, the presence of *Paracoccus denitrificans* and *Clostridium* sp., denitrifying bacteria with the capacity to metabolize sucrose, glucose, starch, lactic acid, and other sugars, has been reported [80]. Also, the genera *Enterobacter* sp. involved in the electrochemical reduction of oxygen and *Escherichia* sp. involved in the reduction of extracellular iron [36]. Urine has been evaluated as domestic waste for its treatment in BES, achieving a degradation of antibiotics such as ampicillin greater than 99%. In these investigations, the most abundant bacterial phyla identified were Proteobacteria, Firmicutes, Bacteroidetes, Chloroflexi, Actinobacteria. The most abundant bacterial genera were *Pseudomonas* sp., *Anaerolineae* sp. known for their cellulolytic capacity and their cellular adhesiveness; and the genera *Nitrospira* sp. and *Methyloversatilis* sp. associated with the metabolism of nitrogenous compounds (NH4/NO3) [76]. Xiao et al. reported the presence of the bacterial genera *Desulfovibrio*, *Lysinibacillus*, *Clostridium* and *Lachnoclostridium* in an MFC directly inoculated with wastewater, achieving greater efficiency in energy production of 543.75 mW/m^2 and a removal of nutrients such as total nitrogen with 76.15%, ammoniacal nitrogen with 83.23%, and a COD reduction of 75.59%, compared to that achieved with a defined inoculum or consortium. The authors indicate a higher microbial diversity in the MFC operated with wastewater [103].

In the case of industrial wastewater, bioelectrochemical treatments of wastewater from the food industry have been reported in the presence of *Geobacter* sp. and *Petrimonas* sp. known for their action to oxidize or ferment substrates such as brewery waste [56], or the metabolic activity of *Pseudomonas* sp. to degrade azo-type dyes derived from the textile industry or even for the production of pigments [78]. Some archaea of the order Methanomicrobiales and Methanosarcinales have also been identified in wastewater from the meat industry that could establish syntrophic interactions with acetate-oxidizing microorganisms and also develop exoelectrogenic activity [4]. In an investigation published by Marshal et al., in a BES with an inoculum from brewery wastewater, there were described the genera *Acetobacterium* sp., *Sulfurospirillum* sp. and *Desulfovibrio* sp. participating in functions such as fixation of carbon in the electrode, indication of a microaerophilic ecosystem and in the possible expression of protein hydrogenases, formate dehydrogenase and proton-to-hydrogen reducing cytochromes [109]. In a study describing the electrogenic behavior of controlled microbial consortia, strains of *Methanococcus maripaludis* and *Acetobacterium woodii* were used as model microorganisms for the hydrogenotrophic synthesis of methane and acetate; electromethanogenesis rates of 0.1–0.14 $\mu mol\ cm^{-2}\ h^{-1}$ at −400 mV (vs SHE) and 0.6–0.9 $\mu mol\ cm^{-2}\ h^{-1}$ at −500 mV were observed and acetate formation was observed at rates of 0.21–0.23 $\mu mol\ cm^{-2}\ h^{-1}$ at −400 mV and 0.57–0.74 $\mu mol\ cm^{-2}\ h^{-1}$ at −500 mV, respectively [84]. Some microorganisms such as *Desulfovibrio* sp. and *Thiobacillus* sp., *Azospirillum* sp. and *Mycobacterium* sp. have the ability to metabolize nutrients such as SO_4, PO_4, and NO_3/NO_2, and transform them into simple molecules and gases with less impact on the environment [19, 30, 74, 97]. Some cyanobacteria, photosynthetic microorganisms, have also been related to the production of intermediate compounds such as acetate and lactate, and it is suggested that they could be used for the denitrification of organic matter and co-generation of energy through exoelectrogens in BES [13].

On the other hand, the characterization of microbial communities in bioelectrochemical systems has been subject to studies evaluating the performance of BES that are subjected to different operational conditions. For example, Koók et al. reported that the predominance of the genera *Geobacter* sp., *Azospirillum* sp., *Castellaniella* sp., *Pandoraea* sp., *Treponema* sp., *Clostridium* sp. varies in relation to the external resistance used in an MFC. The authors report a higher yield, but a lower stability of the system when a dynamic system of external resistances is used, which is attributed to the acidification of the biofilm. These results suggest the adaptive evolution of the microbial community in response to the operational conditions of the BES [110]. In a mixed system known as MFC-CW that has been described as an MFC coupled to a constructed wetland, Tao et al. reported that the most abundant phyla were Proteobacteria, Cyanobacteria, Bacteroidetes, Acidobacteria, Chloroflexi, and Nitrospirae, in the case of the predominant genera *Geobacter* sp., *Desulfobulbus* sp., *Bacillus* sp., and *Geothrix* sp. In this mixed system, a nitrate reduction greater than 90% and a power density of 6.09 mW/m^2 were achieved when xylose was used as carbon source; in comparison, only a 10% nitrate reduction and a power density of 2.91 mW/m^2 were achieved when cellulose was used [13]. In another study by Ge et al., using an MFC-CW, reduction values of COD, NO^{3-}-N, total inorganic nitrogen, and total phosphorus of 71.9, 70.1, 63.2, and 91.2%, respectively, were reported; and current and power densities of 47.77 mA/m^2 and 6.74 mW/m^2, respectively [19]. The predominant electrogenic microorganisms found were the Proteobacteria, Actinobacteria, Acidobacteria, Bacteroidetes, and Chloroflexi phyla. The genera *Geobacter* sp. and *Dechloromonas* sp. associated with autotrophic denitrification and phosphate accumulation; the genus *Thiobacillus* sp. related to pyrite-driven autotrophic denitrification and the genera *Desulfovibrio* sp., *Desulfomicrobium* sp., *Desulfobulbus* sp., and *Desulfuromonas* sp. identified as sulfate-reducing bacteria for the oxidation of organic compounds such as biofilm-detached biomass and residual organic pollutants [19]. For the treatment of organosulfur compounds, Elzinga et al. used a type H MFC, and determined an abundance of the families *Halomonadaceae*, *Clostridiaceae*, *Eubacteriaceae*, and *Clostridiaceae*. The coulombic efficiency associated with this system in relation to different compounds was 1.7% for methanethiol, 3.4% for ethanethiol, and 5.0% for propanethiol and dimethyl disulfide [22]. Research in microbial ecology for BES has increased in the recent years. However, there are few reports focused directly on the study of native or autochthonous microbial communities in BES. Examples include the work of Revelo et al., who evaluated the effect of the inoculum in an MFC with anodic and cathodic chambers separated by a salt bridge. Anaerobic sludge from a solid waste treatment plant and sulfidogenic sludge and wastewater from artisanal tanneries were evaluated. The researchers reported reductions of 97.9% in organic matter content and 86.3% in chromium content, as well as a coulombic efficiency of 0.92% [111]. In a recent study, Agudelo-Escobar et al. reported the electrogenic capacity of native microorganisms in complex communities originating from wastewater from the coffee agroindustry, in an open cathode MFC with an unconventional design, achieving a reduction in organic matter of up to 70% and a power density of 21.16 W/m^3. The authors reported the presence of the genera *Clostridium* sp., *Cohnella* sp., *Aneurinibacillus* sp., *Candida* sp., *Gluconacetobacter* sp., *Clostridium* sp., *Acetobacter* sp., *Bacillus* sp., *Weissella* sp., *Leuconostoc* sp., and *Lactobacillus* sp., microorganisms known mainly in anaerobic or microaerophilic ecosystems, associated with the reduction of organic compounds and that they suggest are related to the electrochemical performance of the system [112]. Do et al. indicate that the bacterial community at the anode is mainly affected by the type of substrates used and ultimately this influences current generation [113]. It has been shown that current may be sensitive not only to COD concentrations, but also to the presence of aerobic and electrogenic bacteria, as well as nutrient ratios in water samples [114]. The success of the implementation of a BES system for the treatment of wastewater, the co-generation of energy and the generation of high value-added products, will be conditioned by the type and origin of the substrate used and by the choice of microorganism or microbial community, which will be responsible for carrying out the metabolic transformation and use of organic matter in the system. **Figure 8** proposes a consolidated scheme that groups the main components identified in the wastewater with the intermediate metabolites and final products that can be obtained through the main metabolic activities identified in the electrogenic microorganisms in the BES.

What is proposed is a route to identify the best bioelectrochemical alternative. To do this, the purpose of the implementation must be established, that is, to identify whether the main objective is bioremediation or treatment of a specific wastewater or the obtaining of biofuels, energy generation, or obtaining high value-added

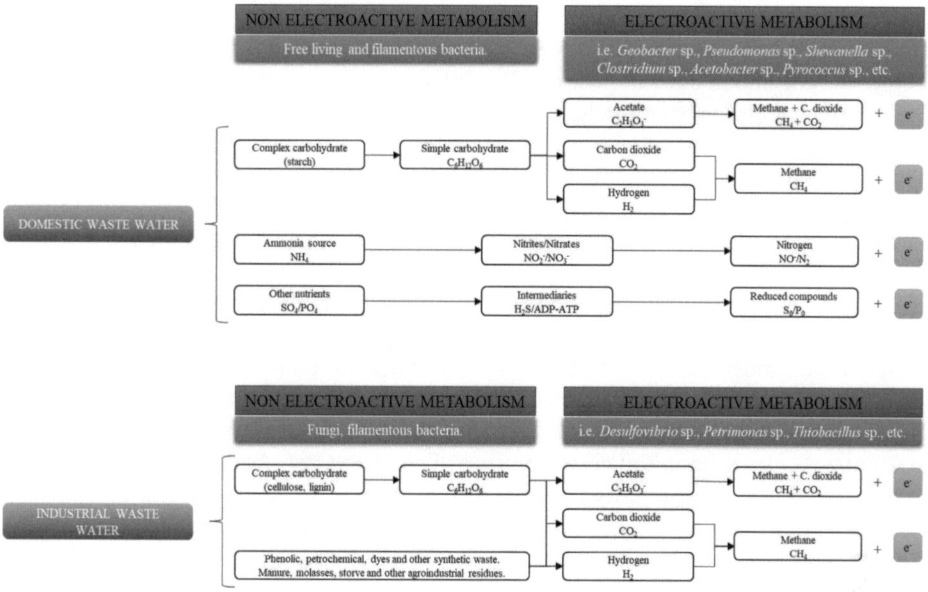

Figure 8.
Unified scheme unveiling wastewater transformation by electrogenic microorganisms.

metabolites. The origin and composition of the wastewater must be identified, to establish what type of substances will be present in the substrate and thus choose the adapted microbial community with the metabolic capacity to transform that type of substrate. For example, for the treatment of domestic wastewater, the use of an MFC or a combined system such as the MFC-CW could be implemented. For industrial wastewater, depending on their origin, MECs or METs could be implemented coupled to gas capture systems, in which the use of microbial communities with metabolic capacity for the successive transformation of organic substrates, until reaching the production of biofuels such as H_2 and CH_4.

7. Perspectives of electroactive microorganisms

Microorganisms and different microbial communities play a fundamental role in oxidation-reduction processes for the transformation of organic matter, energy production, and obtaining compounds of interest in bioelectrochemical systems. Some main metabolic pathways have been identified that are directly associated with the degradation of organic matter for the recovery of electrical energy or the obtaining of biofuels for some specific microorganisms. However, in diverse biological systems, such as microbial communities, in which there are ecological successions and different metabolic reactions that are essential to achieve the transformation of complex organic and inorganic compounds, as well as the transformation of highly polluting substances such as heavy metals, pesticides, and other contaminants; the combination of metabolic pathways that; they do not necessarily obey bioelectrochemical processes; have not been clearly established. Although knowledge of the microbial communities responsible for the biochemical functioning of BES has been progressively deepening in relation to technological advances in molecular biology and genetic engineering, the population and ecological study of microbial communities is constantly updated [45, 54, 59, 113], and research directly focused on the ecological study of complex microbial communities in bioelectrochemical systems is scarce.

Scientific advances have made it possible to identify which microorganisms are present in these systems and the possible metabolic mechanisms that favor their ability to transform organic matter and electrochemical performance. However, research focuses mainly on the design, configuration, and electrochemical performance of these systems. The detailed study of the changes in the diversity and abundance of microbial communities in BES provide fundamental information to determine the role that each microorganism plays in the microbial community and establish their participation in the oxidation-reduction processes of organic matter, the energy production, the electron transport mechanism, and the formation of biofilms and their interaction with the substrate and other microorganisms in BES systems; there is fundamental information for the implementation of bioelectrochemical systems as biotechnological alternatives for wastewater treatment and co-generation of clean energy. As a perspective, it is necessary to deepen the investigation of the biological interactions of microbial communities to analyze the electrochemical potential with application in the optimization of bioelectrochemical systems. For example, for the implementation of BES as biotechnological tools such as biosensors based on microbial fuel cells known as economic *in situ* sensing technologies of wastewater quality [113, 114].

There is evidence of microorganisms' capability to generate electricity and biofuels from complex organic matters such as cellulose, pectin, pesticides, dyes. While many studies delve into the treatment of domestic wastewater, industrial wastewater, particularly of agricultural origin, could be considered as unconventional substrates for biotechnological processes like BES. Agro-industrial wastewater consists mainly of organic matter and nutrients derived from both plant sources (biomass) and the addition of manure and fertilizers to crops. This composition makes them substrates with high potential for microbial metabolism.

Access to wastewater treatment systems in rural and agricultural areas is primarily limited by the difficulty of reaching these locations. Moreover, the extensive cultivation areas and their geographical separation make it challenging to aggregate the waste. The use of bioelectrochemical systems as an alternative for treating agro-industrial waste shows promise as a sustainable strategy that allows for the implementation of local treatment systems in agricultural zones. Considering the properties and composition of agro-industrial wastewater, as well as the adaptability of electrogenic microorganisms to different environmental conditions and substrates, supports the applicability of these systems in rural areas.

It is necessary to delve deeper into the study and characterization of native microbial communities in agro-industrial wastewater responsible for energy production through the oxidation-reduction of organic matter in bioelectrochemical systems. These communities offer several advantages, such as their adaptation to the composition, their development under ambient conditions, and their stable ecological relationships in these substrates. Compared to the introduction of exogenous microorganisms and microbial consortia into the treatment matrix, native microbial communities do not require adaptation phases to achieve electrogenic metabolism.

In-depth exploration of biostimulus and bioaugmentation techniques holds great promise in the optimization of wastewater bioelectrochemical treatment processes. These techniques involve adapting and enhancing the microorganisms present in the system, leading to a range of benefits in terms of electrogenic potential and overall efficiency of BES [9]. Through biostimulus techniques, the electrogenic activity and performance of the microorganisms can be stimulated and optimized providing suitable growth factors, adjusting environmental parameters, or employing specific additives that enhance microbial activity and promote electron transfer processes. Considering bioaugmentation, it is proposed to evaluate the introduction of selected microorganisms or microbial consortia into the existing native microbial community of the BES. These added microorganisms could improve specific capabilities, such as enhanced electron transfer abilities or the ability to degrade specific complex organic or remove inorganic compounds. By introducing these specialized microorganisms, the electrogenic potential of the system can be enhanced, leading to improved performance in terms of electricity generation or pollutant removal.

Both biostimulus and bioaugmentation techniques offer exciting opportunities for optimizing wastewater bioelectrochemical treatment processes. Through the adaptation and enhancement of microorganisms, these strategies can unlock the full potential of BES, improving their electrogenic performance and overall efficiency. However, further research is needed to understand the specific mechanisms and conditions that lead to optimal results, as well as to ensure the long-term stability and sustainability of these approaches in real-world applications.

Conflict of interest

The authors declare that the research was conducted in the absence of any commercial or financial relationships that could be construed as a potential conflict of interest.

Notes/thanks/other declarations

All authors listed have made a substantial, direct, and intellectual contribution to the work, and approved it for publication.

Author details

Lina María Agudelo-Escobar* and Santiago Erazo Cabrera
Research Group Biotransformación, School of Microbiology, University of Antioquia, Colombia

*Address all correspondence to: lina.agudelo@udea.edu.co

References

[1] UNESCO. Agua y cambio climático. In: Place de Fontenoy. 1st ed. Vol. 167 (3618). 75352 Paris 07 SP, France: United Nations Educational, Scientific and Cultural Organization; 2020. [Online]. Available from: https://unesdoc.unesco.org/ark:/48223/pf0000372876_spa

[2] UNESCO. Las aguas residuales: el recurso desaprovechado. UNESCO 06134 Colombella, Perugia, Italy: Division of Water Sciences; 2017. Available from: https://es.unesco.org/news/son-aguas-residuales-nuevo-oro-negro [Accessed: July 29, 2023]

[3] Qasim S. Wastewater Treatment Plants: Planning, Design and Operation. 2nd ed. New York: Routledge; 2017. DOI: 10.1201/9780203734209

[4] Sotres A, Tey L, Bonmatí A, Viñas M. Microbial community dynamics in continuous microbial fuel cells fed with synthetic wastewater and pig slurry. Bioelectrochemistry. 2016;**111**:70-82. DOI: 10.1016/j.bioelechem.2016.04.007

[5] Aquae. Datos sobre aguas residuales en el mundo—Fundación Aquae. Aguas residuales: datos y usos. 2020. Available from: https://www.fundacionaquae.org/aguas-residuales-datos-usos/?gclid=CjwKCAjwhaaKBhBcEiwA8acsHPhLynhA-iUA4-7AQwbmo2Nf_8OP3NexuaZZLZe16ekfcPevBN5D8RoCnIoQAvD_BwE [Accessed: September 22, 2021]

[6] Gul M, Ahmad K. Bioelectrochemical systems: Sustainable bio-energy powerhouses. Biosensors and Bioelectronics. 2019;**142**(August):111576. DOI: 10.1016/j.bios.2019.111576

[7] Mahmoudi A, Mousavi SA, Darvishi P. Effect of ammonium and COD concentrations on the performance of fixed-bed air-cathode microbial fuel cells treating reject water. International Journal of Hydrogen Energy. 2020;**45**(7):4887-4896. DOI: 10.1016/j.ijhydene.2019.12.081

[8] Palanisamy G, Jung H, Sadhasivam T, Kurkuri M, Kim S, Roh S. A comprehensive review on microbial fuel cell technologies: Processes, utilization, and advanced developments in electrodes and membranes. Journal of Cleaner Production. 2019;**221**:598-621. DOI: 10.1016/j.jclepro.2019.02.172

[9] Zhang P et al. Accelerating the startup of microbial fuel cells by facile microbial acclimation. Bioresource Technology Reports. 2019;**8**(November):100347. DOI: 10.1016/j.biteb.2019.100347

[10] Kumar A et al. The ins and outs of microorganism-electrode electron transfer reactions. Nature Reviews Chemistry. 2017;**1**(24):1-13. DOI: 10.1038/s41570-017-0024

[11] Logan B, Rossi R, Ragab A, Saikaly P. Electroactive microorganisms in bioelectrochemical systems. Nature Reviews Microbiology. 2019;**17**(5):307-319. DOI: 10.1038/s41579-019-0173-x

[12] Xu F et al. Electricity production and evolution of microbial community in the constructed wetland-microbial fuel cell. Chemical Engineering Journal. 2018;**339** (December 2017):479-486. DOI: 10.1016/j.cej.2018.02.003

[13] Tao M et al. Enhanced denitrification and power generation of municipal wastewater treatment plants (WWTPs) effluents with biomass in microbial fuel cell coupled with constructed wetland. Science of the Total Environment. 2020;**709**:136159. DOI: 10.1016/j.scitotenv.2019.136159

[14] Jiang X et al. Comprehensive comparison of bacterial communities in a membrane-free bioelectrochemical system for removing different mononitrophenols from wastewater. Bioresource Technology. 2016;**216**:645-652. DOI: 10.1016/j.biortech.2016.06.005

[15] Zhang L, Fu G, Zhang Z. Electricity generation and microbial community in long-running microbial fuel cell for high-salinity mustard tuber wastewater treatment. Bioelectrochemistry. 2019;**126**:20-28. DOI: 10.1016/j.bioelechem.2018.11.002

[16] Yong XY et al. An integrated aerobic-anaerobic strategy for performance enhancement of *Pseudomonas aeruginosa*-inoculated microbial fuel cell. Bioresource Technology. 2017;**241**:1191-1196. DOI: 10.1016/j.biortech.2017.06.050

[17] Leiva E, Leiva-Aravena E, Rodríguez C, Serrano J, Vargas I. Arsenic removal mediated by acidic pH neutralization and iron precipitation in microbial fuel cells. Science of the Total Environment. 2018;**645**:471-481. DOI: 10.1016/j.scitotenv.2018.06.378

[18] Wang H et al. Bioelectricity generation from the decolorization of reactive blue 19 by using microbial fuel cell. Journal of Environmental Management. 2019;**248**(March):109310. DOI: 10.1016/j.jenvman.2019.109310

[19] Ge X et al. Bioenergy generation and simultaneous nitrate and phosphorus removal in a pyrite-based constructed wetland-microbial fuel cell. Bioresource Technology. 2019, 2020;**296**(September):122350. DOI: 10.1016/j.biortech.2019.122350

[20] Wu Y et al. Copper removal and microbial community analysis in single-chamber microbial fuel cell. Bioresource Technology. 2018;**253**:372-377. DOI: 10.1016/j.biortech.2018.01.046

[21] Kumar R, Chiranjeevi P, Sukrampal, Patil SA. Integrated drip hydroponics-microbial fuel cell system for wastewater treatment and resource recovery. Bioresource Technology Reports. 2020;**9**:100-392. DOI: 10.1016/j.biteb.2020.100392

[22] Elzinga M, Liu D, Klok JBM, Roman P, Buisman CJN, ter Heijne A. Microbial reduction of organosulfur compounds at cathodes in bioelectrochemical systems. Environmental Science and Ecotechnology. 2020;**1**(December 2019):100009. DOI: 10.1016/j.ese.2020.100009

[23] Zhou Y et al. Relationship between electrogenic performance and physiological change of four wetland plants in constructed wetland-microbial fuel cells during non-growing seasons. Journal of Environmental Sciences. 2018;**70**:54-62. DOI: 10.1016/j.jes.2017.11.008

[24] Santoro C, Arbizzani C, Erable B, Ieropoulos I. Microbial fuel cells: From fundamentals to applications. A review. Journal of Power Sources. 2017;**356**:225-244. DOI: 10.1016/j.jpowsour.2017.03.109

[25] Saratale G et al. A comprehensive overview on electro-active biofilms, role of exo-electrogens and their microbial niches in microbial fuel cells (MFCs). Chemosphere. 2017;**178**:534-547. DOI: 10.1016/j.chemosphere.2017.03.066

[26] Almatouq A, Babatunde AO, Khajah M, Webster G, Alfodari M. Microbial community structure of anode electrodes in microbial fuel cells and microbial electrolysis cells. Journal of Water Process Engineering. 2020;**34**(December 2019):101140. DOI: 10.1016/j.jwpe.2020.101140

[27] Cai W, Lesnik KL, Wade MJ, Heidrich ES, Wang Y, Liu H. Incorporating microbial community data with machine learning techniques to predict feed substrates in microbial fuel cells. Biosensors and Bioelectronics. 2019;**133**(March):64-71. DOI: 10.1016/j.bios.2019.03.021

[28] dos Passos V et al. Hydrogen and electrical energy co-generation by a cooperative fermentation system comprising Clostridium and microbial fuel cell inoculated with port drainage sediment. Bioresource Technology. 2019; **277**(November 2018):94-103. DOI: 10.1016/j.biortech.2019.01.031

[29] Ishii S, Suzuki S, Tenney A, Norden-Krichmar T, Nealson K, Bretschger O. Microbial metabolic networks in a complex electrogenic biofilm recovered from a stimulus-induced metatranscriptomics approach. Scientific Reports. 2015;**5**(September):1-14. DOI: 10.1038/srep14840

[30] Miran W, Jang J, Nawaz M, Shahzad A, Lee DS. Sulfate-reducing mixed communities with the ability to generate bioelectricity and degrade textile diazo dye in microbial fuel cells. Journal of Hazardous Materials. 2018;**352**:70-79. DOI: 10.1016/j.jhazmat.2018.03.027

[31] Ghimire U, Gude VG. Accomplishing a N-E-W (nutrient-energy-water) synergy in a bioelectrochemical nitritation-anammox process. Scientific Reports. 2019;**9**(1):1-13. DOI: 10.1038/s41598-019-45620-2

[32] Arkatkar A, Mungray AK, Sharma P. Effect of treatment on electron transfer mechanism in microbial fuel cell. Energy Sources, Part A: Recovery, Utilization and Environmental Effects. 2019;**45**(2): 3843-3858. DOI: 10.1080/15567036.2019.1668878

[33] Allam F, Elnouby M, Sabry SA, El-Khatib KM, El-Badan DE. Optimization of factors affecting current generation, biofilm formation and rhamnolipid production by electroactive *Pseudomonas aeruginosa* FA17. International Journal of Hydrogen Energy. 2021;**46**(20):11419-11432. DOI: 10.1016/j.ijhydene.2020.08.070

[34] Marks S, Makinia J, Fernandez-Morales FJ. Performance of microbial fuel cells operated under anoxic conditions. Applied Energy. 2019;**250** (October 2018):1-6. DOI: 10.1016/j.apenergy.2019.05.043

[35] Bosire E, Blank L, Rosenbaum M. Strain- and substrate-dependent redox mediator and electricity production by *Pseudomonas aeruginosa*. Applied and Environmental Microbiology. 2016;**82** (16):5026-5038. DOI: 10.1128/AEM.01342-16

[36] Park Y et al. Response of microbial community structure to pre-acclimation strategies in microbial fuel cells for domestic wastewater treatment. Bioresource Technology. 2017;**233**:176-183. DOI: 10.1016/j.biortech.2017.02.101

[37] Fazal T et al. Simultaneous production of bioelectricity and biogas from chicken droppings and dairy industry wastewater employing bioelectrochemical system. Fuel. 2019; **256**(February):115902. DOI: 10.1016/j.fuel.2019.115902

[38] Ren ZJ. Running on gas. Nature Energy. 2017;**2**(June):1-2. DOI: 10.1038/nenergy.2017.93

[39] Jang J, Jeon BW, Kim YH. Bioelectrochemical conversion of CO2 to value added product formate using engineered Methylobacterium extorquens. Scientific Reports. 2018;**8**(1): 1-7. DOI: 10.1038/s41598-018-23924-z

[40] Potter M. Electrical effects accompanying the decomposition of organic compounds. Proceedings of the Royal Society of London. 1911;**84**:260-276

[41] Zhao Z, Zhang Y, Wang L, Quan X. Potential for direct interspecies electron transfer in an electric-anaerobic system to increase methane production from sludge digestion. Scientific Reports. 2015;**5**(February):1-12. DOI: 10.1038/srep11094

[42] Arkatkar A, Mungray AK, Sharma P. Study of electrochemical activity zone of *Pseudomonas aeruginosa* in microbial fuel cell. Process Biochemistry. 2021;**101** (October 2020):213-217. DOI: 10.1016/j.procbio.2020.11.020

[43] Estevez-Canales M, Pinto D, Coradin T, Laberty-Robert C, Esteve-Núñez A. Silica immobilization of Geobacter sulfurreducens for constructing ready-to-use artificial bioelectrodes. Microbial Biotechnology. 2018;**11**(1):39-49. DOI: 10.1111/1751-7915.12561

[44] Wang VB et al. Metabolite-enabled mutualistic interaction between Shewanella oneidensis and Escherichia coli in a co-culture using an electrode as electron acceptor. Scientific Reports. 2015;**5**(May):1-11. DOI: 10.1038/srep11222

[45] Rehman Z, Fortunato L, Cheng T, Leiknes TO. Metagenomic analysis of sludge and early-stage biofilm communities of a submerged membrane bioreactor. Science of the Total Environment. 2020;**701**:134682. DOI: 10.1016/j.scitotenv.2019.134682

[46] Modestra JA, Reddy CN, Krishna KV, Min B, Mohan SV. Regulated surface potential impacts bioelectrogenic activity, interfacial electron transfer and microbial dynamics in microbial fuel cell. Renewable Energy. 2020;**149**:424-434. DOI: 10.1016/j.renene.2019.12.018

[47] Deng Q et al. Performance and functional microbial communities of denitrification process of a novel MFC-granular sludge coupling system. Bioresource Technology. 2020;**306** (January):123173. DOI: 10.1016/j.biortech.2020.123173

[48] Kumar S, Kumar V, Kumar R, Malyan S, Pugazhendhi A. Microbial fuel cells as a sustainable platform technology for bioenergy, biosensing, environmental monitoring, and other low power device applications. Fuel. 2019;**255**(February):115682. DOI: 10.1016/j.fuel.2019.115682

[49] Sun G, Kang K, Qiu L, Guo X, Zhu M. Electrochemical performance and microbial community analysis in air cathode microbial fuel cells fuelled with pyroligneous liquor. Bioelectrochemistry. 2019;**126**:12-19. DOI: 10.1016/j.bioelechem.2018.11.006

[50] Paitier A, Godain A, Lyon D, Haddour N, Vogel TM, Monier JM. Microbial fuel cell anodic microbial population dynamics during MFC start-up. Biosensors and Bioelectronics. 2017; **92**:357-363. DOI: 10.1016/j.bios.2016.10.096

[51] Chen J et al. Bacterial community shift and antibiotics resistant genes analysis in response to biodegradation of oxytetracycline in dual graphene modified bioelectrode microbial fuel cell. Bioresource Technology. 2019;**276** (December 2018):236-243. DOI: 10.1016/j.biortech.2019.01.006

[52] Bagchi S, Behera M. Bioaugmentation using *Pseudomonas aeruginosa* with an approach of intermittent aeration for enhanced

power generation in ceramic MFC. Sustainable Energy Technologies and Assessments. 2021;**45**(January):101138. DOI: 10.1016/j.seta.2021.101138

[53] McAnulty M, Poosarla VG, Kim KY, Jasso-Chávez R, Logan BE, Wood TK. Electricity from methane by reversing methanogenesis. Nature Communications. 2017;**8**(May):15419. DOI: 10.1038/ncomms15419

[54] Chen J et al. Bacterial community structure and gene function prediction in response to long-term running of dual graphene modified bioelectrode bioelectrochemical systems. Bioresource Technology. 2020;**309**(April):123398. DOI: 10.1016/j.biortech.2020.123398

[55] Long X, Cao X, Song H, Nishimura O, Li X. Characterization of electricity generation and microbial community structure over long-term operation of a microbial fuel cell. Bioresource Technology. 2019;**285**(April):121395. DOI: 10.1016/j.biortech.2019.121395

[56] Asefi B et al. Characterization of electricity production and microbial community of food waste-fed microbial fuel cells. Process Safety and Environmental Protection. 2019; **125**:83-91. DOI: 10.1016/j.psep.2019.03.016

[57] Madsen C, TerAvest M. NADH dehydrogenases Nuo and Nqr1 contribute to extracellular electron transfer by Shewanella oneidensis MR-1 in bioelectrochemical systems. Scientific Reports. 2019;**9**(1):1-6. DOI: 10.1038/s41598-019-51452-x

[58] Rago L et al. A study of microbial communities on terracotta separator and on biocathode of air breathing microbial fuel cells. Bioelectrochemistry. 2018;**120**: 18-26. DOI: 10.1016/j.bioelechem.2017.11.005

[59] Liu J et al. Enhanced mechanism of extracellular electron transfer between semiconducting minerals anatase and *Pseudomonas aeruginosa* PAO1 in euphotic zone. Bioelectrochemistry. 2021;**141**:107849. DOI: 10.1016/j.bioelechem.2021.107849

[60] Yang C et al. Carbon dots-fed Shewanella oneidensis MR-1 for bioelectricity enhancement. Nature Communications. 2020;**11**(1):1-11. DOI: 10.1038/s41467-020-14866-0

[61] Zhang M, Ma Z, Zhao N, Zhang K, Song H. Increased power generation from cylindrical microbial fuel cell inoculated with *P. aeruginosa*. Biosensors and Bioelectronics. 2019;**141**(May): 111394. DOI: 10.1016/j.bios.2019.111394

[62] Qiao YJ, Qiao Y, Zou L, Wu XS, Liu JH. Biofilm promoted current generation of *Pseudomonas aeruginosa* microbial fuel cell via improving the interfacial redox reaction of phenazines. Bioelectrochemistry. 2017;**117**:34-39. DOI: 10.1016/j.bioelechem.2017.04.003

[63] Patel R, Deb D. Parametrized control-oriented mathematical model and adaptive backstepping control of a single chamber single population microbial fuel cell. Journal of Power Sources. 2018;**396**(June):599-605. DOI: 10.1016/j.jpowsour.2018.06.064

[64] Hassan H, Jin B, Donner E, Vasileiadis S, Saint C, Dai S. Microbial community and bioelectrochemical activities in MFC for degrading phenol and producing electricity: Microbial consortia could make differences. Chemical Engineering Journal. 2018;**332** (September 2017):647-657. DOI: 10.1016/j.cej.2017.09.114

[65] Wang HYA, Yang CS, Lin CH. A rapid quantification method for energy conversion efficiency of microbial fuel

cells. Journal of the Taiwan Institute of Chemical Engineers. 2020;**109**:124-128. DOI: 10.1016/j.jtice.2020.02.001

[66] Hirose A, Kasai T, Aoki M, Umemura T, Watanabe K, Kouzuma A. Electrochemically active bacteria sense electrode potentials for regulating catabolic pathways. Nature Communications. 2018;**9**(1):1-10. DOI: 10.1038/s41467-018-03416-4

[67] Wang H et al. Cascade degradation of organic matters in brewery wastewater using a continuous stirred microbial electrochemical reactor and analysis of microbial communities. Scientific Reports. 2016;**6**(73):1-12. DOI: 10.1038/srep27023

[68] Chen J, Zhang L, Hu Y, Huang W, Niu Z, Sun J. Bacterial community shift and incurred performance in response to in situ microbial self-assembly graphene and polarity reversion in microbial fuel cell. Bioresource Technology. 2017;**241**: 220-227. DOI: 10.1016/j.biortech.2017.05.123

[69] Islam MA, Ethiraj B, Cheng CK, Yousuf A, Khan MMR. An Insight of synergy between *Pseudomonas aeruginosa* and *Klebsiella variicola* in a microbial fuel cell. ACS Sustainable Chemistry & Engineering. 2018;**6**(3):4130-4137. DOI: 10.1021/acssuschemeng.7b04556

[70] Wang F et al. Synchronously electricity generation and degradation of biogas slurry using microbial fuel cell. Renewable Energy. 2019;**142**:158-166. DOI: 10.1016/j.renene.2019.04.063

[71] Chen D, Yang K, Wei L, Wang H. Microbial community and metabolism activity in a bioelectrochemical denitrification system under long-term presence of p-nitrophenol. Bioresource Technology. 2016;**218**:189-195. DOI: 10.1016/j.biortech.2016.06.081

[72] Zhang S, You J, Chen H, Ye J, Cheng Z, Chen J. Gaseous toluene, ethylbenzene, and xylene mixture removal in a microbial fuel cell: Performance, biofilm characteristics, and mechanisms. Chemical Engineering Journal. 2020;**386**:123916. DOI: 10.1016/j.cej.2019.123916

[73] Fang H et al. Exploring bacterial communities and biodegradation genes in activated sludge from pesticide wastewater treatment plants via metagenomic analysis. Environmental Pollution. 2018;**243**:1206-1216. DOI: 10.1016/j.envpol.2018.09.080

[74] Hou R, Luo X, Liu C, Zhou L, Wen J, Yuan Y. Enhanced degradation of triphenyl phosphate (TPHP) in bioelectrochemical systems: Kinetics, pathway and degradation mechanisms. Environmental Pollution. 2019;**254**: 113040. DOI: 10.1016/j.envpol.2019.113040

[75] Shen J, Du Z, Li J, Cheng F. Co-metabolism for enhanced phenol degradation and bioelectricity generation in microbial fuel cell. Bioelectrochemistry. 2020;**134**:107527. DOI: 10.1016/j.bioelechem.2020.107527

[76] Obata O, Salar-Garcia MJ, Greenman J, Kurt H, Chandran K, Ieropoulos I. Development of efficient electroactive biofilm in urine-fed microbial fuel cell cascades for bioelectricity generation. Journal of Environmental Management. 2020;**258**(September 2019):109992. DOI: 10.1016/j.jenvman.2019.109992

[77] Cai J, Zheng P, Xing Y, Qaisar M. Effect of electricity on microbial community of microbial fuel cell simultaneously treating sulfide and nitrate. Journal of Power Sources. 2015; **281**:27-33. DOI: 10.1016/j.jpowsour.2015.01.165

[78] Ilamathi R, Merline Sheela A, Nagendra Gandhi N. Comparative evaluation of Pseudomonas species in single chamber microbial fuel cell with manganese coated cathode for reactive azo dye removal. In: International Biodeterioration & Biodegradation. Vol. 144. 2019. pp. 104744. DOI: 10.1016/j.ibiod.2019.104744

[79] Hsiang A, Chun Y, Cheng L. A rapid quantification method for energy conversion efficiency of microbial fuel cells. Journal of the Taiwan Institute of Chemical Engineers. 2020;**000**:1-5. DOI: 10.1016/j.jtice.2020.02.001

[80] Liu R et al. Microbial community dynamics in a pilot-scale MFC-AA/O system treating domestic sewage. Bioresource Technology. 2017;**241**: 439-447. DOI: 10.1016/j.biortech.2017.05.122

[81] Xin X, Hong J, Liu Y. Insights into microbial community profiles associated with electric energy production in microbial fuel cells fed with food waste hydrolysate. Science of the Total Environment. 2019;**670**:50-58. DOI: 10.1016/j.scitotenv.2019.03.213

[82] Munjal M, Tiwari B, Lalwani S, Sharma M, Singh G, Sharma RK. An insight of bioelectricity production in mediator less microbial fuel cell using mesoporous Cobalt Ferrite anode. International Journal of Hydrogen Energy. 2020;**45**(22):12525-12534. DOI: 10.1016/j.ijhydene.2020.02.184

[83] Salar-García MJ, Ortiz-Martínez VM, Gajda I, Greenman J, Hernández-Fernández FJ, Ieropoulos IA. Electricity production from human urine in ceramic microbial fuel cells with alternative non-fluorinated polymer binders for cathode construction. Separation and Purification Technology. 2017;**187**:436-442. DOI: 10.1016/j.seppur.2017.06.025

[84] Deutzmann JS, Spormann AM. Enhanced microbial electrosynthesis by using defined co-cultures. ISME Journal. 2017;**11**(3):704-714. DOI: 10.1038/ismej.2016.149

[85] Toczyłowska-Mamińska R et al. Evolving microbial communities in cellulose-fed microbial fuel cell. Energies. 2018;**11**(1):1-12. DOI: 10.3390/en11010124

[86] Wu Y, Wang L, Jin M, Zhang K. Simultaneous copper removal and electricity production and microbial community in microbial fuel cells with different cathode catalysts. Bioresour Technol. 2020;**305**(January):123166. DOI: 10.1016/j.biortech.2020.123166

[87] Liang P, Duan R, Jiang Y, Zhang X, Qiu Y, Huang X. One-year operation of 1000-L modularized microbial fuel cell for municipal wastewater treatment. Water Research. 2018;**141**:1-8. DOI: 10.1016/j.watres.2018.04.066

[88] Bajracharya S, Srikanth S, Mohanakrishna G, Zacharia R, Strik DP, Pant D. Biotransformation of carbon dioxide in bioelectrochemical systems: State of the art and future prospects. Journal of Power Sources. 2017;**356**:256-273. DOI: 10.1016/j.jpowsour.2017.04.024

[89] Kumar SS et al. Microbial fuel cells (MFCs) for bioelectrochemical treatment of different wastewater streams. Fuel. 2019;**254**(February): 115526. DOI: 10.1016/j.fuel.2019.05.109

[90] Obata O, Greenman J, Kurt H, Chandran K, Ieropoulos I. Resilience and limitations of MFC anodic community when exposed to antibacterial agents. Bioelectrochemistry. 2020;**134**:107500. DOI: 10.1016/j.bioelechem.2020.107500

[91] Sciarria T, Arioli S, Gargari G, Mora D, Adani F. Monitoring microbial

communities' dynamics during the start-up of microbial fuel cells by high-throughput screening techniques. Biotechnology Reports. 2019;**21**(2018): e00310. DOI: 10.1016/j.btre.2019.e00310

[92] Ishii S, Suzuki S, Tenney A, Nealson K, Bretschger O. Comparative metatranscriptomics reveals extracellular electron transfer pathways conferring microbial adaptivity to surface redox potential changes. ISME Journal. 2018;**12**(12):2844-2863. DOI: 10.1038/s41396-018-0238-2

[93] Chen H, Dong F, Minteer SD. The progress and outlook of bioelectrocatalysis for the production of chemicals, fuels and materials. Nature Catalysis. 2020;**3**(3):225-244. DOI: 10.1038/s41929-019-0408-2

[94] Ciniciato GPMK et al. Investigating the association between photosynthetic efficiency and generation of biophotoelectricity in autotrophic microbial fuel cells. Scientific Reports. 2016;**6**(July):1-10. DOI: 10.1038/srep31193

[95] Ishii S et al. Microbial population and functional dynamics associated with surface potential and carbon metabolism. ISME Journal. 2014;**8**(5):963-978. DOI: 10.1038/ismej.2013.217

[96] Rabaey K, Rozendal R. Microbial electrosynthesis—Revisiting the electrical route for microbial production. Nature Reviews Microbiology. 2010;**8**(10):706-716. DOI: 10.1038/nrmicro2422

[97] Jin X, Guo F, Liu Z, Liu Y, Liu H. Enhancing the electricity generation and nitrate removal of microbial fuel cells with a novel denitrifying exoelectrogenic strain EB-1. Frontiers in Microbiology. 2018;**9**(NOV):1-10. DOI: 10.3389/fmicb.2018.02633

[98] Li M et al. Microbial fuel cell (MFC) power performance improvement through enhanced microbial electrogenicity. Biotechnology Advances. 2018;**36**(4):1316-1327. DOI: 10.1016/j.biotechadv.2018.04.010

[99] Jones A, Buie C. Continuous shear stress alters metabolism, mass-transport, and growth in electroactive biofilms independent of surface substrate transport. Scientific Reports. 2019;**9**(1): 1-8. DOI: 10.1038/s41598-019-39267-2

[100] Guo J, Cheng J, Li B, Wang J, Chu P. Performance and microbial community in the biocathode of microbial fuel cells under different dissolved oxygen concentrations. Journal of Electroanalytical Chemistry. 2019;**833**: 433-440. DOI: 10.1016/j.jelechem.2018.12.015

[101] Haavisto JM, Lakaniemi AM, Puhakka JA. Storing of exoelectrogenic anolyte for efficient microbial fuel cell recovery. Environmental Technology. United Kingdom. 2019;**40**(11):1467-1475. DOI: 10.1080/09593330.2017.1423395

[102] Zhao N, Jiang Y, Alvarado M, Treu L, Angelidaki I, Zhang Y. Electricity generation and microbial communities in microbial fuel cell powered by macroalgal biomass. Bioelectrochemistry. 2018;**123**: 145-149. DOI: 10.1016/j.bioelechem.2018.05.002

[103] Xia T et al. Power generation and microbial community analysis in microbial fuel cells: A promising system to treat organic acid fermentation wastewater. Bioresource Technology. 2019;**284**:72-79. DOI: 10.1016/j.biortech.2019.03.119

[104] Li C, Ren H, Xu M, Cao J. Study on anaerobic ammonium oxidation process coupled with denitrification microbial fuel cells (MFCs) and its microbial community analysis. Bioresource

Technology. 2015;**175**:545-552. DOI: 10.1016/j.biortech.2014.10.156

[105] Cui MH, Cui D, Gao L, Cheng HY, Wang AJ. Analysis of electrode microbial communities in an up-flow bioelectrochemical system treating azo dye wastewater. Electrochimica Acta. 2016;**220**:252-257. DOI: 10.1016/j.electacta.2016.10.121

[106] Li T, Guo F, Lin Y, Li Y, Wu G. Metagenomic analysis of quorum sensing systems in activated sludge and membrane biofilm of a full-scale membrane bioreactor. Journal of Water Process Engineering. 2019;**32**(June):1-11. DOI: 10.1016/j.jwpe.2019.100952

[107] Santoro C et al. Ceramic microbial fuel cells stack: Power generation in standard and supercapacitive mode. Scientific Reports. 2018;**8**(1):1-12. DOI: 10.1038/s41598-018-21404-y

[108] Gajda I, Greenman J, Ieropoulos I. Microbial fuel cell stack performance enhancement through carbon veil anode modification with activated carbon powder. Applied Energy. 2020;**262** (December 2019):114475. DOI: 10.1016/j.apenergy.2019.114475

[109] Marshall CW et al. Metabolic reconstruction and modeling microbial electrosynthesis. Scientific Reports. 2017;**7**(1):1-12. DOI: 10.1038/s41598-017-08877-z

[110] Koók L, Nemestóthy N, Bélafi-Bakó K, Bakonyi P. Investigating the specific role of external load on the performance versus stability trade-off in microbial fuel cells. Bioresource Technology. 2020; **309**(February):123313. DOI: 10.1016/j.biortech.2020.123313

[111] Revelo DM, Hurtado NH, Ruiz JO, López S. Uso de microorganismos nativos en la remoción simultánea de materia orgánica y cr(VI) en una celda de combustible microbiana de biocátodo (CCM). Informacion Tecnologica. 2015; **26**(6):77-88. DOI: 10.4067/S0718-07642015000600010

[112] Agudelo-Escobar L, Erazo S, Avignone-Rossa C. A Bioelectrochemical system for waste industrial coffee waste water. Frontiers in Chemical Engineering. 2022;**4**(February):1-16. DOI: 10.3389/fceng.2022.814987

[113] Do MH et al. Microbial fuel cell-based biosensor for online monitoring wastewater quality: A critical review. Science of the Total Environment. 2020; **712**:135612. DOI: 10.1016/j.scitotenv.2019.135612

[114] Xiao N, Selvaganapathy PR, Wu R, Huang JJ. Influence of wastewater microbial community on the performance of miniaturized microbial fuel cell biosensor. Bioresource Technology. 2020;**302**:122777. DOI: 10.1016/j.biortech.2020.122777